MARINE FORECASTING

PREDICTABILITY AND MODELLING IN OCEAN HYDRODYNAMICS

FURTHER TITLES IN THIS SERIES

Elsevier Oceanography Series, 25

MARINE FORECASTING

Predictability and Modelling in Ocean Hydrodynamics

PROCEEDINGS OF THE 10th INTERNATIONAL LIÈGE COLLOQUIUM
ON OCEAN HYDRODYNAMICS

Edited by

JACQUES C.J. NIHOUL

Professor of Ocean Hydrodynamics,
University of Liège,
Liège, Belgium

ELSEVIER SCIENTIFIC PUBLISHING COMPANY
Amsterdam — Oxford — New York   1979

ELSEVIER SCIENTIFIC PUBLISHING COMPANY
335 Jan van Galenstraat
P.O. Box 211, 1000 AE Amsterdam, The Netherlands

Distributors for the United States and Canada:

ELSEVIER/NORTH-HOLLAND INC.
52, Vanderbilt Avenue
New York, N.Y. 10017

Library of Congress Cataloging in Publication Data

International Liege Colloquium on Ocean Hydrodynamics,
    10th, 1978.
    Marine forecasting.

    (Elsevier oceanography series ; 25)
    Bibliography: p.
    Includes index.
    1. Oceanography--Mathematical models--Congresses.
2. Hydrodynamics--Mathematical models--Congresses.
I. Nihoul, Jacques C. J. II. Title.
GC200.I57 1978   551.4'7'00184   79-11360
ISBN 0-444-41797-4

ISBN 0-444-41797-4 (Vol. 25)
ISBN 0-444-41623-4 (Series)

Printed in The Netherlands

FOREWORD

   The International Liège Colloquia on Ocean Hydrodynamics are organized annually. Their topics differ from one year to another and
try to address, as much as possible, recent problems and incentive
new subjects in physical oceanography.

   Assembling a group of active and eminent scientists from different countries and often different disciplines, they provide a forum
for discussion and foster a mutually beneficial exchange of information opening on to a survey of major recent discoveries, essential
mechanisms, impelling question-marks and valuable suggestions for
future research.

   Basic studies of atmospheric processes continuously feed a science
called Meteorology and a public service called Meteorological Forecasting. For a long time, ocean sciences have remained more descrip-
tive in nature, more concerned with the understanding of the basic
processes and mathematical models were often designed with the main
purpose of elucidating particular aspects of the ocean dynamics.

   However, the rapid advancement, in the recent years, of both the
physical sciences of the ocean and the mathematical techniques of
marine modelling have made possible the development, in the field of
marine hydrodynamics and air-sea interactions, of prognostic models
serving a new science and initiating a public service : Marine
Forecasting.

   The papers presented at the Tenth International Liège Colloquium
on Ocean Hydrodynamics report fundamental or applied research and
they address such different fields as storm surges, mixing in the
upper ocean layers, surface waves, cyclogenesis and other air-sea or
sea-air interactions. Their unity resides in a common approach,
seeking a better understanding (by modellers and users) of the scientific maturity and of the incentive new prospects of Marine Forecasting.

                                         Jacques C.J. NIHOUL.

The Scientific Organizing Committee of the Tenth International Liège Colloquium on Ocean Hydrodynamics and all the participants wish to express their gratitude to the Belgian Minister of Education, the National Science Foundation of Belgium, the University of Liège and the Office of Naval Research for their most valuable support.

LIST OF PARTICIPANTS

ADAM,Y., Dr., Ministère de la Santé Publique et de l'Environnement, Belgium.

ARANUVACHAPUN,S., Dr., Mekong Project, United Nations, Bangkok, Thailand.

BACKHAUS,J.O., Mr., Deutsches Hydrographisches Institut, Hamburg, W. Germany.

BAH,A., Ir., Université de Liège, Belgium.

BELHOMME,G., Ir., Université de Liège, Belgium.

BERGER,A., Dr., Université Catholique de Louvain, Belgium.

BERNARD,E., Dr., Institut Royal Météorologique, Bruxelles, Belgium.

BESSERO,G., Ir., Service Hydrographique et Océanographique de la Marine, Brest, France.

BUDGELL,W.P., Mr., Ocean & Aquatic Sciences, Burlington, Canada.

CANEILL,J.Y., Ir., ENSTA, Laboratoire de Mécanique des Fluides, Paris, France

CAVANIE,A., Dr., CNEXO/COB, Brest, France.

CHABERT d'HIERES,G., Ir., Institut de Mécanique, Grenoble, France.

DE GREEF,E., Mr., Institut Royal Météorologique, Bruxelles, Belgium.

DE KOK, Mr., Rijkswaterstaat, Rijswijck, The Netherlands.

DELECLUSE,P., Melle, M.H.N., Laboratoire d'Océanographie Physique, Paris, France.

DISTECHE,A., Prof., Dr., Université de Liège, Belgium.

DONELAN,M., Dr., Canada Centre for Inland Waters, Burlington, Canada.

DOWLEY,A., Mr., University College, Dublin, Ireland.

DUNN-CHRISTENSEN,J.T., Dr., Meteorologisk Institut, Copenhagen, Denmark.

ELLIOTT,A.J., Dr., SACLANT ASW Research Centre, La Spezia, Italy.

EWING,J.A., Mr., I.O.S., Wormley, U.K.

FEIN,J., Dr., CDRS, National Science Foundation, Washington D.C., U.S.A.

FISCHER,G., Prof., Dr., Meteorologisches Institut, Universität Hamburg, W. Germany.

FRANKIGNOUL,C.J., Dr., Massachusetts Institute of Technology, Cambridge, U.S.A.

FRASSETTO,r.., Prof., Laboratorio per lo Studio della Dinamica delle Grandi Masse, Venezia, Italy.

FRITZNER,H.E., Mr., Norsk Hydro, Oslo, Norway.

GERRITSEN,H., Ir., Technische Hogeschool Twente, The Netherlands.

GRAF,W.H., Prof., Ecole Polytechnique Fédérale, Lausanne, Switzerland.

HAUGUEL,A., Ir., E.D.F., Chatou, France.

HEAPS,N.S., Dr., IOS, Bidston Observatory, U.K.

HECQ,P., Ir., Université de Liège, Belgium.

HENKE,I.M., Mrs., Institut für Meereskunde, Universität Kiel, W. Germany.

HUA,B.L., Melle, M.H.N., Laboratoire d'Océanographie Physique, Paris, France.

JAUNET,J.P., Ir., Bureau VERITAS, Paris, France.

JONES,J.E., Mr., IOS, Bidston Observatory, U.K.

JONES,S., Dr., University of Southampton, U.K.

KAHMA,K., Mr., Institute of Marine Research, Helsinki, Finland.

KITAYGORODSKIY,S.A., Prof., Dr., Academy of Sciences of the U.S.S.R., Moscow, U.S.S.R., and Institute of Marine Research, Helsinki, Finland.

LEJEUNE,A., Dr., Université de Liège, Belgium.

LOFFET,A., Ir., Université de Liège, Belgium.

MAC MAHON,B., Mr., Imperial College, Civil Engineering Dept., London, U.K.

MAGAARD,L., Prof., Dr., University of Hawaii, Honolulu, U.S.A.

MELSON, L.B., Ir., U.S. Navy Sciences & Technical Group Europe, München, W. Germany.

MESQUITA,A.R. de, Prof., University of Sao Paulo, Brazil.

MICHAUX,T., Ir., Université de Liège, Belgium.

MILLER,B.L., Dr., National Maritime Institute, Teddington, U.K.

MIQUEL,J., Ir., E.D.F., Chatou, France.

MÜLLER,P., Dr., Institut für Geophysik, Universität Hamburg, W. Germany.

NAATZ,O.W., Mr., Fachbereich See, Fachhochschule Hamburg, W. Germany.

NASMYTH,P.W., Dr., Institute of Ocean Sciences, Sidney, Canada.

NIHOUL,J.C.J., Prof., Dr., Université de Liège, Belgium.

NIZET,J.L., Mr., Université de Liège, Belgium.

O'BRIEN,J.J., Prof., Dr., Florida State University, Tallahassee, U.S.A.

O'KANE,J.P., Dr., University College, Dublin, Ireland.

OZER,J., Ir., Université de Liège, Belgium.

PELLEAU,R., Ir., ELF-AQUITAINE, Pau, France.

PICHOT,G., Ir., Ministère de la Santé Publique et de l'Environnement, Belgium.

RAMMING,H.G., Dr., Universität Hamburg, W. Germany.

REID,R.O., Prof.,Dr., Texas A&M University, College Station, U.S.A.

ROISIN,B., Mr., Université de Liège, Belgium.

RONDAY,F.C., Dr., Université de Liège, Belgium.

ROOVERS,P., Ir., Waterbouwkundig Laboratorium, Borgerhout, Belgium.

ROSENTHAL,W., Dr., Institut für Geophysik, Universität Hamburg, W. Germany.

RUNFOLA,Y., Mr., Université de Liège, Belgium.

SCHÄFER,P., Mr., K.F.K.I., Hamburg, W. Germany.

SCHAYES,G., Dr., Université Catholique de Louvain, Belgium.

SETHURAMAN, S., Dr., Brookhaven National Laboratory, Upton, U.S.A.

SHONTING,D.H., Prof., Naval Underwater Systems Center, Newport, U.S.A.

SMITZ,J., Ir., Université de Liège, Belgium.

SPLIID,H., Dr., IMSOR, Technical University of Denmark, Lyngby, Denmark.

THACKER,W.C., Dr., NOAA/AOML Sea-Air Laboratory, Miami, U.S.A.

THOMASSET,F., Ir., IRIA LABORIA, Le Chesnay, France.

TIMMERMANN,H., Ir., KNMI, De Bilt, The Netherlands.

TWITCHELL, P.F., Dr., Office of Naval Research, Boston, U.S.A.

VAN HAMME,J.L., Dr., Institut Royal Météorologique, Bruxelles, Belgium.

VINCENT,C.L., Dr., U.S.A. Engineer Waterways Experiment Station, Vicksburg, U.S.A.

VOOGT,J., Ir., Rijkswaterstaat, s'Gravenhage, The Netherlands.

WANG,D.P., Dr., Chesapeake Bay Institute, The Johns Hopkins University Baltimore, U.S.A.

WILLEBRAND,J., Dr., Princeton University, U.S.A.

WORTHINGTON,B.A., Dr., Hydraulics Research Station, Wallingford, U.K.

CONTENTS

REVIEW OF THE THEORIES OF WIND-MIXED LAYER DEEPENING

S.A. KITAIGORODSKII

PP Shirshov Institute of Oceanology, Academy of Sciences, Moscow
(U.S.S.R.).

English version prepared from the original manuscript in Russian by

Jacques C.J. NIHOUL and A. LOFFET
Mécanique des Fluides Géophysiques, Université de Liège, Sart Tilman B6, Liège
(Belgium).

ABSTRACT

   One considers here the time evolution of the oceanic surface boundary layer in relation with the synoptic variability of atmospheric
processes. Attention is restricted to situations where the major res-
ponsability for the short-period variability of the vertical structure of the surface boundary layer lies on the local thermal and dynamic interactions between the atmosphere and the ocean and on the
internal thermocline - supported transfer processes. Emphasis is
laid on theoretical and experimental results which can be interpreted
by means of simple one-dimensional vertical mixing models.

INTRODUCTION

   The description of the dynamic of wind mixing in oceanic surface

layers (e.g. Kitaigorodskii, 1970) is based on the assumption that

the main sources of turbulent energy are

i) the breaking of wind waves which produces turbulence in a relati-

   vely thin surface layer (having a thickness of the order of the

   amplitude of the breaking waves) which extends into the fluid by

   turbulent energy diffusion effects (Kitaigorodskii and Miropolskii,

   1967 ; Kalatskiy, 1974) ;

ii)the velocity shear associated with drift currents responsible for

   turbulent energy production throughout the turbulent layer and,

   primarily, in those parts of it where the velocity shear is large.

   In oceanic surface layers, the two mechanisms can act simultaneous-

ly. However, in laboratory conditions, it is possible to explore

each of them individually.

To study the wind wave breaking effect, the initial stirring of the thin surface layer can be simulated by means of a vertically oscillating grid placed in the vicinity of the fluid surface (Turner, 1973 ; Linden, 1975). The mixing caused by drift currents can be modelled by experiments in which a constant stress is applied at the surface of the fluid (Kato and Phillips, 1969 ; Kantha et al, 1977).

The laboratory experiments (Turner, 1973 ; Linden, 1975 ; Kato and Phillips, 1969 ; Kantha et al, 1977 ; Moore and Long, 1971) explicitly show that all the mechanisms of turbulence production create a thin region of large vertical density gradient in the initially continuously stratified fluid. This region, referred to as the "turbulent entrainment layer", normally lies below a well-mixed layer, the so-called "upper homogeneous layer". Beneath the turbulent entrainment layer, lies a relatively unperturbed region of the fluid in which internal waves and irregular irrotational perturbations may exist. In laboratory test conditions, the intensity of the fluctuations below the turbulent entrainment layer is found rather insignificant and such motions do not appear to contribute to the vertical momentum, heat and energy transfer processes.

When a steady stress acts on the free surface, a layer of considerable velocity shear (of thickness δ) is formed at the top of the mixed layer. If one excepts the very beginning of the entrainment process, the thickness of the shear layer is always much smaller than the depth D of the mixed layer ($\delta \ll D$). Large mean velocity gradients are also observed in the turbulent entrainment layer (Kato and Phillips, 1969 ; Kantha et al, 1977 ; Moore and Long, 1971) and they may extend to the lower part of the mixed layer (Moore and Long, 1971).

At very large values of the Richardson number (based on the variation of density accross the turbulent entrainment layer) a certain amount of heat and momentum transfer in the core of the entrainment layer can be attributed to molecular diffusion (Kantha et al, 1977 ; Crapper and Linden, 1974 ; Wolanski and Brush, 1975 ; Phillips, 1977). However, in cases of well-developed turbulence in the mixed layer, the molecular effects in the turbulent entrainment layer are obviously negligible. (Molecular diffusion can only play a role in the one-centimeter thick layer of water immediately below the surface).

In situ observations show that the thickness h of the turbulent entrainment layer reaches several meters in storm conditions. The ratio $\frac{h}{D}$ is then of the order of 10^{-1}. Detailed measurements made in laboratory test conditions, (Crapper and Linden, 1974 ; Wolanski and Brush, 1975) show that $\frac{h}{D}$ does not depend on the density variation accross the turbulent entrainment layer (provided the density jump is large enough). Beside, it has become evident that with increasing Peclet number (Pe = $\frac{wD}{\lambda}$ where w is the root mean square of the horizontal fluctuating velocity at the upper boundary of the entrainment layer and λ the molecular diffusivity of heat or salt) $\frac{h}{D}$ decreases and tends to a constant value $\sim 1.5 \ 10^{-1}$. Measurements by Moore and Long (1971), in experiments where turbulence was generated by a velocity shear, lead to $\frac{h}{D} \sim 0.8 \ 10^{-1}$. Finally, laboratory experiments by Wolanski and Bush (1975) also showed that $\frac{h}{D} \sim 0(10^{-1})$ and is independent of the Richardson number (Ri = $\frac{g \, \Delta \rho \, D}{\rho_o w^2}$), where g is the acceleration of gravity and $\Delta \rho$ the density difference accross the entrainment layer.

In modelling the deepening process of the upper homogeneous layer, in the ocean as well as in laboratory experiments, one may thus assume[*]

$$\frac{h}{D} \sim 10^{-1} \qquad ; \qquad \frac{\delta}{D} \sim 10^{-1}$$

EQUATIONS DESCRIBING THE EFFECT OF WIND MIXING ON THE DEEPENING OF
THE UPPER HOMOGENEOUS LAYER IN A STRATIFIED FLUID

The basic features of an oceanic wind-mixed layer can be simulated by one-dimensional models, disregarding advection, horizontal diffusion and large scale vertical motions. It will be assumed here, for simplicity, that the water density is a function of temperature only (the introduction of variations of salinity or horizontal non-homogeneity is not a major difficulty). It will be further assumed that the short-wave radiation is absorbed at the sea surface. A simple technique to account for the volume absorption of solar radiation has been described by Kraus and Turner (1967) and Denman (1973). The corrections introduced thereby have been found to be not very significant since the thickness of the effective absorption layer is, on the average, about one order of magnitude smaller than D (Denman,1973).

---

[*]This assumption provides a good approximation in modelling local one-dimensional vertical mixing processes but may not be applicable to the study of the evolution of the seasonal thermocline (Kitaigorodskii and Miropolskii, 1970). The analysis of the whole year development of the temperature field in the active layer of the ocean (200 - 400 m) must take into account the universal temperature profiles below the upper homogeneous layer. These profiles were found first by Kitaigorodskii and Miropolskii (1970) and were confirmed later by numerous observations of the vertical distributions of temperature and salinity in many parts of the ocean (Moore and Long, 1971 ; Miropolskii et al, 1970 ; Nesterov and Kalatskiy, 1975 ; Reshetova and Chalikov, 1977).

With these assumptions, the equations describing the non-steady, one-dimensional vertical heat, momentum and turbulent energy transfers in a stratified rotating fluid can be written

$$\frac{\partial \Theta}{\partial t} = - \frac{\partial s}{\partial z} \qquad (1)$$

$$\frac{\partial \underset{\sim}{u}}{\partial t} = - \frac{\partial \underset{\sim}{\tau}}{\partial z} - f \underset{\sim}{e}_z \wedge \underset{\sim}{u} \qquad (2)$$

$$\frac{\partial e}{\partial t} = - \underset{\sim}{\tau} \cdot \frac{\partial \underset{\sim}{u}}{\partial z} - g \beta s - \varepsilon - \frac{\partial M}{\partial z} \qquad (3)$$

where Θ , $\underset{\sim}{u}$ and e denote respectively the mean temperature, the mean horizontal velocity and the mean turbulent energy and where s, $\underset{\sim}{\tau}$ and M are the corresponding fluxes (normalized with respect to the mean thermal capacity $\rho_o c_p$ and the mean density ρ_o respectively). f is equal to twice the vertical component of the earth's rotation vector, g is the acceleration of gravity, β the thermal expansion coefficient and ε is the rate of turbulent energy dissipation. The frame of reference is sinistrorsum and such that the x-axis is in the direction of the surface wind and the z-axis is vertical pointing downwards.

At the upper boundary of the mixed layer $(z = 0)$, one must prescribe the fluxes. The fluxes depend on the atmospheric conditions and they are normally parameterized in terms of the meteorological data. In general, they are functions of time. However, in the following, the discussions will be restricted to the steady case, for the sake of simplicity.

If ϕ stands for any of the variables Θ , u, v, e, one defines

$$\overline{\phi} = \frac{1}{D} \int_o^D \phi \, dz \qquad ; \qquad \overset{\sim}{\phi} = \frac{1}{h} \int_D^{D+h} \phi \, dz \qquad (4)(5)$$

$$\phi_o = \phi(0,t) \qquad ; \qquad \phi_- = \phi(D,t) \qquad ; \qquad \phi_+ = \phi(D + h,t)$$
$$(6)(7)(8)$$

Obviously, one has

$$\| \overset{\sim}{\underset{\sim}{u}} \| \leq \| \underset{\sim}{u}_- \| \leq \| \overline{\underset{\sim}{u}} \| \qquad ; \qquad \overset{\sim}{\Theta} < \Theta_- \leq \overline{\Theta}$$

$$\tilde{e} \leq \overline{e} \qquad \qquad ; \qquad \tilde{q} \leq \overline{q}$$

Integrating eqs. 1 - 3 over the upper homogeneous layer and the turbulent entrainment layer, one derives a system of equations for the depth-averaged variables $\overline{\phi}$ and $\overset{\sim}{\phi}$.

Combining these equations and neglecting small terms of relative magnitude $\frac{h}{D}$ (in the hypothesis of a "thin interface" $\frac{h}{D} \ll 1$)* , one obtains, after some calculations,

$$\frac{d}{dt} (\overline{\Theta} D) = s_o - s_+ + \frac{dD}{dt} \Theta_+ \tag{9}$$

$$\frac{d}{dt} (\overline{u} D) + f \underset{\sim}{e}_z \Lambda \overline{u} D = \underset{\sim}{\tau}_o - \underset{\sim}{\tau}_+ + \frac{dD}{dt} \overline{u}_+ \tag{10}$$

$$\frac{d}{dt} (\overline{e} D) + g \beta \overline{s} D = M_o + \Pi_D + \Pi_h - \overline{\varepsilon} D - \widetilde{\varepsilon} h - M_+ + \frac{dD}{dt} e_+ \tag{11}$$

where

$$\Pi_D = - \int_o^D \underset{\sim}{\tau} \cdot \frac{\partial u}{\partial z} \, dz \tag{12}$$

$$\Pi_h = - \int_D^{D+h} \underset{\sim}{\tau} \cdot \frac{\partial u}{\partial z} \, dz \tag{13}$$

The calculation of Π_D can be most easily done with the assumption that the velocity shear in the upper homogeneous layer is concentrated in the constant stress layer δ . Then

$$\Pi_D \sim \Pi_\delta = - \int_o^\delta \underset{\sim}{\tau} \cdot \frac{\partial u}{\partial z} \, dz \sim \underset{\sim}{\tau}_o \cdot (\underset{\sim}{u}_o - \underset{\sim}{u}_\delta) \tag{14}$$

where $\underset{\sim}{u}_\delta$ is the velocity at the lower boundary of the constant stress layer of thickness δ .

From eq.(2) and its scalar product by $\underset{\sim}{u}$, one gets, after some rearrangement and neglecting small terms involving h

$$\underset{\sim}{\tau}_- = \underset{\sim}{\tau}_+ + \frac{dD}{dt} (\underset{\sim}{u}_- - \underset{\sim}{u}_+) \tag{15}$$

$$\Pi_h \sim \frac{1}{2} \frac{dD}{dt} \| \underset{\sim}{u}_- - \underset{\sim}{u}_+ \|^2 + \underset{\sim}{\tau}_+ \cdot (\underset{\sim}{u}_- - \underset{\sim}{u}_+) \tag{16}$$

It can be shown that the turbulent energy production in the upper homogeneous layer and in the turbulent entrainment layer is not very sensitive to the detailed velocity distribution in the main part of the upper homogeneous layer.. In a first approach, it seems thus reasonable to make the so-called "slab model approximation" where the vertical velocity distribution is assumed homogeneous for $\delta \le z \le D$ so that

---

*Even, in the hypothesis $\frac{h}{D} \ll 1$, such simplification is difficult to justify because the remaining terms can partially cancel each other and sum up to be comparatively small. It must be regarded as a first approximation liable to revision. The term $\widetilde{\varepsilon} h$ is retained in the absence of a clear-cut evaluation of the respective orders of magnitude of $\overline{\varepsilon}$ and $\widetilde{\varepsilon}$.

$$\underset{\sim}{u}_- = \overline{\underset{\sim}{u}} = \underset{\sim}{u}_\delta \tag{17}$$

In this particular case, one can write

$$\Pi_D = \Pi_\delta = \underset{\sim}{\tau}_o \cdot (\underset{\sim}{u}_o - \overline{\underset{\sim}{u}}) = \tau_o (u_o - \overline{u}) \tag{18}$$

$$\Pi_h = \frac{1}{2} \frac{dD}{dt} \| \overline{\underset{\sim}{u}} - \underset{\sim}{u}_+ \|^2 + \underset{\sim}{\tau}_+ \cdot (\overline{\underset{\sim}{u}} - \underset{\sim}{u}_+) \tag{19}$$

$$\Pi_{D-\delta} = 0 \tag{20}$$

Velocity shear layers are thus taken into account as velocity jumps $(\underset{\sim}{u}_o - \overline{\underset{\sim}{u}})$ and $(\overline{\underset{\sim}{u}} - \underset{\sim}{u}_+)$ in thin layers of thickness $\delta \ll D$ and $h \ll D$, respectively.

There is some experimental evidence that one can assume

$$u_o - \overline{u} \sim \alpha \, \tau_o^{1/2} \tag{21}$$

where α is an empirical constant.

Then, if one sets

$$G_\delta = M_o + \Pi_\delta = M_o + \alpha \, \tau_o^{3/2} \tag{22}$$

and restrict attention to the case of steady fluxes at the air-sea interface, G_δ does not depend on time and may be used successfully as one of the external characteristic parameters ($G_\delta \, \delta^{1/3}$ is the velocity scale) of turbulence in the wind mixed layer of the ocean.

Along the same line, one may assume that the temperature is uniform in the upper homogeneous layer. Then, integrating eq.(1) in the mixed layer and over the turbulent entrainment layer, one gets

$$s(z) = s_o - \frac{s_o - s_-}{D} z \tag{23}$$

$$s_- = s_+ + \frac{dD}{dt} (\overline{\theta} - \theta_+) \tag{24}$$

Hence

$$g \, \beta \, \overline{s} \, D = \frac{1}{2} \left[g \, \beta \, s_o D + g \, \beta \, s_+ D + g \, \beta \, \frac{dD}{dt} (\overline{\theta} - \theta_+) D \right] \tag{25}$$

The time scale of turbulent energy dissipation $\frac{\overline{e}}{\varepsilon}$ does not exceed a few minutes whereas the deepening of the mixed layer has a characteristic time of several hours. One may thus regard the turbulence as adapting itself instantaneously to the modifications of the mixed layer and following the "slow" evolution of the latter. At the scale of turbulence, this slow evolution is not noticeable and the turbulent energy may be regarded as quasi steady, i.e.

$$\frac{\partial}{\partial t} (\overline{e} \, D) \sim \overline{e} \, \frac{dD}{dt} \tag{26}$$

In this case, one can usually assume

$$\bar{e} = c_o \tau_o \tag{27}$$

where c_o is a constant of order unity ($c_o \sim 2 - 3$, in the atmospheric boundary layer) or, more generally

$$\bar{e} = c_\delta G_\delta^{2/3} \tag{28}$$

which is valid also in the case of turbulence generated by a turbulent energy flux M_o in the absence of momentum flux.

The values of the variables at the lower boundary of the turbulent entrainment layer depend on the characteristics of the layer of fluid below. These can be affected by different factors : molecular diffusion (Mellor and Durbin, 1975), internal waves generated inside the turbulent entrainment layer, but also below it, by turbulent disturbances at the bottom of the mixed layer and radiating momentum and energy away, turbulent diffusion caused by the instability and breaking of the internal waves or other mechanisms not directly related with the local wind mixing process (Garnich, 1975).

The effect of both molecular and turbulent diffusion below the turbulent entrainment layer is not perceivable at the time scale of the mixed layer deepening (of the order of a day) and it may accordingly be neglected.

The contribution, to the energy balance, of internal waves excited by the upper turbulent layer is difficult to evaluate (Linden, 1975 ; Thorpe, 1973 ; Kantha, 1977).

In the absence of vertical density stratification below the turbulent entrainment layer $\frac{\partial \Theta}{\partial z} = 0$ for $z \geq D + h$) , the fluid there may be regarded as being at rest. The internal waves generated by the upper turbulent layer do not propagate downwards and, in the non-viscous approximation which is actually made here, the fluid motion reduces to weak irrotational fluctuations which are rapidly attenuated as they move deeper from the bottom of the upper homogeneous layer and which contribute very little to e_+ and not at all to τ_+ , s_+ and M_+ .

In that case, one may take

$$e_+ = u_+ = \tau_+ = M_+ = s_+ = 0 \tag{29}$$

This assumption is not quite valid when the fluid below the turbulent entrainment layer is stratified but, in a first approximation, when the density variation accross the turbulent entrainment layer is large, perturbations below it can be regarded as insignificant and eq. (29) can be used.

The temperature at the lower boundary of the turbulent entrainment layer can be written quite generally

$$\theta_+ = \theta_{oo} - \gamma D \tag{30}$$

where θ_{oo} is a constant and γ the temperature gradient below the turbulent entrainment layer.

With the approximations described above (18, 19, 21, 22, 24, 25, 26, 28, 29, 30), the basic equations (9)-(11) can be written

$$\frac{d}{dt} \left(\Delta\theta D - \frac{1}{2} \gamma D^2 \right) = s_o \tag{31}$$

$$\frac{d}{dt} (D\bar{u}) - f \bar{v} D = \tau_o \tag{32}$$

$$\frac{d}{dt} (D\bar{v}) + f \bar{u} D = 0 \tag{33}$$

$$\frac{dD}{dt} \left[\frac{g\beta\Delta\theta D}{2} + c_\delta G_\delta^{2/3} \right] = G_\delta + \frac{1}{2} \frac{dD}{dt} \left\| \bar{\underset{\sim}{u}} \right\|^2 - (\bar{\varepsilon} D + \tilde{\varepsilon} h) - \frac{Dg\beta s_o}{2} \tag{34}$$

where

$$\Delta\theta = \bar{\theta} - \theta_+ \tag{35}$$

Similarly the entrainment fluxes reduce to[*]

$$\tau_- = \frac{dD}{dt} \underset{\sim}{u}_- \tag{36}$$

$$s_- = \frac{dD}{dt} \Delta\theta \tag{37}$$

$$M_- = \tilde{\varepsilon} h + \frac{dD}{dt} e_- - \Pi_h \tag{38}$$

$$= \tilde{\varepsilon} h + \frac{dD}{dt} \left(e_- - \frac{1}{2} \left\| \underset{\sim}{u} \right\|^2 \right)$$

---

[*] Attention is restricted here to the case $\frac{dD}{dt} > 0$. As a result of eq. (29), eqs. (36)-(38) cannot be used when $dD/dt < 0$ ("reverse entrainment"). As suggested by Krauss and Turner (1967), one should write

$$s_- = 0 \quad , \quad \tau_- = 0 \quad \text{for} \quad \frac{dD}{dt} \leq 0 \tag{39},(40)$$

When one considers the upper homogeneous layer throughout a long period when strong storms may give way to perfectly still weather, it is important to describe the reverse entrainment process (Garnich, 1975 ; Kosnizev et al, 1976).

It is important to note that, for $h \rightarrow 0$, $\tilde{\varepsilon}h$ has a finite but small value. Then when the turbulent kinetic energy e_- is smaller than the mean flow kinetic energy $\frac{1}{2} \| u \|^2$, the flux of turbulent energy can be directed from the turbulent entrainment layer into the upper homogeneous layer, contributing to the turbulent mixing in that layer. This is confirmed by experiments in the laboratory (Kato and Phillips, 1969 ; Kantha et al, 1977).

Eqs.(31)-(33) can be integrated easily since s_o and t_o are known functions of time. In particular, when they are constant, one gets

$$\Delta\Theta D - \frac{1}{2} \gamma D^2 = \Delta\Theta(o)D(o) - \frac{1}{2} \gamma D^2(o) + s_o t \tag{41}$$

$$D\bar{u} = \left(\frac{\tau_o}{f} + D(o)\bar{v}(o)\right) \sin ft + D(o)\bar{u}(o) \cos ft \tag{42}$$

$$D\bar{v} = - D(o)\bar{u}(o) \sin ft + \left(\frac{\tau_o}{f} + D(o)\bar{v}(o)\right) \cos ft - \frac{\tau_o}{f} \tag{43}$$

where
$D(o)$, $\bar{u}(o)$, $\bar{v}(o)$ are the initial values.

To close the system of equations (31)-(34), one needs an estimate of the integrated viscous dissipation in the turbulent energy balance equation. A question arises : what portion of G_δ and Π_h is used for mixing and what portion dissipates into heat ? This is probably the most important problem in one-dimensional modelling of mixed layer deepening and the models can be classified according to their particular parameterization of the integral energy dissipation.

Let[*]

$$\bar{\varepsilon}D + \tilde{\varepsilon}h = (1 - \Phi_1)G_\delta + (1 - \Phi_2)\Pi_h \tag{44}$$

where Φ_1 and Φ_2 are non-dimensional functions depending on the external parameters of the problem. Their dependence will be examined later, using similarity theory together with experimental data obtained in well controlled external conditions.

However, by simple inspection of eq.(34) and (44), one can get some information about the functions Φ_1 and Φ_2 .

If the deepening of the mixed layer is produced only by the diffusion of turbulent energy down from the surface into the stratified fluid $(M_o \neq 0$, $\tau_o = 0$, $\Pi_h = 0)$, then obviously

[*] Attention is restricted here to $s_o \geq 0$. When the ocean surface is cooling $\Pi_q = \frac{g\beta s_o D}{2} < 0$ acts as a complementary source of energy and (44) must generalized to

$$\bar{\varepsilon}D + \tilde{\varepsilon}h = (1 - \phi_1)G_\delta + (1 - \phi_2)\Pi_h + (1 - \phi_3)|\Pi_q| \tag{45}$$

$$0 \leq \Phi_1 \leq 1 \tag{46}$$

On the other hand, in the case of shear generated turbulence $(M_o \ll \Pi_D , \Pi_h ; G_\delta \sim \Pi_D = \Pi_\delta)$ deepening of the upper homogeneous layer is possible only if

$$(1 - \Phi_1)G_\delta + (1 - \Phi_2)\Pi_h < G_\delta + \Pi_h - \Pi_q \tag{47}$$

where

$$\Pi_q = \frac{g \beta s_o D}{2} \geq 0 \tag{48}$$

Hence

$$\frac{\Phi_1}{\Phi_2} > \frac{\Pi_q}{G_\delta \Phi_2} - \frac{\Pi_h}{G_\delta} \tag{49}$$

If one considers, to begin with, a situation where the effects of rotation and surface heating can be neglected, in the asymptotic case $\frac{h}{D} \to 0$, there is one non-dimensional number characterizing the problem, the overall Richardson number defined as

$$R_{iG} = \frac{g \beta \Delta \Theta D}{G_\delta^{2/3}} \tag{50}$$

The functions Φ_1 and Φ_2 should then be functions of R_{iG} only.

In the following, one shall write for simplicity u, v, ... *instead of* $\bar{u}$, $\bar{v}$, ...

FUNDAMENTAL RESULTS OF LABORATORY EXPERIMENTS

In laboratory studies of the entrainment process in a stratified fluid, two essentially different mechanisms were used to generate the turbulence.

In the first type of experiments, (e.g. Turner, 1973 ; Thorpe, 1973 ; Long, 1974) turbulence is produced by an oscillating grid placed at a given depth Z_1 below the free surface in the tank. The depth of the mixed layer is

$$D = Z_1 + Z_D \tag{51}$$

where Z_D is the distance from the grid to the turbulent entrainment layer.

The most detailed studies of entrainment in this case have been performed in a fluid consisting of two homogeneous layers of different density. The results have been interpreted, as a rule, using a relationship of the form

$$\frac{1}{\omega_o} \frac{dD}{dt} = C \left(\frac{\omega_o^2 \rho_o}{g \, \Delta \, \rho}\right)^n \qquad (52)$$

where C is some appropriate dimensional factor independent of the grid oscillation frequency ω_o and of the density jump $\Delta \rho$ accross the turbulent entrainment layer (ρ_o is the mean density in the mixed layer).

The exponent n is a function of the Peclet number which tends asymptotically to the value 1.5 for large Peclet numbers (Turner, 1973 ; Crapper and Linden, 1974 ; Wolanski, 1972).

Hence, in the asymptotic case

$$\frac{1}{\omega_o} \frac{dD}{dt} = C \frac{\omega_o^3}{\left(g \frac{\Delta\rho}{\rho_o}\right)^{3/2}} \qquad (53)$$

Long (1974), using experimental values of ω_o and $g \frac{\Delta\rho}{\rho_o}$, suggested, on the basis of dimensional arguments, the following form for C

$$C = z_o^{5/2} \, \Psi\left(\frac{z_o}{D}, \frac{z_1}{D}\right) \qquad (54)$$

where z_o is the amplitude of the grid oscillations kept constant during the experiment, Ψ is some non-dimensional function tending to a constant value $\Psi(o,o)$ for $\frac{z_o}{D} \to 0$ and $\frac{z_1}{D} \to 0$.

The flux of turbulent energy must be proportional to the third power of ω_o and z_o, i.e.

$$M_o = C_1 (\omega_o z_o)^3 \qquad (55)$$

where C_1 is a non-dimensional constant[*].

Combining (51), (53), (54) and (55), one obtains

$$\frac{1}{M_o^{1/3}} \frac{dD}{dt} = C_2 \, R_{iM}^{-3/2} \qquad (56)$$

where C_2 is another non-dimensional constant defined by

$$C_2 = \frac{\Psi}{C_1^{4/3}} \left(\frac{z_1 + z_o}{z_o}\right)^{3/2} \qquad (57)$$

and where the overall Richardson number R_{iM} is defined by

$$R_{iM} = \frac{g\beta\Delta\theta D}{M_o^{2/3}} \qquad (58)$$

The results can also be presented using a "turbulent" Richardson number (Turner, 1973)

---

[*]At least, C_1 is a constant for each particular geometry of the grid.

$$R_{iT} = \frac{g\beta\Delta\theta\ell}{w^2} \tag{59}$$

where ℓ is the turbulent integral length scale in the vicinity of the upper boundary of the turbulent entrainment layer and w the root mean square turbulent velocity, as before.

The variations of the non-dimensional entrainment velocity $\tilde{w}_e = \frac{1}{w}\frac{dD}{dt}$ plotted as a function of the Richardson number R_{iT} according to Turner's experimental data (Turner, 1968, 1973) show that, for $R_{iT} \geq 5$, one can write with a good approximation

$$\tilde{w}_e = \frac{1}{w}\frac{dD}{dt} = C_3\, R_{iT}^{-3/2} \tag{60}$$

where $C_3 \sim 2$.

In all Turner's experiments however (Turner, 1968, 1973 ; Thompson and Turner, 1975), Z_1, Z_D, Z_o were fixed $(Z_1 = Z_D = 9$ cm ; $Z_o = 1$ cm). Computing C_3 from the measurements made by Linden (1975),$(Z_1 = 6$ cm, $Z_D = 7$ cm), one finds $C_3 \sim 0.7, 0.6$.

On the other hand, if the results are interpreted in terms of R_{iM}, both sets of experiments give similar values for the constant $C_2 C_1^{4/3}$ ($\sim 1.85\ 10^{-3}$) and this is an argument in favor of formula (56) where C_2 may be regarded as a universal constant.

Linden (1975) extended its study to the case where there is a constant density gradient below the turbulent entrainment layer and he found an appreciable decrease in the entrainment rate which he interpreted by the loss of energy due to radiating internal waves.

Wolanski and Brush (1975) presented experimental evidence that the entrainment rate depends mainly on the overall Richardson number R_{iM} and that the influence of the non-dimensional time $(g\beta\gamma)^{1/2}t$ is negligible. Since the Brunt-Väisälä frequency $N = (g\beta\gamma)^{1/2}$ in the fluid below the turbulent entrainment layer is one of the main factors determining internal waves radiation, one can conclude that, in this particular experiment, this effect was very small[*].

One of the most important mechanisms of turbulence generation by wind mixing is related to the presence of a mean drift current.

---

[*] Wolanski and Brush (1975) also showed that the entrainment rate depended on the Peclet number. This was later explained by Phillips (1977) who determined that molecular transfer processes play a role in the ocean when $R_{i*} = \frac{g\Delta\rho D}{\rho_o \tau_o} \geq 10^8$ if $\Delta\rho$ is due to a salinity gradient and when $R_{i*} \geq 10^7$ if $\Delta\rho$ is due to a temperature gradient whereas in laboratory experiments (Kato and Phillips, 1969 ; Kantha et al, 1977) it occurs for $R_{i*} \geq 10^3$.

Laboratory experiments in which shear generated turbulence is responsible for entrainment are thus very important and, in this respect,
the work of Kato and Phillips can be considered as a classical one.
To avoid the influence of the limited length of the channel and the
generation of secondary flows, a circular tank was designed where,
at the fluid surface, a constant stress was applied by a rotating
screen. In the top layer, the fluid velocity $u_o(t)$ was set such
that τ_o remains essentially constant in the course of the experiment.

In the case of a linear continuous density stratification, Kato
and Phillips (1969) found

$$\frac{u_e}{u_*} \equiv \frac{1}{\tau_o^{1/2}} \frac{dD}{dt} = 2.5 \ R_{i*}^{-1} \tag{61}$$

where $\quad R_{i*}' = \dfrac{g\beta\Delta\theta D}{\tau_o}$ \hfill (62)

In the case of two superposed layers of fluid, the dependence of
$\frac{u_e}{u_*}$ on R_{i*} was more variable but, for a large range of Richardson
numbers R_{i*} $(R_{i*} \leq 2 \ 10^2)$, one can write a similar formula

$$\frac{u_e}{u_*} = 8.2 \ R_{i*}^{-1} \tag{63}$$

The main difference between eq.(61) and (63) was attributed by
Phillips (1977) to internal waves energy radiation and by Garnich and
Kitaigorodskii (1976, 1977) to the different turbulent energy produc-
tion in the entrainment layer.

If the initial density profile is linear, eq.(61) yields

$$D = \tau_o^{1/2} \left(\frac{15t}{N}\right)^{1/3} \tag{64}$$

The laboratory experiments discussed above have been performed without heat flux accross the sea surface $(s_o = 0)$ and without rotation $(f = 0)$. In this case, eqs.(31)-(34) give, using (44) and zero
initial conditions for the velocity

$$\Delta\theta D - \frac{1}{2} \gamma D^2 = \Delta\theta(o)D(o) - \frac{1}{2} \gamma D^2(o) \tag{65}$$

$$Du = \tau_o t \tag{66}$$

$$\frac{dD}{dt} \left(\frac{g\beta\Delta\theta D}{2} + C_\delta G_\delta^{2/3} - \phi_2 \frac{u^2}{2}\right) = \phi_1 G_\delta \tag{67}$$

Krauss and Turner (1967), restricting attention to the particular
case

$$C_\delta = 0 \quad ; \quad \Phi_2 = 0 \quad ; \quad \Phi_1 = \frac{\tau_o^{3/2}}{G_\delta} \quad ; \quad D(o) = 0$$

derived the formula

$$D(t) = \tau_o^{1/2} \left(\frac{12t}{N}\right)^{1/3} \tag{68}$$

This result seems in very close agreement with (64) but one may argue that this is nothing but a fortuitous coïncidence. Indeed, in the experiments by Kato and Phillips (1969), the fundamental role in the entrainment process was played by the turbulent energy production in the turbulent entrainment layer and not by the term $\Phi_1 G_\delta$.

In fact, with the hypotheses of Krauss and Turner one gets, in the case of two superposed layers

$$\frac{1}{\tau_o^{1/2}} \frac{dD}{dt} = 2 \; R_{i*}^{-1} \tag{69}$$

which should be compared to eq.(63).

The hypothesis that the term $\Phi_1 G_\delta = \tau_o^{3/2}$ plays the main role in the process of mixed layer deepening in the presence of a velocity shear cannot therefore be considered as sustained by laboratory experiments.

An alternative approach was proposed by Pollard et al (1973) who assumed $\Phi_1 = 0$, $C_\delta = 0$, $\Phi_2 = 1$. In this case, when $D(o) = 0$, eqs.(65)-(67) yield

$$D(t) = 2^{1/4} \tau_o \; N^{-1/2} t^{1/2} \tag{70}$$

which differs from Kato and Phillips' result (eq.64).

The authors argued that eq.(70) was in good agreement with experimental results but their attempt is not very convincing because the final choice between (64) and (70) lies on the verification of the dependence of D , not only on time t but also on N and this is a problem which is still to be solved. Moreover, at the beginning of the deepening process (t $\sim$ 10 - 60 sec), experimental results (Kato and Phillips, 1969) lie far away from the curve represented by eq.(70). For a fluid of two superposed layers, the assumptions of Pollard et al (1973) lead to the following formula for the drag coefficient

$$C_f = \left(\frac{u_*}{u}\right)^2 = \frac{\tau_o}{u^2} = R_{i*}^{-1} \tag{71}$$

which is often very different from the observed values.

It is clear that the most important similarity criterium in the analysis of the mixed layer deepening in the ocean is the bulk

Richardson number (R_{iG} *or* R_{i*} *or* R_{iM}). *In spite of enormous dif-*
ferences between in situ and laboratory experiments, the Richardson
number simularity holds and the theory of the oceanic mixed layer can
and must rely on the experimental data obtained by Kato and Phillips
(1969) and Kantha et al (1977).

THEORETICAL MODELLING OF THE DEEPENING OF THE UPPER HOMOGENEOUS
LAYER BY WIND MIXING

(I) Two superposed homogeneous layers of different densities, no rotation, no heat flux and steady momentum and turbulent energy fluxes
accross the air-sea interface.

In this case, setting $\gamma = 0$ in eq.(65), one has

$$g\beta\Delta\theta D = g\beta\Delta\theta(o)D(o) = \text{const.} \tag{72}$$

and the overall Richardson number R_{iG} , and thus Φ_1 and Φ_2 ,
are independent of time.

When $\gamma = 0$, it is reasonable to neglect momentum and energy
losses due to internal waves radiation. Then, the functions Φ_1 and
Φ_2 can be determined as follows (Garnich and Kitaigorodskii, 1977).

Using $G_\delta^{1/3}$ as a velocity scale and $D(o)$ as a length scale,
one can write eq.(67) in the non-dimensional form :

$$\frac{d\xi}{d\eta}\left(\frac{1}{2}R_{iG} + C_\delta - C_* \frac{\eta^2}{2\xi^2}\Phi_2\right) = \Phi_1 \tag{73}$$

where

$$\xi = \frac{D}{D(o)} \quad ; \quad \eta = \frac{G_\delta^{1/3}t}{D(o)} \quad ; \quad C_* = \frac{\tau_o^2}{G_\delta^{4/3}} \tag{74)(75)(76}$$

The solution of eq.(73) satisfying the condition $\xi = 1$ at $\eta = 0$
is readily found to be

$$\frac{\xi}{\eta} = \frac{\Phi_1(1 + \mu)}{R_{iG} + 2C_\delta}\Delta(\mu_1\xi) \tag{77}$$

where

$$\mu = \left(\frac{(R_{iG} + 2C_\delta)C_*\Phi_2}{\Phi_1^2} + 1\right)^{1/2} \tag{78}$$

$$\Delta(\mu_1\xi) = \frac{\xi^\mu - \frac{1 - \mu}{\mu + 1}}{\xi^\mu - 1} \tag{79}$$

and

$$\tilde{u}_e = G_\delta^{-1/3} \frac{dD}{dt} = \frac{d\xi}{d\eta} = \frac{\Phi_1}{R_{iG} + 2C_\delta} B(\xi,\mu) \qquad (80)$$

where

$$B(\xi,\mu) = \frac{\left[(\mu + 1)\xi^\mu + \mu - 1\right]^2}{(\mu + 1)\xi^{2\mu} + 2(\mu^2 - 1)\xi^\mu + 1 - \mu} \qquad (81)$$

The limiting value

$$\tilde{u}_e^{(1)} = \lim_{\xi \to 1} \frac{d\xi}{d\eta} = \frac{2\Phi_1}{R_{iG} + 2C_*} \qquad (82)$$

is independent of Φ_2 .

The same result is obtained directly in the model where deepening of the mixed layer is determined by turbulent energy diffusion from the upper portion of the mixed layer, turbulence being generated by a vertically oscillating grid. (In this case $\tau_o = 0$, $C_* = 0$, $G_\delta = M_o$, $R_{iG} = R_{iM}$) .

For sufficiently large values of R_{iG} ($R_{iG} \gg 2C_\delta$) , combining (56) and (82), one gets

$$\Phi_1 = \frac{\dot{C}_2}{2 R_{iG}^{1/2}} \qquad (83)$$

For small values of R_{iG} , the entrainment velocity must tend to a constant independent of R_{iG} .

The simplest interpolation formula between these two asymptotic cases is

$$\Phi_1 = \frac{\alpha_1}{(1 + \alpha_2 R_{iG})^{1/2}} \qquad (84)$$

where α_1 and α_2 are two non-dimensional universal constants.

To determine Φ_2 , one examines the final period of the entrainment process ($\xi \gg 1$) and assumes that, in the same time, $\mu \gg 1$. Then, from eqs. (78) and (80), one gets

$$\frac{d\xi}{d\eta} = \tilde{u}_e^{(2)} = \left(\frac{C_* \Phi_2}{R_{iG} + 2C_\delta}\right)^{1/2} \qquad (85)$$

This formula obviously applies in the case of two superposed layers with velocity shear when turbulent energy production in the turbulent entrainment layer is mainly responsible for the entrainment process as in the laboratory experiment of Kantha et al (1977). In this case, for sufficiently large Richardson number, the non-dimensional entrainment velocity is proportional to R_{i*}^{-1} (eq.63) and thus to $R_{iG}^{-1\ *}$.

Thus, for $R_{iG} \gg 2\ C_\delta$

$$\Phi_2 \sim \frac{n_2}{R_{iG}} \tag{86}$$

where $n_2 \sim \dfrac{64}{\alpha^{2/3}}$.

In the general case, when the contribution of M_o to G_δ must be taken into account, the dependence of n_2 on α and C_* must be considered. The value of C_* must then be estimated from the momentum and energy transfers from wind to drift currents and waves (Garnich and Kitaigorodskii, 1977).

On the model of eq.(84), one can write a more general relation for Φ_2 which is valid for a wide range of Richardson number R_{iG} , i.e.

$$\Phi_2 = \frac{\alpha_3}{1 + \alpha_4\ R_{iG}} \tag{87}$$

where α_3 and α_4 are non-dimensional constants.

The values of α_1, α_2, α_3 and α_4 can be determined by a careful inspection of the experimental data (Turner, 1973 ; Kantha et al, 1977).

The empirical expressions

$$\alpha_1 \sim 1.24\ C_\delta^{3/2} \ ; \ \alpha_2 \sim 0.5\ C_\delta^{-1} \ ; \ \alpha_3 \sim 0.9 \ ; \ \alpha_4 \sim 1.3\ 10^{-2} C_*^{-1/2} \tag{88}$$

are found in very good agreement with the laboratory data[**] (Garnich and Kitaigorodskii, 1977, 1978 ; Kitaigorodskii, 1977).

In the experiments by Kantha et al (1977), the essential contribution to G_δ is Π_δ . Thus, from eqs.(22) and (76)

$$C_* = \frac{\tau_o^2}{G_\delta^{4/3}} \sim \frac{\tau_o^2}{\Pi_\delta^{4/3}} = \alpha^{-4/3} \quad (\alpha \sim 8 - 20) \tag{89}$$

To estimate the relative importance of M_o due to wave breaking in real situations, the value of C_* must be varied in numerical computations such that $C_* \lesssim \alpha^{-4/3}$.

---

[*] $M_o \sim u_*^3 \ll \Pi_\delta = \alpha\ u_*^3$ $(\alpha \sim 10)$. Hence $G_\delta \sim \alpha\ u_*^3$ and $R_{iG} = R_{i*}\ \alpha^{-2/3}$.

[**] Knowing the α_i 's , it is possible to verify the hypothesis $\mu \gg 1$. One finds that it is satisfied over most of the range of interesting Richardson numbers.

Because of the similarity between laboratory experiments and in situ conditions, one can argue that the empirical expressions (88) can be applied to the analysis of wind mixing in the ocean. They will be used in the following.

THEORETICAL MODELLING OF THE UPPER HOMOGENEOUS LAYER BY WIND MIXING

(II) <u>Continuously stratified fluid, no rotation, no heat flux and steady momentum and turbulent energy fluxes accross the air-sea interface</u>

With a constant temperature gradient below the turbulent entrainment layer, eq.(65) shows that the Richardson number $R_{iG} = \dfrac{g\beta\Delta\Theta D}{G_\delta^{2/3}}$ varies with time as follows

$$R_{iG} = R_{iG}(o) + \frac{g\beta\gamma D^2(o)}{2\ G_\delta^{2/3}}\ (\xi^2 - 1) \tag{90}$$

The difference $R_{iG}(o) - \dfrac{g\beta\gamma D^2(o)}{2\ G_\delta^{2/3}}$ characterizes the deviation of the initial buoyancy jump $\dfrac{g\Delta\rho(o)}{\rho_o}$ from the value $\dfrac{N^2 D^2(o)}{2}$ which can be interpreted as the result of mixing a fluid with a constant density gradient down to a depth $D(o)$. One shall assume here

$$R_{iG}(o) \sim \frac{g\beta\gamma D^2(o)}{2\ G_\delta^{2/3}} \tag{91}$$

Thus $\quad R_{iG} = R_{iG}(o)\ \xi^2 \tag{92}$

and, from eq.(73),

$$\frac{d\eta}{d\xi} = \frac{1}{\lambda_1}\ (\xi^2 + \lambda_3 - \lambda_2\ \frac{\eta^2}{\xi^2}) \tag{93}$$

where

$$\lambda_1 = \frac{2\Phi_1}{R_{iG}(o)}\quad ;\quad \lambda_2 = \frac{C_* \Phi_2}{R_{iG}(o)}\quad ;\quad \lambda_3 = \frac{2C_\delta}{R_{iG}(o)} \tag{94}$$

It is illuminating to consider two asymptotic cases of eqs.(84) and (87), i.e.

(i) Φ_1 and Φ_2 constant $\tag{95}$

(ii) $\Phi_1 \sim \dfrac{\alpha_1}{\alpha_2^{1/2} R_{iG}^{1/2}}\quad ;\quad \Phi_2 \sim \dfrac{\alpha_3}{\alpha_4 R_{iG}} \tag{96}$

which correspond respectively to small values of the Richardson number in the initial stage of development of the mixed layer in a continuously stratified fluid and to large values of the Richardson number in the final stage of the deepening process.

(i) In the first case, λ_1, λ_2 and λ_3 are constants. Changing variables from η to $f(\xi)$ defined by

$$\eta = \frac{\lambda_1}{\lambda_2} \xi^2 \frac{f'(\xi)}{f(\xi)} \qquad (97)$$

one gets the linear equation

$$f'' + \frac{2}{\xi} f' - \frac{\lambda_2}{\lambda_1^2} (1 + \frac{\lambda_3}{\xi^2}) f = 0 \qquad (98)$$

The solution of eq.(98) which satisfies the initial condition $\eta = 0$ at $\xi = 1$ is readily found to be

$$\eta = \frac{\lambda_1}{\lambda_2} \xi^2 \left[\frac{\nu - 1}{\xi} + a \frac{A I_{\nu+1}(a\xi) - K_{\nu+1}(a\xi)}{A I_\nu(a\xi) + K_\nu(a\xi)} \right] \qquad (99)$$

where

$$\nu = \frac{1}{2} (1 + 4 \frac{\lambda_3 \lambda_2}{\lambda_1^2}) \quad ; \quad a = \frac{\lambda_2^{1/2}}{\lambda_1} \qquad (100)(101)$$

$$A = \frac{K_{\nu+1}(a) - \frac{\nu-1}{a} K_\nu(a)}{I_{\nu+1}(a) + \frac{\nu-1}{a} I_\nu(a)} \qquad (102)$$

One notes that for small values of $\frac{\lambda_2}{\lambda_1^2}$, corresponding to the case of "purely diffusive" entrainment considered by Krauss and Turner (1967), one has

$$a \ll 1 \quad , \quad a(\xi - 1) \ll 1 \quad \nu - \frac{1}{2} \ll 1$$

and

$$\xi \sim (\lambda_1 \eta)^{1/3} \qquad (103)$$

On the other hand, for large values of $\frac{\lambda_2}{\lambda_1^2}$ such that $a\xi \gg 1$, one obtains using the asymptotic expansions of I_ν and K_ν

$$\xi \sim (\lambda_2 \eta)^{1/2} \qquad (104)$$

This case corresponds to the model of Pollard et al (1973) where the main mixing source is the turbulent energy production in the turbulent entrainment layer (Π_h) .

(ii) In the second case, λ_1 and λ_2 can be written

$$\lambda_1 = \frac{\tilde{\lambda}_1}{\xi} \quad ; \quad \lambda_2 = \frac{\tilde{\lambda}_2}{\xi^2} \qquad (105)(106)$$

where $\tilde{\lambda}_1$ and $\tilde{\lambda}_2$ are constants.

Substituting in eq.(93), one gets

$$\frac{d\eta}{d\xi} = \frac{\xi}{\tilde{\lambda}_1} (\xi^2 + \lambda_3 - \tilde{\lambda}_2 \, \eta^2/\xi^4) \tag{107}$$

The change of variables

$$\eta = \frac{\tilde{\lambda}_1}{\tilde{\lambda}_2} \xi^3 \frac{f'(\xi)}{f(\xi)} \tag{108}$$

leads again to a linear differential equation of which the solution has the form (99) with a defined as before and

$$\nu = \left\{ 1 + \frac{\tilde{\lambda}_2 \lambda_3}{\tilde{\lambda}_1^2} \right\}^{1/2} \tag{109}$$

For $a\xi \gg 1$, the solution takes the asymptotic form

$$\xi = (\tilde{\lambda}_2^{1/2} \eta)^{1/3} \tag{110}$$

and

$$\frac{d\xi}{d\eta} = \frac{1}{3} \frac{\tilde{\lambda}_2^{1/2}}{\xi^2} = \frac{1}{3} \left(\frac{c_* \alpha_3}{\alpha_4} \right)^{1/2} R_{iG}^{-1} \tag{111}$$

i.e. the non-dimensional entrainment velocity is proportional to the inverse of the overall Richardson number in agreement with the observations in the laboratory (cf.eq.(61)).

In the general case, eq.(73) can be solved numerically, using eqs. (84) and (87) for Φ_1 and Φ_2 . The results can be compared with the actual measurements made by Kato and Phillips (1969).

Fig.1 shows the depth D of the mixed layer in a linearly stratified fluid as a function of time. The dots represent the observed values in two typical situations differing by the magnitudes of the density gradient and of the surface stress. The continuous curves are the predictions of the model in the same conditions. Although the trend is well reproduced, the theoretical results seem to give systematically larger values of the depth D at the beginning of the entrainment process. This may indicate that the turbulent diffusion part of the entrainment (described by the function Φ_1) is overestimated.

The calibration of the model (the determination of the α_i's) was partly made with the results of laboratory experiments where there was no stratification below the turbulent entrainment layer, i.e.

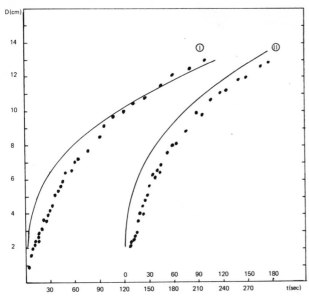

Fig. 1. Depth of the mixed layer in a stratified fluid as a function of time. The dots represent the observed values in two typical situations. The plain curves are drawn from the predictions of the model.

$\gamma = 0$. (Kantha et al, 1977). The fact that the model remains applicable in the case $\gamma \neq 0$ is essentially due to the similar set-up of the two types of experiments (Kato and Phillips, 1969 ; Kantha et al, 1977) where, in particular, the relationship between the surface stress and the drift current was the same. In other words, the experimental values of C_* were quite similar ($C_* \sim \alpha^{-4/3} \sim 0.06 - 0.02$).

In real ocean situations, as mentioned before, one should expect values of C_* slightly less than in the laboratory, as a result of the existence of a turbulent energy flux M_0 associated with wave breaking. The determination of C_* in situ is however still an open question.

The results of the numerical computation of the non-dimensional entrainment velocity are in good agreement with the experimental data (fig. 2).

The models described above (both in the case $\gamma = 0$ and in the case $\gamma \neq 0$) are of course limited to initial periods of mixing sufficiently smaller than f^{-1} . This is actually the case for many in situ measurements.

Fig. 3 shows a comparison between predicted and observed values (Kullenberg, 1977) of the non-dimensional entrainment velocity.

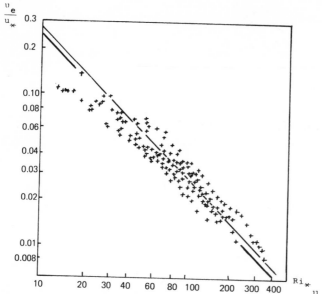

Fig. 2. The non-dimensional entrainment rate $\frac{u_e}{u_*}$ as a function of the overall Richardson number R_{i*} for a linearly stratified fluid. The continuous curve corresponds to the solution of the model. Two short lines in the upper and lower part of the figure correspond to Kato and Phillips' approximation.

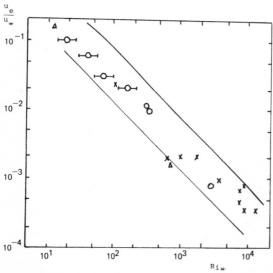

Fig. 3. The non-dimensional entrainment rate $\frac{u_e}{u_*}$ as a function of the Richardson number R_{i*}. The upper curve corresponds to the theoretical predictions in the case $\gamma = 0$. The lower curve corresponds to the theoretical predictions in the case $\gamma \neq 0$, constant. The experimental points (x, 0, Δ) are taken from the measurements made by Kullenberg in the Baltic Sea.

The two continuous lines correspond to the two cases $\gamma = 0$, $\gamma \neq 0$. The experimental points suggest that, in natural conditions, the actual stratification is somehow intermediate between the two cases. Indeed, at the initial moment, an important density jump is created accross the turbulent entrainment layer while a continuous density gradient exists below it.

The excellent agreement shown in fig. 3 between in situ observations and theoretical predictions in an argument in favor of using the present model, calibrated with laboratory data, to the study of wind mixing and entrainment in seas and lakes.

THEORETICAL MODELLING OF THE UPPER HOMOGENEOUS LAYER BY WIND MIXING

(III) The influence of the rotation of the earth ; no heat flux and steady momentum and turbulent energy fluxes accross the air-sea interface.

Substituting the first integrals (42) and (43) in eq.(34), neglecting small terms and changing variables to ξ and η as before (eqs. 74, 75, 76), one obtains

$$\frac{d\xi}{d\eta} \left[\frac{1}{2} R_i G + C_\delta - \frac{C_*}{f_*^2 \xi^2} (1 - \cos f_* \eta) \Phi_2 \right] = \Phi_1 \tag{112}$$

where

$$f_* = \frac{f \, D(o)}{G_\delta^{1/3}} \tag{113}$$

The functions Φ_1 and Φ_2 have the same meaning as before. Φ_1 is associated with the vertical diffusion of turbulent energy and should be rather insensitive to the earth's rotation. One shall thus assume that it has the same form as in the non rotating case (eq. 84). Φ_2 represents the portion of the turbulent energy Π_h which is dissipated. With Π_h strongly influenced by the earth's rotation $\left(\Pi_h = \frac{C_*}{f_*^2 \xi^2} (1 - \cos f \, t) \frac{dD}{dt} \right)$ the question arises whether Φ_2 can still be parameterized as before. Conserving the same parameterization of Φ_2 has the advantage that the solutions of eq.(112) tend asymptotically to the solutions obtained in the non-rotating case and found in good agreement with the laboratory experiments (Garnich and Kitaigorodskii, 1977, 1978). In both cases $\gamma = 0$ and $\gamma \neq 0$, the solutions of eq.(112) are found to be practically the same as the solutions of eq.(73) for $ft \lesssim \Pi/3$.

Thus, the assumption will be made here that eq.(87) is applicable when the effects of the earth's rotation are no longer negligible.

For $ft > \Pi/3$, Π_h decreases and this reflects on the deepening process as shown on figs 4 and 5.

In the case $\gamma = 0$ (fig.4), the earth's rotation produces a deviation from the linear law (77) valid for $ft << 1$. In the case $\gamma \neq 0$, the rotation has a radical effect. The entrainment rate decreases and the depth of the mixed layer appears to tend to some maximum value.

The maximum depth D_* the mixed layer can reach under the influence of Π_h alone ($\Phi_1 = 0$) can be estimated using eqs.(92) and (96). One finds, for $ft = \Pi$

$$D_* = \left(\frac{16\ \alpha_3}{\alpha_4 c_*^{1/2}}\right)^{1/6} \frac{\tau_o^{1/2}}{(N^2\Omega)^{1/3}} \sim 3\ \frac{u_*}{(N^2\Omega)^{1/3}} \tag{114}$$

The non-dimensional depth $\tilde{D} = \dfrac{D}{D_*}$ is plotted on fig.6 for three different values of ft , as a function of $\dfrac{N}{f}$ (Garnich and Kitaigorodskii, 1978). One can see that a relationship $D \sim C\dfrac{u_*}{(N^2\Omega)^{1/3}}$ is well verified in the range of values of $\dfrac{N}{f}$ most often encountered. The value of the coefficient C can be slightly larger than 1 and this may be attributed to the diffused energy flux (the influence of Φ_1). The variations, however, are small and the value

$$D_{st} \equiv D(ft=\Pi) = 1.2\ D_* \sim 3.6\ \frac{u_*}{(N^2\Omega)^{1/3}} \tag{115}$$

can be used as a characteristic depth of penetration of wind mixing in a rotating stratified fluid.

Eq.(115) can be compared with the results of Pollard et al (1973)

$$D_{st} \sim \frac{8^{1/4}}{3}\ D_*(\frac{N}{f})^{1/6} \tag{116}$$

and by Phillips (1977)

$$D_{st} \sim \frac{1}{3}\ D_*(\frac{N}{f})^{1/3} \tag{117}$$

The comparison is shown on fig.7. The three formulae are found to give very close values when, typically, $f \sim 10^{-4}$, $N \sim 10^{-2}$.

To find a definitive answer to the problem of choosing between the different parameterizations of the turbulent energy balance in the upper homogeneous layer (Phillips, 1977 ; Pollard et al, 1973 ;

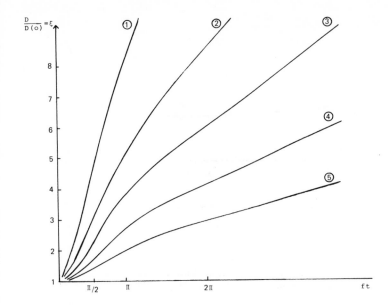

Fig. 4. Theoretical curves ξ(ft) according to the model in the case
γ = 0. The curves correspond to the values

① $R_{i*}(o) = 30$; ② $R_{i*}(o) = 50$; ③ $R_{i*}(o) = 70$

④ $R_{i*}(o) = 100$; ⑤ $R_{i*}(o) = 140$.

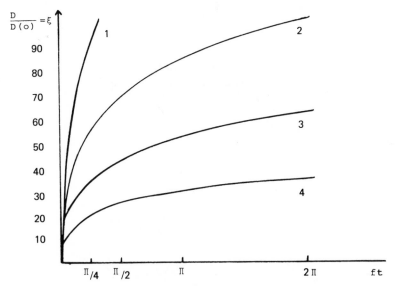

Fig. 5. Theoretical curves ξ(ft) according to the model in the case
γ ≠ 0. The curves correspond to the values

① $R_{i*}(o) = 4.4 \ 10^{-3}$; ② $R_{i*}(o) = 2.2 \ 10^{-2}$

③ $R_{i*}(o) = 8.8 \ 10^{-2}$; ④ $R_{i*}(o) = 4.4 \ 10^{-1}$

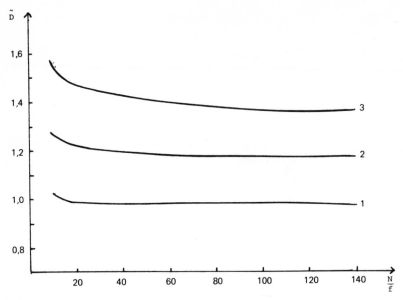

Fig. 6. The non-dimensional thickness of the mixed layer $\tilde{D} = \dfrac{D}{D_*}$ as a function of the non-dimensional ratio $\dfrac{N}{f}$ for three values of ft .

① ft = $\dfrac{\pi}{2}$; ② ft = π ; ③ ft = 2π

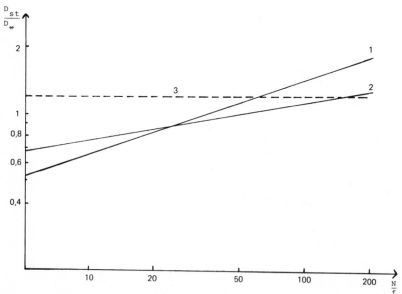

Fig. 7. The non-dimensional thickness of the wind mixed layer $\dfrac{D_{st}}{D_*} = \dfrac{D(ft=\pi)}{D_*}$ as a function of the non-dimensional ratio $\dfrac{N}{f}$

1. Formula (116)
2. Formula (117)
3. (Dashed curve ---) Formula (115)

Garnich and Kitaigorodskii, 1977, 1978) and to determine which one
provides the best estimate for the characteristic depth of the oceanic
wind mixed layer, many careful experiments must be performed both in
the laboratory and in nature, for a broad range of values of $\frac{N}{f}$.
It would be interesting to use,as a basis,the stock of oceanic observations as it was done previously by Kitaigorodskii and Filushkin
(Kitaigorodskii, 1960 ; Kitaigorodskii and Filushkin, 1964).

In this study, a different parameterization of the energy dissipation was used, i.e.

$$\bar{\epsilon}D + \tilde{\epsilon}h = G_\delta \ \Psi_1 \ \left(R_{i*} \ , \ \frac{fD}{u_*} \right)$$
(118)

and this is not, in general, equivalent to the parameterization used
in this paper.

In a subsequent publication (Resnyanskiy, 1975) a simpler form of
eq.(118) was used, i.e.

$$\Psi_1 \ \left(R_{i*} \ , \ \frac{fD}{u_*} \right) \sim \frac{fD}{u_*}$$
(119)

The thickness of the mixed layer was derived from the simple energy
balance $G_\delta = \bar{\epsilon}D + \tilde{\epsilon}h$ which of course gives the well-known Rossby-
Montgomery formula (e.g. Kitaigorodskii, 1970)

$$D_{st} \sim \frac{u_*}{f}$$
(120)

In situ experimental data (Kitaigorodskii, 1960 ; Kitaigorodskii and
Filushkin, 1964) do not rule out eq.(120) but the arbitrariness of
the approximation (119) must be emphasized. On the other hand, as
shown above, the dependence of the energy dissipation on the Richardson
number must be taken into account to reproduce satisfactorily the
laboratory experiments and it has a primary importance in modelling
the deepening of the mixed layer.

THEORETICAL MODELLING OF THE DEEPENING OF THE MIXED LAYER BY WIND-
MIXING

(IV) Influence of the earth's rotation and of the heating of the
ocean surface ; steady heat momentum and turbulent energy fluxes
accross the air-sea interface.

Eq.(41) can be written

$$R_{iG} = R_{iG}(o) + R_1(\xi^2-1) + R_2\eta$$
(121)

where ξ and η are defined as before (eqs. 74 and 75) and where

$$R_1 = \frac{\beta g \gamma D_o^2}{2 G_\delta^{2/3}} \quad ; \quad R_2 = \frac{\beta g \, s_o \, D_o}{G_\delta} \equiv \frac{D_o}{L_*} \qquad (122)(123)$$

L_* denoting the Monim - Obukhov length scale

$$L_* = \frac{G_\delta}{g \beta s_o} \qquad (124)$$

From eqs.(34) and (44), one gets, using eq.(121),

$$\frac{d\xi}{d\eta} \left[\frac{1}{2} R_{iG}(o) + \frac{1}{2} R_1 (\xi^2 - 1) + \frac{1}{2} R_2 \eta + C_\delta - \frac{C_*}{f_*^2 \xi^2} (1 - \cos f_* \eta) \Phi_2 \right]$$

$$= \Phi_1 - \frac{R_2}{2} \xi \qquad (125)$$

The functions Φ_1 and Φ_2 are defined by eqs.(84) and (87) where R_{iG} is given by eq.(121). The functions Φ_1 and Φ_2 are thus assumed independent of the earth rotation and functions of the stratification and of the surface heat flux through the expression of R_{iG} .

Since both ξ and R_{iG} increase with time, the right-hand side of eq.(125) is a decreasing function of time which may vanish for some critical values η_{cr} and ξ_{cr} .

Restricting attention to small times ($f_* \eta \equiv ft \ll 1$) , Garnich (1978) has shown that, in the case where ϕ_1 and ϕ_2 are constant, an analytical solution of eq.(125) could be found. The critical depth is then given by

$$D_{cr} = 2 \Phi_1 L_* \qquad (126)$$

and the critical entrainment velocity $\lim_{\xi \to \xi_{cr}} \frac{d\xi}{d\eta}$ is different from zero.

For times larger than the critical time, the solution becomes multivalued and to describe the mixing process additional hypotheses are required.[*]

The typical wind-mixing length scale D_* can be computed from the solution of eq.(125) at $ft \equiv f_* \eta = \Pi$, one finds

$$D_* = 2 \left(\frac{\alpha_3}{\alpha_4 C_*^{1/2}} \right)^{1/2} \frac{u_*}{f \, R_{i*}} \qquad (127)$$

When the surface heat flux is zero, the temperature difference accross the turbulent entrainment layer is created by the stratification below and one can write

_____

[*]For instance, one can imagine that for $t > t_{cr}$ the vertical structure of the temperature field is destroyed and a new wind-mixed layer is generated near the surface.

$$R_{i*} = \frac{N^2 D^2}{2\ u_*^2} \tag{128}$$

Substituting in eq.(127), one finds

$$D_* = \left(\frac{16\alpha_3}{\alpha_4 c_*^{1/2}}\right)^{1/6} \frac{u_*}{(N^2 f)^{1/3}} \tag{114'}$$

identical to eq.(114).

On the contrary, if one assumes that the temperature jump $\Delta\theta$ is mainly due to the surface flux one can write, according to eq.(41) where $\gamma = 0$, $\Delta\theta(o) = 0$,

$$\Delta\theta D \sim s_o t \tag{129}$$

In the fundamental stages of the deepening process, the mixing of the entrained fluid is mainly due to the turbulent energy production in the turbulent entrainment layer and one has, approximately

$$\frac{g\beta\Delta\theta D}{2} \frac{dD}{dt} \sim \phi_2 \Pi_h = \phi_2 \frac{dD}{dt} \frac{\|\underset{\sim}{u}\|^2}{2} \tag{130}$$

For small values of ft, using eq.(42) and (43), one gets then

$$R_{iG} \sim \frac{c_* G_\delta^{2/3} t^2}{D^2} \phi_2 \tag{131}$$

and in the asymptotic case (96)

$$D \sim \left(\frac{c_* \alpha_3}{\alpha_4}\right)^{1/2} G_\delta^{1/3} t\ R_{iG}^{-1} \tag{132}$$

$$= \left(\frac{c_* \alpha_3}{\alpha_4}\right)^{1/2} L_*$$

Comparison of formula (132) with observations in the ocean (Kitaigorodskii, 1960) has confirmed a relationship of the form $D \sim C\ L_*$ but the coefficient determined from the experimental data was found different. The difference can be attributed partly to the contribution of the stratification γ to the temperature jump accross the turbulent entrainment layer during the spring-summer heating period when the observations were made.

The solution of eq.(125) has been computed numerically for different values of the surface heat flux. The results are presented on figs. 8 and 9. (The value $s_o = 10^{-4}$ °C m/sec corresponds to a <u>maximum</u> value of the heat flux in mid latitudes during a summer day).

Fig. 10 shows the computed values of $\tilde{D} = \frac{D}{D_*}$ at three different times $ft = \frac{\Pi}{2}$, Π and 2Π as a function of the non-dimensional

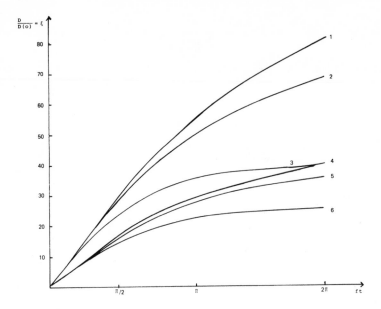

Fig. 8. Theoretical curve ξ(ft) for different values of the surface
heat flux s_O in the case $\gamma = 0$.
The curves 1, 2, 3 were calculated for $Ri_{*}(o) = 50$, $u_{*} = 1,5\ 10^{-2}$m/sec,
$D(o) = 1$m and $s_O = 0$; 2.10^{-5} and 10^{-4}°C m/sec, respectively.
The curves 4, 5, 6 were calculated for $Ri_{*}(o) = 100$, $u_{*} = 1,5\ 10^{-2}$m/sec
$D(o) = 1$m and $s_O = 0$; 2.10^{-5} and 10^{-4}°C m/sec, respectively.

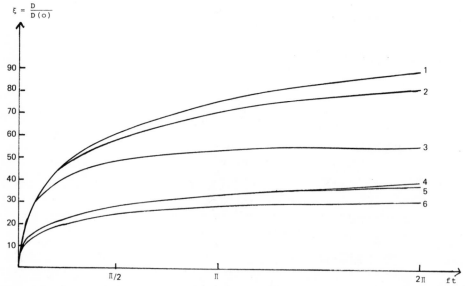

Fig. 9. Theoretical curve ξ(ft) for different values of the surface
heat flux s_O in the case $\gamma \neq 0$.
The curves 1, 2, 3 were calculated for $N^2 = g\beta\gamma = 5.10^{-6}$sec^{-2} and
$s_O = 0$; 2.10^{-5} and 10^{-4}°C m/sec respectively.
The curves 4, 5, 6 were calculated for $N^2 = 5.10^{-5}$sec^{-2} and $s_O = 0$;
2.10^{-5} and 10^{-4}°C m/sec respectively.

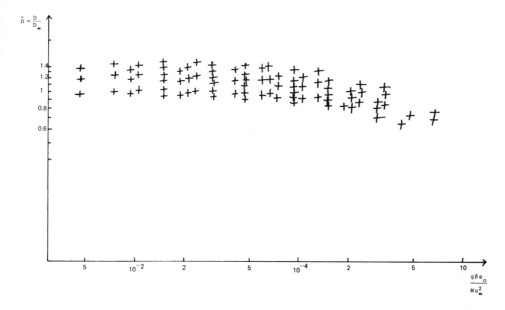

Fig. 10. The non-dimensional thickness of the mixed layer $\tilde{D} = \dfrac{D(t)}{D_*}$ as a function of the non-dimensional parameter $\dfrac{g\beta s_o}{N^2 u_*^2}$ for three values of $ft(\Pi/2, \Pi, 2\Pi)$ and five different $\dfrac{g\beta s_o}{N^2 u_*^2}$ values of N. $N^2 = 10^{-6} sec^{-2}$; $N^2 = 2.10^{-6} sec^{-2}$; $N^2 = 5.10^{-6} sec^{-2}$; $N^2 = 5.10^5 sec^{-2}$ ($u_* = 1{,}5\ 10^{-2}$ m/sec , D(o) = 1m).

parameter $\dfrac{g\beta s_o}{Nu_*^2}$.

Although there are fairly important variations with the surface heat flux in the representation used in figs. 8 and 9, one can see on fig. 10 that for $\dfrac{g\beta s_o}{N u_*^2} < 10^{-4}$, - a condition which is usually realized, in mid latitude, during the spring summer heating period - , a short range forecast of the thickness of the upper homogeneous layer under the effect of wind mixing can be based on eq.(115) provided the initial stratification is known.

CONCLUSIONS

In the present review, attention has been restricted to a discussions of previous studies contributing to a better understanding of the dynamics of vertical mixing processes and the problem of modelling the upper ocean layer. One has voluntarily left aside the question of large scale well-organized motions such as, Langmuir circulation, for instance.

Observations clearly show that large scale vertical motions (like

Langmuir vortices) are closely related to the upper homogeneous layer and the question remains to be solved whether the thickness of the mixed layer can be determined by such organized patterns or whether these structures depend on the vertical density profile built up by turbulence in the upper homogeneous layer.

ACKNOWLEDGMENT

This review is mainly based on the results reported in (Garnich and Kitaigorodskii, 1977, 1978 ; Kitaigorodskii, 1977). The author is much indebted to Dr. N. Garnich for permission to present here also some of the graphs and results of his PhD dissertation (Garnich, 1978).

REFERENCES

Crapper, P.F., Linden, P.F., 1974. The structure of turbulent density interfaces. J. Fluid Mech., 65:45-64.

Denman, K.L., 1973. A time-dependent model of the upper ocean. J. Phys. Oceanogr., 3:173-184.

Garnich, N.G., 1975. A model of the continuous evolution of the seasonal thermocline. Oceanologia, 15:233-238.

Garnich, N.G., 1978. The theory of wind-mixed layer. Ph.D. Dissertation, Laboratory of Physics of Atmosphere-Ocean Interactions, Institute of Oceanology, Academy of Science, USSR.

Garnich, N.G., Kitaigorodskii, S.A., 1977. On the rate of mixed layer deepening. Izv. Acad. of Sciences USSR, Physics of Atmosphere and Ocean, 13:1287-1296.

Garnich, N.G., Kitaigorodskii, S.A., 1978. To the theory of wind mixed layer deepening in the Ocean. Izv. Acad. of Sciences USSR, Physics of Atmosphere and Ocean, 14:1287-1296.

Kantha, L.H., 1977. Note on the role of internal waves in thermocline erosion. Chap. 10a. In: E.B. Kraus (Editor), Modelling and prediction of the upper layers of the ocean. Pergamon Press.

Kantha, L.H., Phillips, O.M., Azad, R.S., 1977. On turbulent entrainment at a stable density interface. J. Fluid Mech., 79:753-768.

Kato, H., Phillips, O.M., 1969. On the penetration of a turbulent layer into stratified fluid. J. Fluid Mech., 37:643-655.

Kitaigorodskii, S.A., 1960. On the computation of the thickness of the wind-mixing layer in the Ocean. Izv. Acad. of Sciences USSR, Geophysical ser., 3:425-431 (English edition pp. 284-287).

Kitaigorodskii, S.A., 1970. Physics of air-sea interaction. Leningrad, Gidromet. Izdatel'stvo (English edition 1973, Jerusalem Israel Progr. Scien. Transl.).

Kitaigorodskii, S.A., 1977. Oceanic surface boundary layer. Report of the JOC/SCOR Joint Study Conference on general circulation models of the ocean and their relation to climate (Helsinki, 23-27 may 1977) 1, Geneva.

Kitaigorodskii, S.A., Filushkin, B.N., 1964. Application of the similarity theory to the analysis of the observations in the upper ocean. Oceanological Studies, 13, Izd-vo., "Nauka", Moskva.

Kosnizev, V.K., Kuftarkov, Yu.M., Felsenbaum, A.I., 1976. One-dimensional asymptotic model of the upper ocean. Proceedings of Acad. of Sciences. USSR (DAN), 1:70-72.

Kraus, E.B. and Turner, J.B., 1967. A one-dimensional model of the seasonal thermocline : II. The general theory and its consequences. Tellus, 19:98-106.

Kullenberg, G., 1977. Entrainment velocity in natural stratified vertical shear flow. Estuarine and coastal marine science, 5:329-338.

Linden, P.F., 1975. The deepening of a mixed layer in stratified fluid. J. Fluid Mech., 71:385-405.

Long, R.B., 1974. Turbulence and mixing processes in stratified fluids Lectures, Tech. Rep. N°6 (serie C). The Johns Hopkins University.

Mellor, G.L. and Durbin, P.A., 1975. The structure and dynamics of the ocean surface mixed layer. J. Phys. Cceanogr., 5:718-728.

Moore, M.J., Long, R.ᴴ., 1971. An experimental investigation of turbulent stratified shearing flow. J. Fluid Mech., 49:635-656.

Phillips, O.M., 1977. Entrainment. Chap. 7. In: E.B. Kraus (Editor), Modelling and Prediction of the upper layers in the ocean. Pergamon Press.

Pollard, R.T., Rhines, P.B., Thompson, R.O.R.Y., 1973. The deepening of the wind-mixed layer. Geoph. Fluid Dyn., 3:381-404.

Resnyanskiy, Yu.D., 1975. Parametrization of the integral turbulent energy dissipation in the upper quasi-homogeneous layer of the ocean. Izv. Atmospheric and Oceanic Physics, 11:726-733 (English edition pp. 453-457).

Thompson, S.M. and Turner, J.S., 1975. Mixing across an interface due to turbulence generated by an oscillating grid. J. Fluid Mech., 67:349-368.

Thorpe, S.A., 1973. Turbulence in stably stratified fluids a review of laboratory experiments. Boundary-layer meteorology, 5:95-119.

Turner, J.S., 1968. The influence of molecular diffusivity on turbulent entrainment across a density interface. J. Fluid Mech., 33:639-656.

Turner, J.S., 1973. Buoyancy effects in fluids. Cambridge Univ. Press.

Wolanski, B.J., 1972. Turbulent entrainment across stable density-stratified liquids and suspensions. Ph. D. Dissertation. The Johns Hopkins University.

Wolanski, B.J., Brush, L.M., 1975. Turbulent entrainment across stable density step structures. Tellus, 27:259-268.

LARGE SCALE AIR-SEA INTERACTIONS AND CLIMATE PREDICTABILITY

CLAUDE FRANKIGNOUL

Department of Meteorology, Massachusetts Institute of Technology, Cambridge, MA

ABSTRACT

After reviewing recent empirical studies of short-term climate predictability
based on observations of the sea surface temperature (SST), the physical processes
that govern the generation and decay of large scale SST anomalies are discussed.
Using a slab model of the oceanic mixed layer, we find that large scale mid-
latitude SST anomalies can be described as a first-order autoregressive process
in regions of small mean current, as suggested by Frankignoul and Hasselmann
(1977). The SST anomalies are continuously generated by the natural variability
of the air-sea fluxes. Short time scale variations in the local heat exchanges
seem dominant, although mixed-layer depth variations are important during certain
seasons. Temperature advection plays a large role in some regions, and mesoscale
eddies mainly contribute a small scale noise. The decay of the SST anomalies
can be represented by a linear negative feedback, and seems largely controlled
by their back-interaction on the atmosphere. The importance of the feedback
processes for climate predictability is stressed, as well as the seasonal
variabilities in the SST anomaly dynamics.

INTRODUCTION

In recent years, there has been growing concern about the impact of climate
changes on man's agricultural, economic and social activities, leading to
increasing climate research efforts (see GARP report No. 16, 1975). On monthly
to decadal time scales, the surface layers of the ocean are believed to play a
prominent role in climate variations, through exchanges of heat, moisture and
momentum at the air-sea interface. Indeed, there is increasing evidence of large-
scale relationships between monthly or seasonal anomalies in the sea surface
temperature (SST) and the atmospheric circulation (e.g., Namias, 1969; Bjerknes,
1969; Ratcliffe and Murray, 1970).

The dominant SST anomaly patterns are large, nearly ocean-wide, and have a
persistence time of the order of seasons. This persistence reflects the large
thermal and mechanical inertia of the upper ocean, and suggests that the ocean
is likely to contribute most to the predictability of the coupled ocean-
atmosphere system. Empirical climate forecasts based on the observed distribution
of SST anomalies have been made, reaching some degree of success (Namias, 1978).
However, higher predictive skills are linked to our understanding of the physics
of these large scale air-sea interactions. The effect of SST anomalies on the

atmosphere is difficult to model simply, and there are complex teleconnections between tropical and extratropical systems (Bjerknes, 1969). Much information comes from simulations with general circulation models (GCM) of the atmosphere. In these numerical experiments, a fixed SST anomaly is prescribed, and its effects on the atmosphere circulation analyzed. In mid-latitudes where the natural variability of the atmosphere is largest, the effects of SST anomalies seem mainly local, although some non-local effects have been found (Kutzbach et al., 1977). In the tropics, the response of atmospheric models is larger (e.g., Shukla, 1975).

To investigate the influence of SST anomalies on climate changes, the SST field must be allowed to vary. It has often been suggested that SST anomalies arise through positive ocean-atmosphere feedback processes, by which the presence of a SST anomaly modifies the atmospheric circulation in such a manner that the anomaly is strengthened (e.g., Namias, 1963). However, the empirical evidence suggests that it is the atmosphere that is driving the ocean, rather than vice-versa (Davis, 1976; Trenberth, 1975; Haworth, 1978). This is consistent with recent theoretical studies by Salmon and Hendershott (1976), and Frankignoul and Hasselmann (1977), who suggested that SST anomalies arise spontaneously in response to the natural variability of the weather fluctuations. The climate predictability problem is therefore twofold, since the evolution of the SST anomaly field and its back interaction on the atmosphere must be predicted. In their numerical experiment, Salmon and Hendershott found that the atmosphere was too noisy to be much affected by the SST anomalies on the time scales over which the anomalies are themselves predictable. However, their ocean model was very simple (a copper plate ocean). The observations suggest that climate predictability on time scales of months exists only during certain seasons (Davis, 1978). Hence, more refined models that can simulate the seasonal variability of the atmosphere and the ocean are needed. To avoid excessive computational costs, only the essential physical mechanisms should be retained in these models.

The main emphasis of this paper is on extratropical air-sea interactions. In section 2, we briefly review recent empirical studies of short-term climate predictability based on observations of SST anomalies. Encouraging results have been obtained, but it is difficult to distinguish between competing predictive models. As discussed by Davis (1977) and Hasselmann (1978), the most useful selection of predictors will be based on theoretical concepts. This introduces naturally section 3, where the physical processes that govern the evolution of large scale SST anomalies are discussed, using a slab model of the oceanic mixed layer. It is found that in typical central ocean conditions, the rate of change of the SST anomalies is described by a first-order autoregressive process, as suggested by Hasselmann (1976), and Frankignoul and Hasselmann (1977). The upper ocean responds as an integrator of the short-time scales changes in the

air-sea fluxes. Random walk SST fluctuations with linearly growing variance
develop, until negative feedback processes lead to a statistically steady state.
The need to investigate further the feedback processes is stressed in the
conclusions, because of their important role in climate predictability.

SEA SURFACE TEMPERATURE AND SHORT-TERM CLIMATE PREDICTABILITY

In the absence of satisfactory dynamical models of the large scale interactions
between anomalies in the weather, SST, sea ice, snow cover, etc., historical
observations are best suited for attempting climate forecasts. If there are
recurrent patterns of developments other than these occurring by chance, future
developments may be in part predictable from past and present observations. For
sufficiently small excursions about an equilibrium mean climatic state, linear
relations can be used for describing climatic changes (e.g. Leith, 1975;
Hasselmann, 1976). Then, to predict a variable y from N observations z_i of the
same or other variables at prior times (the predictors), one can write

$$\hat{y} = \sum_{i=1}^{N} a_i \, z_i \tag{1}$$

where $\hat{y}$ denotes the predicted value of y. The optimum linear prediction is the
one which minimizes the mean square error $<\varepsilon>=<(y-\hat{y})^2>$, where the angle braces
indicate ensemble average. The solution is given by

$$a_i = \sum_{j=1}^{N} <z_i \, z_j>^{-1} <z_i \, y> \tag{2}$$

The quality of the prediction is generally characterized by the skill

$$S = 1 - \frac{<\varepsilon>}{<y^2>} \tag{3}$$

which represents the fraction of variance that can be predicted. As first
discussed by Lorenz (1956), the performance of statistical estimators is
influenced by the fact that the mean products in (2) are not known but must be
estimated from finite data sets. This introduces artificial predictive skill
that increases with increasing number of predictors N, and decreasing length of
the data set used to estimate the mean products (see also Davis, 1976). It is
therefore necessary to limit N before carrying out the analysis. Davis (1977)
and Hasselmann (1978) discuss in detail why it is much preferable to select the
number of predictors by an a priori criteria, rather than by coefficient
screening. They also point out the difficulty in distinguishing between different
competing models on purely empirical grounds. The best data selection techniques
will therefore be based on theoretical studies, and Hasselmann concludes that
"the credibility of a model will have to rest to a large part also on the
intrinsic credibility of the physics of the model".

As mentioned in the introduction, numerous case studies of the co-occurrence of SST and atmospheric anomaly patterns have been reported in the literature. To use SST as climate predictor, however, some correlation must be observed between SST anomalies and subsequent weather anomaly patterns. Some early evidence for such correlations is reported by Namias (1969). Little attention had been given to the other lag correlations, until Davis (1976) made the first systematic analysis of the relationship between SST and sea level pressure (SLP) over the North Pacific. Using statistical estimators of the form (1), Davis found that the only significant connection between SST and SLP anomalies is one where SLP leads SST in time by a lag of zero to 2 or 3 months, and that there was no predictive skill for predicting SLP from SST or SLP. Subsequently, Namias (1976) showed convincing evidence that summer SST anomalies in the Aleutian low are significantly correlated with SLP (or mid-tropospheric height) over the North Pacific and with air temperature and precipitation anomalies in the United States (downstream) in the following fall. Davis (1978) showed that the difference in the results occurs because he had used year-round data, whereas Namias only considered the transition from summer to fall. Using seasonally stratified statistics, Davis found that up to 20% of the variance of autumn and winter SLP anomalies could be predicted from prior observations of either SST or SLP, while there was no predictive skill for SLP in the other seasons.

These studies suggest that useful climate forecasts based on the observed SST anomaly field can be made, at least for some seasons. The highly abnormal winter of 1976-1977 was characterized in the United States by severe cold over the central and eastern parts of the country, and drought over the West. Figure 1 reproduces the forecast made by Namias (1978) with a very simple prediction model; the winter SST anomaly pattern was estimated from the fall pattern using persistence and advection around the North Pacific gyre, and from this the US climate was predicted on the basis of teleconnections observed in the past. The general patterns of air temperature and precipitation was reasonably anticipated, but the extreme severity of the winter was not foretold. Interestingly, Davis (1978) notes the inability of his statistical model to predict the North Pacific SLP during the same winter.

Similar correlations between SST anomalies and subsequent weather patterns might be found in other oceans. Ratcliffe and Murray (1970) discuss some evidence of a relationship between large scale SST anomalies south of Newfoundland and weather patterns over Europe during the following month, mainly during the fall. Near the equator, the air-sea interactions are stronger. For example, Newell and Weare (1976) found that the tropical SST anomalies in the Pacific were leading the globally averaged temperature in the tropical troposphere by 6 months. More recently, Hasselmann and Barnett (1978) were able to predict SST and atmospheric anomalies in the tropical Pacific up to about 7 months in advance.

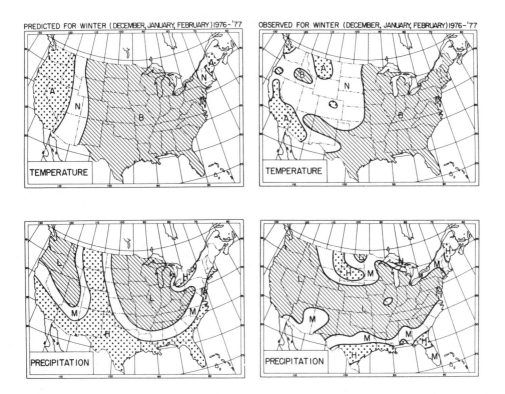

Fig. 1. Predicted and observed temperature and precipitation patterns for winter 1976-77 made on 1 December 1976. Shaded areas denote the three categories - below (B), near (N) and above normal (A) whose range are determined as terciles from a 30-year climatic record (from Namias, 1978).

MODELLING SEA SURFACE TEMPERATURE ANOMALIES

Because of wind mixing and turbulent convection, the upper layer of the ocean is nearly homogeneous. Large seasonal changes are observed in the temperature and depth of the mixed layer, but there is practically no seasonal variability below the largest depth it reaches in late winter. Thus, it seems sufficient to consider the near surface layers for understanding the SST variability over periods of a few years.

The heat exchanges and the wind stress at the sea surface fluctuate with a dominant time scale of a few days, mainly on the synoptic scale of motion, reflecting the rapid changes in the weather. The weather has limited predictability, hence can be represented as a stochastic process on the climatic time scales. Although the oceanic mixed-layer responds only weakly - because of its inertia -

to these daily fluctuations, the integral effect of the stochastic atmospheric forcing is large. Frankignoul and Hasselmann (1977) have shown that random walk SST fluctuations with linearly growing variance develop, until negative feedback processes lead to a statistically steady state. Using a copper plate model of the oceanic mixed layer, they suggested that the SST anomalies T' can be approximated by a first-order autoregressive (Markov) process obeying the equation

$$\frac{\partial T'}{\partial t} = \frac{v'}{h} - \lambda T' \tag{4}$$

where v' represents the anomalous heat exchanges of short characteristic time scale τ_x, h the (constant) mixed-layer depth and λ the negative feedback processes. The SST anomaly spectrum $F_T(\omega)$ is then given by

$$F_T(\omega) = \frac{1}{h^2} \frac{F_V(o)}{\omega^2 + \lambda^2} \tag{5}$$

where $F_V(o)$ denotes the (approximately) constant power density of the forcing at low frequencies (the white noise level). With $\lambda \sim$ (a few month)$^{-1}$, the model (4) reproduces the main statistical features and orders of magnitudes of the SST anomalies observed in the mid-latitudes central oceans (see also Reynolds, 1978; Frankignoul, 1978).

In this section, we consider a more realistic mixed layer model and discuss the different mechanisms contributing to the generation and decay of SST anomalies. It is found that equation (4) describes indeed, to a reasonable approximation, the rate of change of SST anomalies in regions far from strong currents and large mesoscale eddy activity.

A slab model of the mixed-layer

We represent the upper oceanic layers by a well-mixed layer of depth h and uniform temperature θ overlying a stratified region of temperature $\theta_S(z)$. If the subscript t denotes turbulent fluctuations and the symbol $\wedge$ a Reynolds averaging, the thermodynamics energy equation can be written

$$\frac{\partial}{\partial t} \hat{\theta} + \nabla \cdot (\hat{\underline{u}}\, \hat{\theta} + \widehat{\underline{u}_t\, \theta_t}) + \frac{\partial}{\partial z}(\hat{w}\, \hat{\theta} + \widehat{w_t\, \theta_t}) = \frac{\hat{q}}{\rho C} \tag{6}$$

where u and w denote the horizontal and vertical velocity, ρ the water density, C its specific heat, q the heat sources due to surface heat exchanges and radiation effects, and $\nabla = (\frac{\partial}{\partial x}, \frac{\partial}{\partial y})$. Integration of (6) from the surface (at z = 0) to just below the mixed layer yields

$$h \frac{\partial}{\partial T} T + h\underset{\sim}{U}.\nabla T + \Delta\theta \, w_e = \chi h \nabla^2 T + \frac{Q}{\rho C} \tag{7}$$

where T and $\underset{\sim}{U}$ denote averaged values over the mixed-layer. Here we have used the incompressibility condition $\nabla \cdot \underset{\sim}{u} + \frac{\partial w}{\partial z} = 0$, and assumed zero fluxes below the mixed-layer. The entrainment velocity w_e is defined by

$$w_e = \Lambda \, (\frac{\partial h}{\partial t} + \nabla \cdot h\underset{\sim}{U}) \tag{8}$$

Following Kraus and Turner (1967), we have introduced the function $\Lambda = 1$ for $\frac{\partial h}{\partial t} + \nabla \cdot h\underset{\sim}{U} > 0$ and zero otherwise, because the mixed-layer temperature does not change by detrainment. In (7), $\Delta\theta = T - \theta_s$ is the temperature jump at the bottom of the mixed-layer, and Q the total heat flux across the air-sea interface. We have introduced a constant horizontal diffusivity coefficient χ to parameterize horizontal mixing. Without advection, (7) and (8) reduce to the mixed-layer equations used by Kraus and Turner (1967), and others.

We decompose each variable into a (seasonally varying) mean denoted by an over-bar, and a random fluctuation or anomaly denoted by a prime. The SST anomaly obeys the equation

$$(1 + D') \frac{dT'}{dt} = \frac{Q'}{\rho\overline{Ch}} - D' \frac{d\overline{T}}{dt} + \overline{D' \frac{dT'}{dt}} - \underset{\sim}{U}' \cdot \nabla\overline{T} - \underset{\sim}{U}' \cdot \nabla T'$$

$$\qquad\quad (a) \qquad\qquad (b) \qquad\quad (c) \qquad\quad (d) \qquad\quad (e)$$

$$+ \overline{\underset{\sim}{U}' \cdot \nabla T'} - D' \underset{\sim}{U}' \cdot \nabla\overline{T} + \overline{D' \underset{\sim}{U}' \cdot \nabla\overline{T}} - D' \underset{\sim}{U}' \cdot \nabla T' + \overline{D' \underset{\sim}{U}' \cdot \nabla T'} - \Delta\theta w_e' \tag{9}$$

$$\qquad (f) \qquad\qquad (g) \qquad\qquad (h) \qquad\qquad (i) \qquad\qquad (j) \qquad\qquad (k)$$

$$- \Delta\theta' \overline{w_e} - \Delta\theta' w_e' + \overline{\Delta\theta' w_e'} + \chi(\nabla^2 T' + D' \nabla^2\overline{T} + D' \nabla^2 T' - \overline{D' \nabla^2 T'})$$

$$\qquad (l) \qquad\quad (m) \qquad\quad (n) \qquad\quad (o) \qquad\quad (p) \qquad\quad (q) \qquad\quad (r)$$

where $\frac{d}{dt} = \frac{\partial}{\partial t} + \underset{\sim}{U} \cdot \nabla$ is the time derivative following the mean notion, $D' = h'/\overline{h}$ the ratio of anomaly to mean mixed-layer depth, and $W_e = w_e/\overline{h}$. Although the magnitude of the different terms in (9) depends on scale, geographical position and season, some simplifications can be made for typical central ocean conditions. We shall consider large scale SST anomalies, say of wavenumber $k \le 2\pi/1000 \text{ km}^{-1}$.

The rate of change of the mixed-layer depth depends on the energy input by the wind, and the rate of turbulent energy production and dissipation in the mixed-layer (e.g., Niiler, 1977). This equation will not be written here. It is sufficient to remark that the characteristic time scale of w_e' is normally the one τ_x of the wind forcing. From (8), one sees that the dominant time scale of h' (or D') will be determined by the rate of energy dissipation. It has not been calculated, but experiments with one-dimensional mixed-layer models generally suggest that it is only slightly larger than τ_x (i.e. dissipation occurs rapidly).

Approximate equation for the rate of change of SST anomalies

a. *SST anomaly forcing*. The fluctuations in the surface heat fluxes, term (a), play an important role in the generation of SST anomalies (e.g., Jacob, 1967; Adem, 1975; Clark, 1972; Frankignoul and Hasselmann, 1977). The main contribution to these anomalies arise from the short time scale fluctuations in the flux of latent heat H_L and, to a lesser extent, sensible heat H_s. Fluctuations in the incoming solar radiation H_R seem mainly important during summer, and the back radiation H_B contributes essentially a small negative feedback (see below). Frankignoul (1978) has suggested that stochastic forcing by local heat fluxes was most efficient in generating SST anomalies during summer and fall.

Term (b) describes the influence of mixed-layer depth anomalies on the forcing by the mean heat fluxes (which mainly determine $\frac{d\overline{T}}{dt}$). It should be most important during spring and early summer, when $D' \sim O(1)$, and $\frac{d\overline{T}}{dt}$ is large. A case study by Clark (1972) suggests that term (b) is comparable to term (a) during summer and twice as large during spring, but negligible otherwise. On the basis of simple numerical experiments Willebrand (personal communication) has suggested that the fluctuations in the mixed-layer depth -- including contributions from terms (k), (l), and (m) discussed below -- may be as efficient as the local heat exchanges in generating SST anomalies. Further investigations are needed.

The mean products (c), (f), (h), (j), (n), and (r) contribute to SST fluctuations at zero frequency (neglecting seasonal variations) and are not of interest here.

Terms (d) and (e) represent advection by anomalous currents. Namias (1965) has suggested that temperature advection by anomalous ageostrophic wind-driven currents is the main generating mechanism for SST anomalies, but Clark (1972) found that local heat exchanges were more important during summer and fall. However, these studies are based on an empirical formula relating linearly surface drift and geostrophic wind, which corresponds typically to a very small Ekman depth. It seems more consistent with observations to assume that the Ekman transport is (nearly) uniformly distributed over the mixed-layer depth (Gill and Niiler, 1973). This yields $|\underset{\sim}{\tau}|/\rho f \overline{h}$ as typical drift current magnitude, where $\underset{\sim}{\tau}$ is the surface wind stress and f is the Coriolis parameter. The advection terms can then be conveniently compared with the heat exchange term (a) in the spectral domain, knowing the white noise level $F_\tau(o)$ and $F_H(o)$ of the wind stress and heat flux spectra:

$$\frac{(d) + (e)}{(a)} = \frac{F_T(o) \ (\nabla T)^2 \ c^2}{F_H(o) \ f^2} \tag{10}$$

At ocean weathership (OWS) P (50°N, 145°W), one has F_H (o) = 4 x 10^9 W^2 m^{-4}/Hz (latent plus sensible heat flux only), F_τ (o) = 2 x 10^4 N^2 m^{-4}/Hz (one component) (after Fissel et al., 1976). With f = 1.1 x $10^{-4} s^{-1}$, C = 4 x 10^3 J kg^{-1} $°C^{-1}$, ∇T = 5 x 10^{-6} °C m^{-1} (in general, one has $\overline{\nabla T}$ >> $\nabla T'$), the ratio (10) is about 0.2 and suggests that local heat exchanges are more efficient than temperature advection in the North-east Pacific (note that $\overline{\nabla T}$ is minimum in summer and fall, which explains Clark's result). At lower latitudes, F_H (o) generally increases and F_τ (o) decreases, and the ratio (10) should be even smaller. However, in regions of large temperature gradient ($\nabla T \geq 10^{-5}$°C m^{-1} in the vicinity of the Subarctic front, between 40 and 45°N), wind driven currents are likely to become important. Case studies by Daly (1978) suggest comparable effects of local heat exchanges and temperature advection in the North Atlantic. Pettersen et al. (1962) have documented the distribution of heat exchange in relation to the winds and weather pattern of typical oceanic cyclones (see also Simpson, 1969). It can be shown that the SST anomaly patterns induced by heat exchange and advection (assuming, for example, a uniform meridional SST gradient) will often be rather similar (see also Daly, 1978). This may explain why observed SST changes have been reproduced with some success using advection only (e.g., Namias, 1965).

Quasi-geostrophic eddies also contribute to terms (d) and (e). Eddy frequency spectra are red, with maximum variance at periods $\sim$ 0(100d), and most of the eddy energy is in the oceanic mesoscale $\sim$ 0(50 km). Hence, eddies mainly cause a low-frequency, small scale noise, whose magnitude can be decreased by spatially averaging the SST observations. A comparison with the heat exchange term (a) gives:

$$\frac{\text{eddy noise}}{(a)} \approx \frac{E(\omega) \ (\nabla T)^2 \rho^2 c^2 h^{-2}}{F_H \ (o)} \qquad (11)$$

if we neglect the eddy modulation of the mixed-layer depth. Here $E(\omega)$ is the near surface eddy kinetic energy spectrum, which varies strongly with the geographical position. In the MODE region that has rather large eddy activity, one has $E(\omega) \geq 2$ x 10^4 m^2 s^{-2}/Hz at 100 day period (this value corresponds to 500 m depth, hence the inequality sign), F_H (o) = 10^{10} W^2 m^{-4}/Hz (a guess), h = 50 m and ∇T = 3 x 10^{-6} °C m^{-1}. Then the ratio (11) is $\geq$ 1 and the eddy signal is large. Correspondingly, mesoscale eddies should be detectable from their surface signature, as observed by Voorhis et al. (1976). In the central part of the oceans, however, the eddy activity is smaller than in the MODE region and the eddy noise should be smaller (see also Gill, 1975).

In general, advection effects are smaller than local heat exchanges. Consequently, terms (g) and (i) are likely to be small, except perhaps during spring and early summer, when D' $\sim$ 0(1).

Terms (k), (l), and (m) represent the effect of anomalies in the entrainment velocity and temperature jump at the bottom of the mixed layer. During the cooling season (fall and early winter), the entrainment terms can be very large. Camp and Elsberry (1978) have shown that during atmospheric events with large wind increase, the entrainment heat flux is generally more important than the surface heat flux at high latitudes (OWS P), where the wind forcing is large. At lower latitudes, the entrainment heat flux is smaller, but not negligible. The overall influence of these terms is likely to be rather small, except perhaps at high latitudes.

b. *Feedback mechanisms*. The back interaction of SST anomalies on the atmosphere determines the role of the upper ocean in climatic changes. The observations suggest a complex interplay between the large scale atmospheric and oceanic circulations, involving possibly a number of ocean-atmosphere feedback mechanisms (Bjerknes, 1969; Reiter, 1978). Some theoretical support for the existence of a positive feedback has been presented by White and Barnett (1972), and Pedlosky (1975). However, these studies are based on highly idealized models of the coupled ocean-atmosphere circulation, and their relevance difficult to establish.

In the present context, the local interactions are mainly of interest, whereby SST anomalies may modify the atmospheric circulation in their vicinity. Namias (1963) has argued that there is a positive feedback between SST anomalies and weather patterns: warm SST anomalies cause an intensification of the cyclogenesis, and the increased cyclonic activity keeps the water warm, mainly by altering the local heat exchanges. However, Frankignoul and Hasselmann (1977) have suggested that term (a) causes in fact a negative feedback. This is seen by using the bulk formulae for the sensible and latent heat fluxes

$$H_S + H_L = C_H \ (1+B) \rho^a c^a \ (T^a - T) \ |\underset{\sim}{u}^a| \tag{12}$$

where C_H is the bulk transfer coefficient for the sensible heat flux, B the Bowen ratio of latent to sensible heat flux, and the superscript indicates atmospheric variables. If we assume that the atmosphere is not affected by the SST anomalies, a strong linear negative feedback is found, with

$$\lambda_{H_S+H_L} = \frac{\partial}{\partial T'}(-\frac{H_S'+H_L'}{\rho c \overline{h}}) = \frac{\rho^a c^a C_H (1+B) <|u^a|>}{\rho c \overline{h}} \tag{13}$$

For $C_H = 1.5 \times 10^{-3}$, $B = 3$, $<u^a> = 8 \ m \ s^{-1}$, $h = 50 \ m$, we find $\lambda_{H_S+H_L} = (1.3 \ month)^{-1}$. However, the air temperature adjusts itself to a value fairly close to the local SST, hence the negative feedback is not as strong as indicated by (13). In mid-latitudes, this adjustment depends on the radiative-convective

properties of the atmospheric boundary layer and on the large scale eddy flow
causing horizontal advection and vertical heat exchanges. This is difficult to
model analytically. However, some information can be found in the response of
atmospheric GCM to prescribed (and fixed) SST anomalies. Figure 2 suggests that,
on the SST anomaly time scale, $T^{a\prime}$ is approximately proportional to T'. The
turbulent heat flux does not depend linearly on T' because of the exponential
dependence of the saturation vapor pressure on T. In the temperature range of
observed SST anomalies, however, a linear dependence seems appropriate, and a
better estimate of (13) is $\lambda_{H_S+H_L} = (4 \text{ month})^{-1}$.

No firm link has been established between SST anomalies and cloudiness, hence
the contribution of H_R to feedback is not known. The net long wave radiation H_B
into the mixed-layer may be estimated using Efimova's formula (Reed, 1976):

$$H_B = - 0.97 \, \sigma \, T^4 \, (0.254-0.00495e)(1-\gamma C) \tag{14}$$

where σ is the Stefan-Boltzman constant, e the vapor pressure in millibars, C
the fractional cloud cover and γ a constant depending on the type of cloud (varying between 0.9 and 0.25). The temperature dependence of e can be found by
assuming that e is 80% of the saturation water pressure e_s and using the Clausius-
Clapeyron equation

$$\frac{de_s}{dT} = 5.4 \times 10^3 \, \frac{e_s}{T^2} \tag{15}$$

Introducing $T = \bar{T} + T'$ and keeping only the dominant terms, we find that the
long wave radiation causes a linear negative feedback given by

$$\lambda_{H_B} = \frac{\partial}{\partial T'}(\frac{-H_B'}{\rho C \bar{h}}) = \frac{1.4 \times 10^{-14} \, \bar{T}^3}{\bar{h}} (1 - 0.0156e_s + \frac{21.10}{\bar{T}} e_s)(1 - \gamma C) \tag{16}$$

For $\bar{h} = 50$ m and $T = 295°K$, the negative feedback in the absence of clouds is
$\lambda_{H_B} = (21 \text{ month})^{-1}$. Even smaller values are found when taking cloud effects
into account. A discussion of the negative feedback caused by the atmosphere
can also be found in Gill (1978).

Term (m) contributes to feedback during periods of entrainment, since $\Delta\theta' =$
$T' - \theta'_s$. The character of this intermittent feedback due to vertical mixing
appears to be complex. If there is no anomaly below the mixed layer ($\theta'_s = o$),
term (1) represents a negative feedback, with $\lambda_e = \bar{w}_e/h$. In early fall when the
mixed layer begins to deepen, this negative feedback is very strong. From Camp
and Elsberry (1978), we take as characteristic values $w_e = 6 \times 10^{-6}$ m s^{-1}, $\bar{h} =$
35 m, so that $\lambda_e = (2 \text{ month})^{-1}$. Later in the season, w_e decreases, $\bar{h}$ increases

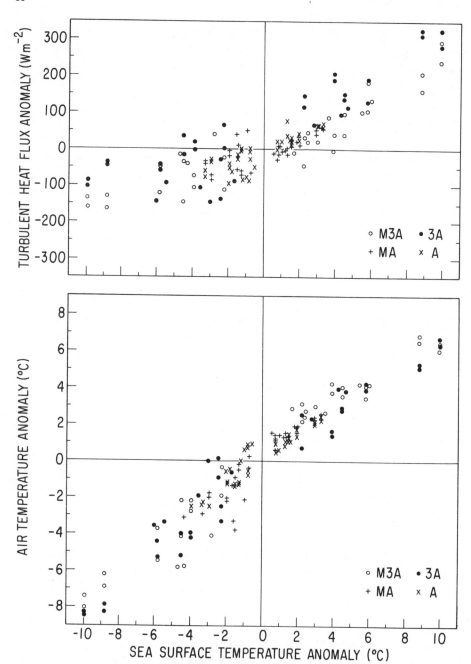

Fig. 2. Scatter plot of 30-day averages of the air temperature anomaly and the sensible plus latent heat flux anomaly into the atmosphere versus imposed SST anomaly at grid points in the GCM experiments by Kutzbach et al. (1977). The different symbols correspond to the 4 experiments described in the paper.

and the negative feedback becomes progressively small. On the other hand, θ'_s will generally contribute an intermittent positive feedback. This effect is likely to be most important for large scale SST anomalies (i.e. not advected away) on time scales of several seasons or more. Indeed, θ'_s will have the same sign as T' during the preceeding winter, thereby favoring the reapparition (or strengthening) of SST anomalies during the fall. This has been observed in the North Pacific (White, private communication). Note that the heat fluxes below the mixed-layer should be taken into consideration for a more detailed study.

Finally, horizontal mixing contributes to SST anomaly decay. If we keep term (o) only (our parameterization is crude anyway) we find

$$\lambda_m = \frac{\partial}{\partial T'} \ (-\chi \nabla^2 T') \tag{17}$$

For $\chi = 10^3$ m^2 s^{-1}, a reasonable value for the MODE region, and wavelength larger than 1000 km, one has $\lambda_m \leq$ (10 month)$^{-1}$. At smaller scales, smaller χ should be used (Csanady, 1973). Hence, horizontal mixing seems to contribute little to SST anomaly dynamics. Our conclusion disagrees with Adem's (1975) suggestion that horizontal mixing is important. However, Adem was using too large a value for χ (= 10^4 m^2 s^{-1}) in his numerical simulations of SST changes.

c. *An approximate equation.* Our analysis suggests that, in a first approximation, the rate of change of SST anomalies can be described by the equation

$$(1 + D') \frac{dT'}{dt} = \frac{Q'}{\rho C \bar{h}} - D' \frac{d\bar{T}}{dt} - \underset{\sim}{U'} . \nabla (\bar{T} + T') - \lambda_M T' \tag{18}$$

where $\lambda_M > o$ represents mixing processes. In high latitudes, the entrainment heat flux should perhaps be added. Combining the forcing and feedback mechanisms, (18) may be written under the form

$$(1 + D') \frac{dT'}{dt} = \frac{v'}{\bar{h}} - \underset{\sim}{U'}_e . \nabla (\bar{T} + T') - \lambda T' \tag{19}$$

where v' represents the short time scale atmospheric forcing (generally dominated by local heat fluxes), $\underset{\sim}{U'}_e$ the eddy velocity and λ the negative feedback processes (probably dominated by $\lambda_{H_S + H_L}$).

Statistical features of the SST anomalies

Equation (19) reduced to equation (4) considered by Frankignoul and Hasselmann (1977) in regions far from strong currents and large eddy activity, if the overall influence of D' in the left-hand side is small (see below). Then, the pre-

dicted SST anomaly spectrum (5) compares generally well with mid-latitudes observations (see in particular Reynolds, 1978). In Fig. 3, SST observations at OWS P
are compared to the model fit (5) using $v' = (H_S+H_L)/\rho C$, $\lambda = (2 \text{ month})^{-1}$ and $\bar{h} =$
25 m. This value of $\bar{h}$ is smaller than the observed mean of 60 m because v' is
underestimated and $\bar{h}$ varies seasonally while entering (5) as a quadratic quantity
(see Frankignoul, 1978).

Fig. 4 reproduces the observed correlation between large scale SST and SLP
anomalies over the North Pacific, and the correlation between SST anomalies and
atmospheric forcing predicted from (4). The agreement is excellent. Here it
was assumed that SLP variations are directly related to v'. Note that SLP is
normally not correlated with air temperature (Willebrand, personal communication)
but is rather coherent with wind speed at low frequencies (Fissel et al., 1976).

As discussed by Frankignoul (1978), a seasonal modulation of the statistical
properties of SST anomalies is expected from (19), since F_v (o), $\bar{h}$, etc. vary
seasonally. Fig. 5 shows seasonal SST spectra at OWS P, and the model fit. The
seasonal variability is well reproduced, except during spring, where much SST
variance is at rather short period. This high-frequency noise is presumably
caused by the term D' in the left-hand side of equation (19), since D' is largest
during that season, $\sim$ O(1). This argument suggests that some high-frequency
variance will be present during summer, but little during fall and winter, as
observed. Note that the characteristic time scale of h' seems to be $\sim$ O(20 days).
The year-round influence of the D' term in (19) should, however, be small, consistent with the observed high-frequency slope of the SST spectrum in Fig. 3,
which is only slightly flatter than ω^{-2}.

The SST anomalies are advected by the mean circulation along the oceanic gyres,
via the derivative term in (19), as observed by Favorite and McLain (1973), and
Davis (1976). In central ocean conditions, the mean currents are generally small
$\lesssim$ O(5 cm s^{-1}) and the spectrum (5) should not be substantially altered, as most
of the SST variance is at very large scales. However, advection may lead to an
apparent increase of the feedback factor λ. Indeed, the SST spectrum at wave-
number $\underset{\sim}{k}$ is given by

$$F_T (\underset{\sim}{k}, \omega) = \frac{F_v(\underset{\sim}{k},o)}{(\omega - \underset{\sim}{k} \cdot \overline{\underset{\sim}{U}})^2 + \lambda^2} \quad ; \qquad \omega \ll \tau_x^{-1} \qquad (20)$$

for uniform mean flow. Using polar coordinates and assuming isotropic forcing
$F_v (\underset{\sim}{k},o) = F_v(k,o)/2\pi k$, we find at low frequencies

$$\int_0^{2\pi} d\theta \; F_T(\underset{\sim}{k},\omega) = \frac{2\pi F_v(\underset{\sim}{k},o)}{\lambda(\lambda^2 + k^2\overline{U}^2)^{\frac{1}{2}}} \quad ; \qquad \omega \ll \lambda \qquad (21)$$

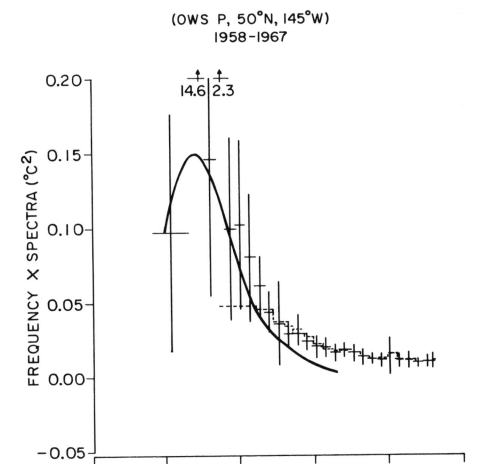

Fig. 3. Variance spectrum of the SST at OWS P, after Fissel et al. (1976). The vertical error bars represent approximately 95% confidence intervals. At the annual and semi-annual periods, the spectral levels are off-scale as indicated by the bold arrows with the numerical values written below the arrows. The continuous curve is the model prediction (from Frankignoul, 1978).

so that the fitted value of λ will increase with k. For $\lambda = (4 \text{ month})^{-1}$, $\bar{U} = 5 \text{ cm s}^{-1}$, the spectral flattening at low frequencies would suggest a feedback factor of $(2 \text{ month})^{-1}$ at 1000 km wavelength, a substantial increase. In real situations, advection will obviously create more complex effects, although they should not be qualitatively different. This could explain in part (together with the effect of entrainment discussed above) why values of λ estimated from observations increase with decreasing SST anomaly scale, and vary with the geo-graphical position (Reynolds, 1978). Fig. 6 illustrates the spatial dependence

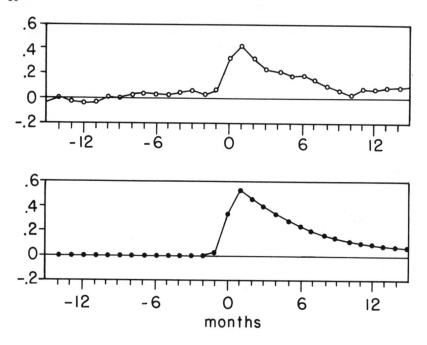

Fig. 4. *Upper panel*: observed correlation between the amplitude of the dominant empirical orthogonal mode of SST and SLP anomalies over the North Pacific, after Davis (1976). *Lower panel*: theoretical correlation curve (from Frankignoul and Hasselmann, 1977).

of λ in the North Pacific, since the SST anomaly scale generally decreases with increasing empirical orthogonal function (EOF) number. Note that the Nyquist wavelength in these data is 1000 km, so that our rough estimate (21) seems quantitatively correct.

CONCLUSIONS

Using a slab model of the oceanic mixed-layer, we have found that large scale mid-latitude SST anomalies can be represented as a first-order autoregressive process in regions with small mean current, as suggested by Frankignoul and Hasselmann (1977). The SST anomalies are continuously generated by the natural variability of the atmospheric fluxes at the air-sea interface. This stochastic forcing seems dominated by the fluctuations in the local heat fluxes, although mixed-layer depth variations are important during certain seasons. In regions with large mean SST gradient, advection by anomalous wind-driven currents also plays a role. The decay of the SST anomalies can be modelled by a linear negative feedback, that seems largely controlled by their back interaction on the atmosphere. The SST anomalies are also advected by the mean oceanic circulation,

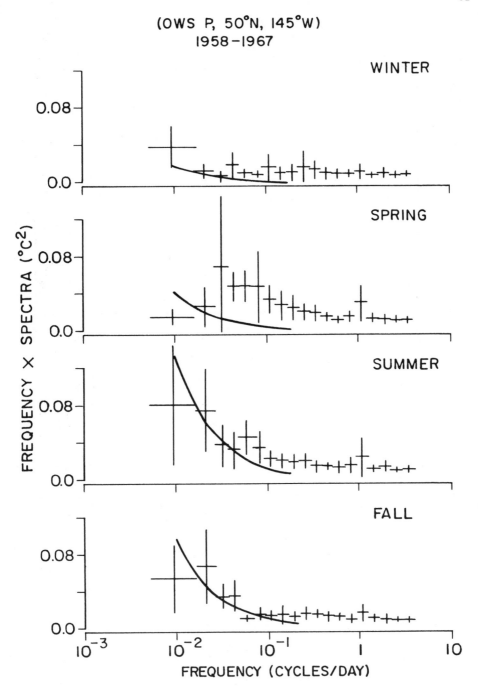

Fig. 5. Observed and predicted variance spectra of SST anomalies at OSW P for each of the seasons, as in Fig. 3 (from Frankignoul, 1978).

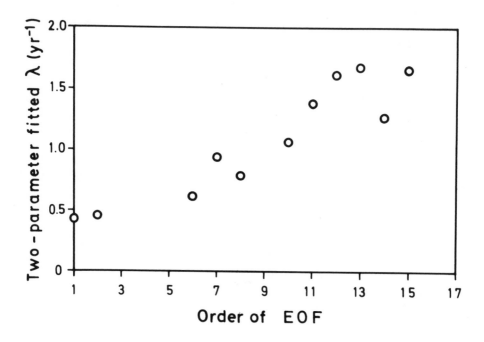

Fig. 6. Fitted value of λ for the EOF of SST anomalies in the North Pacific
that can be represented by the model (4). These values should be multiplied by
2π to compare with values discussed here (from Reynolds, 1978).

and mesoscale eddies contribute essentially a low-frequency, small-scale noise.

On the climatic time scale, the weather forcing is not predictable, hence the
generation of SST anomalies cannot be forecasted. On the other hand, their decay and advection is predictable. Hasselmann (1976) had discussed the predictability of stochastic climate models. In the SST anomaly phase space, the stochastic forcing by the atmosphere induced a diffusion, whereas feedback and
advection cause a propagation. Because of the diffusion, SST prediction will
always entail some degree of statistical uncertainty. For a statistically
stationary SST state, the maximal predictive skill for the SST field may be
significant, but limited to skill parameters of order 0.5.

A substantial part of the SST anomaly predictability is controlled by the
feedback process, and it is therefore of much importance to investigate these
mechanisms further. The strength of the negative feedback varies with the SST
anomaly scale, the geographical position and the season. Since the intensity
of the forcing (the unpredictable part) also varies with season, the SST predictive skill is likely to depend on the time of year.

For climate prediction purposes, the back interaction of the SST anomalies must be understood globally, rather than locally. The problem is formidable, and progress will have to rely mostly on numerical simulations with GCM of the atmosphere having some kind of interacting oceanic surface layer. The observations suggest that the climate predictability varies with the season, and it seems of interest to establish if this is partly caused by the seasonal variability of the SST anomaly field. Perhaps it is not a coincidence that Davis (1978) found some climate predictability for fall and winter, while SST anomalies are largest and most persistent during summer and fall (Fig. 5).

ACKNOWLEDGEMENTS

The GCM data were provided by Dr. R. Chervin who is gratefully acknowledged. Support from the National Science Foundation, office of the International Decade of Oceanic Exploration, under grant OCE 77-28349 is gratefully acknowledged. This is MODE Contribution number 107.

REFERENCES

Adem, J., 1975. Numerical-thermodynamical prediction of monthly ocean temperature. Tellus, 22:541-551.

Bjerknes, J., 1969. Atmospheric teleconnections from the equatorial Pacific. Mon. Wea. Rev., 97:163-172.

Camp, N.T. and Elsberry, R.L., 1978. Oceanic thermal response to strong atmospheric forcing II. The role of one-dimensional processes. J. Phys. Ocean., 8:215-224.

Csanady, G.T., 1973. Turbulent diffusion in the environment. Dordrecht and Boston, D. Reidel Publishing Co., 248 pp.

Daly, W.T., 1978. The response of the North Atlantic sea surface temperature to atmospheric forcing processes. Quart. J. Roy. Met. Soc., 194:363-382.

Davis, R.E., 1976. Predictability of sea surface temperature and sea level pressure anomalies over the North Pacific Ocean. J. Phys. Ocean., 6:249-266.

Davis, R.E., 1977. Techniques for statistical analysis and prediction of geophysical fluid systems. J. Geophys. Astrophys. Fluid Dyn., 8:245-277.

Davis, R.E., 1978. Predictability of sea level pressure anomalies over the North Pacific Ocean. J. Phys. Ocean., 8:233-246.

Favorite, F. and McLain, D.R., 1973. Coherence in transpacific movements of positive and negative anomalies of sea surface temperature, 1953-60. Nature, 244:139-143.

Fissel, D.B., Pond, S. and Miyake, M., 1976. Spectra of atmospheric quantities at ocean weathership P. Atmosphere, 14:77-97.

Frankignoul, C., 1978. Stochastic forcing models of climate variability. Dyn. Atmos. Ocean, to appear.

Frankignoul, C., and Hasselmann, K., 1977. Stochastic climate models, part II. Tellus, 29:284-305.

Gill, A.E., 1975. Evidence for mid-ocean eddies in weather ship records. Deep-Sea Res., 22:647-652.

Gill, A.E., 1978. Remarks on stochastic models of climate. Dyn. Atmos. Ocean, submitted.

Gill, A.E. and Niiler, P.P., 1973. The theory of the seasonal variability in the ocean. Deep-Sea Res., 20:141-177.

Hasselmann, K., 1976. Stochastic climate models, part I. Tellus, 28:473-485.

Hasselmann, K., 1978. Linear statistical models. Dyn. Atmos. Ocean, to appear.

Hasselmann, K., and Barnett, T.P., 1978. Techniques of linear prediction, with application to oceanic and atmospheric fields in the tropical Pacific. To be published.

Haworth, C., 1978. Some relationships between sea surface temperature anomalies and surface pressure anomalies. Quart. J. Roy. Met. Soc., 104:131-146.

Jacob, W.J., 1967. Numerical semiprediction of monthly mean sea surface temperature. J. Geophys. Res., 72:1681-1689.

Kraus, E.B., and Turner, J.S., 1967. A one-dimensional model of the seasonal thermocline: II. Tellus, 19:98-106.

Kutzbach, J.E., Chervin, R.M., and Houghton, D.D., 1977. Response of the NCAR general circulation model to prescribed changes in ocean surface temperature, Part I. J. Atmos. Sci., 34:1200-1213.

Leith, C.E., 1975. Climate response and fluctuation dissipation. J. Atmos. Sci., 32:2022-2026.

Lorenz, E.N., 1956. Empirical orthogonal functions and statistical weather prediction. Report No. 1, Statistical Forecasting Project, Dept. Meteorology, M.I.T.

Namias, J., 1963. Large-scale air-sea interactions over the North Pacific from summer 1962 through the subsequent winter. J. Geophys. Res., 68:6171-6186.

Namias, J., 1965. Macroscopic association between mean monthly sea-surface temperature and the overlying winds. J. Geophys. Res., 70:2307-2318.

Namias, J., 1969. Seasonal interactions between the North Pacific ocean and the atmosphere during the 1960's. Mon. Wea. Rev., 97:173-192.

Namias, J., 1976. Negative ocean-air feedback systems over the North Pacific in the transition from warm to cold seasons. Mon. Wea. Rev., 104:1107-1121.

Namias, J., 1978. Multiple causes of North American abnormal winter 1976-1977. Mon. Wea. Rev., 106:279-295.

Newell, R.E., and Weate, B.C., 1976. Ocean temperature and large scale atmospheric variations. Nature, 262:40-41.

Niiler, P.P., 1977. One-dimensional models of the seasonal thermocline. In: The Sea, E.D. Goldberg (editor), J. Wiley and Son, Inc., pp. 97-115.

Pedlosky, J., 1975. The development on thermal anomalies in a coupled ocean-atmospheric model. J. Atmos. Sci., 32:1501-1514.

Pettersen, S., Bradbury, D.L., and Pederson, K., 1962. The Norwegian cyclone models in relation to heat and cold sources. Geophys. Publn., 24:243-280.

Ratcliffe, R.A.S., and Murray, R., 1970. New lag associations between North Atlantic sea temperature and European pressure applied to long-range weather forecasting. Quart. J. Roy. Met. Soc., 96:226-246.

Reed, R.K., 1976. On estimation of net long-wave radiation from the oceans. J. Geophys. Res., 81:5793-5794.

Reiter, E.R., 1978. The interannual variability of the ocean-atmosphere system. J. Atmos. Sci., 35:349-370.

Reynolds, R.W., 1978. Sea surface temperature in the North Pacific ocean. Tellus, 30:97-103.

Salmon, R., and Hendershott, M.C., 1976. Large-scale air-sea interactions with a simple general circulation model. Tellus, 28:228-242.

Simpson, J., 1969. On some aspects of sea-air interaction in middle latitudes. Deep-Sea Res., Suppl. to vol. 16:233-261.

Shukla, J., 1975. Effect of arabian sea-surface temperature anomaly on Indian summer monsoon: a numerical experiment with the GFDL model. J. Atmos. Sci., 32:503-511.

Tenberth, K.E., 1975. A quasi-biennial standing wave in the Southern Hemisphere and interrelations with sea surface temperature. Quart. J. Roy. Met. Soc., 101: 55-74.

Voorhis, A.D., Schroder, E.H., and Leetmaa, A., 1976. The influence of deep mesoscale eddies on the sea surface temperature in the North Atlantic subtropical convergence. J. Phys. Ocean., 6:953-961.

White, W.B., and Barnett, T.P., 1972. A servomechanism in the ocean-atmosphere system of the mid-latitude North Pacific. J. Phys. Ocean, 4:372-381.

LOW FREQUENCY MOTIONS IN THE NORTH PACIFIC AND THEIR POSSIBLE GENERATION BY
METEOROLOGICAL FORCES

L. MAGAARD

Department of Oceanography, University of Hawaii, Honolulu, Hawaii 96822 (U.S.A.)

ABSTRACT

    Various sets of internal temperature data from the North Pacific Ocean (time
series up to 20 years) have been analyzed with respect to baroclinic Rossby waves.
Fluctuations with periods between one and five years can, to a large extent, be
explained as first order internal Rossby waves. A theoretical model for local
meteorological generation of these Rossby waves has been developed. An application
of this model requires the frequency wave number spectra of wind stress and buoyancy
flux (or temperature) at the sea surface as forcing functions. Such spectra are
presently being determined from observations in the eastern North Pacific.

INTRODUCTION

    Our present understanding of oceanic processes varies greatly from process to

process. For example, much successful work has been done to observe and describe

the sea state, to understand its generation, and to forecast it (e.g. Barnett and

Kenyon, 1975). Much less, however, is known about low frequency processes in the

ocean. We recognize now, mainly from temperature observations, the existence of

significantly strong oceanic fluctuations with time scales of several months to

several years (e.g. White, 1977). But the identification of processes underlying

such fluctuations has just begun. The generation mechanisms of such processes are

still largely unknown, and the prediction of such processes is a very distant goal.

    In this paper we review some of our work, done under the NORPAX project, utilizing

temperature data from the North Pacific to learn more about low frequency oceanic

motion. We have analyzed data by fitting models to them in order to identify

low frequency processes, and we have studied one of the many possible generating

mechanisms of such processes. We consider this a prerequisite for the final

goal of forecasting low frequency processes in the ocean.

DATA ANALYSIS AND MODEL FITTING

    We have at our disposal the following data sets:

1) Monthly mean values of temperature from both hydrographic and XBT casts at

various depths in six 2-degree squares between California and Hawaii. These five-year records are described in Emery and Magaard (1976).

2) Annual mean values of baroclinic potential energy (a bulk measure for the baroclinic temperature fluctuations in the upper 500 m) in the area 20-50N, 145E-130W. These 20-year series have been prepared by White (1977).

3) Objectively analyzed monthly mean values of temperature between San Francisco and Hawaii from 1966-1974. These data have been prepared by Dorman and Saur (1978).

4) The TRANSPAC data. These XBT data have been collected since 1975 under the "Anomaly Dynamics Study" (ADS) program, a subproject of NORPAX. Objectively analyzed monthly maps of temperature in the upper 400 m of the area 30-50N, 160E-130W are currently being prepared by Bernstein and White (Scripps Institution of Oceanography). The records go back to the beginning of 1975. Maps from the first 3½ years are now available. We hope that the TRANSPAC program can be continued over several more years.

We have started our work on data set 1) by developing a model consisting of a random field of free baroclinic Rossby waves. For this model we assumed vanishing mean flow, an assumption that is justified for the area of data set 1). Using a cross spectral fit of the Rossby wave model to the data we have shown that for wave periods from one to two years, 62 to 78% (average 70%) of the observed variances and covariances of the internal temperature field can be interpreted as first order baroclinic Rossby waves traveling at distinct NW directions (one direction per frequency) with wave lengths between 1200 and 1700 km. The group velocities of these waves are directed almost exactly westward and have magnitudes of about 4.5 cm s^{-1}. For smaller wave periods (from the cutoff period of five months to about nine months) the model can interpret only 38 to 73% (average 52%) of the variances and covariances, and the angular range of propagation directions of these shorter waves widens to isotropic propagation in the western half plane. The results of this study are described in Emery and Magaard (1976). In a subsequent paper, Magaard and Price (1977) fitted a more general model, containing the Rossby waves as a special case, to the same data. The result was that, for wave periods from one to two years, the Rossby waves led to the best fit again, which reinforced the idea that baroclinic Rossby waves play a central role in the low frequency fluctuations of the North Pacific.

Data set 2) is presently under study by Price and Magaard. From preliminary results it appears that Rossby waves may not be limited to the restricted area of data set 1) but also may play a role in larger portions of the North Pacific. Moreover, even fluctuations with periods larger than two years (up to about five years) seem to be dominated by Rossby waves.

The analysis of the TRANSPAC data requires a model that considers the North Pacific Current which dominates the area of these data. Such a model was developed

by Kang and Magaard (1978). Their work shows that the influence of the current
on the dispersion features of the Rossby waves is small for the barotropic shear
mode, but is significant for the baroclinic shear mode. A model fit to the TRANSPAC
data is now in preparation. A preliminary study has shown that Rossby waves
appear to be of importance in this case, too. We conclude that even though much
of the nature of the low frequency motions is still unknown, baroclinic Rossby
waves do represent a significant portion of such motion.

GENERATION OF ROSSBY WAVES

After part of the low frequency motion has been identified as Rossby waves
the obvious questions are, of course: why are these waves there? Where do they
come from? We have studied one of the possible generating mechanisms, local
meteorological forcing by low frequency fluctuations of barometric pressure,
wind stress, and buoyancy flux at the sea surface (Magaard, 1977). We have shown
that in this context the barometric pressure is negligible. Wind stress and
buoyancy flux produce divergent horizontal flow (wind-driven and thermohaline
circulation, respectively) in surface boundary layers (viscous and diffusive
layers, respectively). The divergence of this flow leads to vertical pumping
at the lower edge of the surface boundary layers; the pumping then generates
internal Rossby waves in the continuously stratified interior of our model ocean.
A viscous bottom layer results in a finite response even in case of resonance to
which we have given most attention. A rough numerical estimation shows that
actual fluctuations of wind stress as well as buoyancy flux could generate the
waves.

Emery (University of British Columbia), Gallegos (Texas A&M University) and
Magaard are presently trying to determine actual input functions for the generation
model by Magaard (1977). These functions are frequency wave number spectra of
wind stress and buoyancy flux. Since we cannot get enough data for the determination of the buoyancy flux and since the model provides a connection between
the buoyancy flux and the sea surface temperature, we use the temperature spectra
instead of the buoyancy flux spectra and calculate the former. This study is
based on 11-year time series of wind stress and sea surface temperature from the
area 10-40N, 120-160W, which includes the area of data set 1). We want to test
whether the waves observed by Emery and Magaard (1976) can be generated by means
of the model by Magaard (1977) using the above-mentioned wind stress and sea
surface temperature data that stem from the Marine Deck of the Environmental
Data Service, Asheville, North Carolina, U.S.A. Preliminary results suggest
that the annual fluctuation of the sea surface temperature should generate the
strongest internal Rossby wave response, which contradicts the observational

60

finding that in the actual Rossby wave field the annual signal does not play a preferred role. To what extent local meteorological forcing actually generates Rossby waves remains unsolved at this time. We hope, however, that our further studies will shed more light on that problem.

Another mechanism of internal Rossby wave generation has been proposed: Bryan and Ripa (1978) have done a theoretical study about the generation of Rossby waves by reflection of a meteorologically generated, eastward-traveling surface fluctuation at the North American Pacific coast. While their highly idealized study shows that such a mechanism is possible in principle, it remains open to what extent this mechanism is actually efficient.

DISCUSSION

We expect the continued acquisition of data and the development of advanced methods of analysis will increase our insight into the nature of the observed fluctuations. We believe, however, that it will take much more time and effort to successfully tackle the generation problem because of its complexity. That leaves the final goal of predicting the low frequency fluctuations far away.

ACKNOWLEDGEMENTS

This study has been supported by the Office of Naval Research under the North Pacific Experiment of the International Decade of Ocean Exploration; this support is gratefully acknowledged. Hawaii Institute of Geophysics contribution no. 923.

REFERENCES

Barnett, T.P. and Kenyon, K.E., 1975. Recent advances in the study of wind waves. Rep. Prog. Phys., 38:667-729.
Bryan, K. and Ripa, P., 1978. The vertical structure of North Pacific temperature anomalies. J. Geophys. Res., 83:2419-2429.
Dorman, C.E. and Saur, J.F.T., 1978. Temperature anomalies between San Francisco and Honolulu, 1966-74, gridded by an objective analysis. J. Phys. Oceanogr., 8:247-257.
Emery, W.J. and Magaard, L., 1976. Baroclinic Rossby waves as inferred from temperature fluctuations in the Eastern Pacific. J. Mar. Res., 34:365-385.
Kang, Y.Q. and Magaard, L., 1978. Stable and unstable Rossby waves in the North Pacific Current as inferred from the mean stratification. Dyn. Atmos. Oceans, in press.
Magaard, L., 1977. On the generation of baroclinic Rossby waves in the ocean by meteorological forces. J. Phys. Oceanogr., 7:359-364.
Magaard, L. and Price, J.M., 1977. Note on the significance of a previous Rossby wave fit to internal temperature fluctuations in the Eastern Pacific. J. Mar. Res., 35:649-651.
White, W.B., 1977. Secular variability in the baroclinic structure of the interior North Pacific from 1950-1970. J. Mar. Res., 35:587-607.

WIND-INDUCED LOW-FREQUENCY OCEANIC VARIABILITY

Jürgen Willebrand[1] and George Philander

Geophysical Fluid Dynamics Program, Princeton University,
Princeton, New Jersey 08540

[1]Present affiliation: Institut für Meereskunde an der Universität Kiel,
2300 Kiel, Germany

ABSTRACT

   The generation of low-frequency current fluctuations in the ocean
by variable winds is reconsidered. Observed spectral characteristics
of the wind field over the ocean are discussed. A simplified analytical
model is used to derive spectral properties of the oceanic response
to broadband atmospheric forcing. The results qualitatively agree
with a more realistic numerical model, and may help to explain some
aspects of deep-sea current meter observations.

INTRODUCTION

   Variability in space and time of all fields which describe the
oceanic state is an ubiquitously observed phenomenon. The following
remarks will be concerned with variability on time scales from a day
to several months, and spatial scales from a few hundred to a few
thousand kilometers. (Of course, oceanic variability is not confined
to those scales).

   A major reason for the current interest in these fluctuations is
their potential ability to affect the large-scale, long-time behaviour
of the atmosphere-ocean system. Variable currents ("eddies") can interact with the general circulation, either extract energy from or feed
energy into the mean currents. Also, they can contribute to the total
oceanic heat transport, which is especially important in regions where
the mean flow is zonal (e.g., antarctic circumpolar current). Here
any poleward heat flux is possible only due to fluctuating currents.

   Fluctuations in the thermal structure of the upper layers of the
ocean also fall into that range of scales. Anomalies in sea-surface
temperature (SST) can influence the atmospheric circulation and most

likely contain the clue towards any progress in the climate prediction problem; see Frankignoul, 1979 (this issue) for a further discussion and references.

One obvious candidate for the generation of oceanic variability is direct forcing by the atmosphere, as the relevant fields of wind stress, atmospheric pressure and buoyancy flux at the air-sea interface are themselves highly variable in space and time. The relative importance of several mechanisms is discussed in Magaard, 1977, and Frankignoul and Müller, 1978. Here we are concerned with some overall aspects of atmospheric generation processes.

The discussion will be led in terms of a stochastic model, and it is appropriate to make a few remarks on the philosophy behind this approach. The term stochastic is to be understood as opposed to deterministic, not as opposed to dynamical. Stochastic and deterministic models are identical with respect to the governing equations of motion They differ in that stochastic models explicitly account for the broad band of scales which is present in atmospheric as well as oceanic motions. There still will be a dominant scale but that can be quite different in both systems even if the dynamical model is linear, depending on the details of the oceanic response. Thus, a stochastic forcing model is - at least in principle - able to account for the observed differences in scale between oceanic and atmospheric fluctuations, the latter having larger space and smaller time scales. A further characteristic of stochastic models is that they are normally evaluated in terms of certain statistical parameters, e.g. probability distributions, correlation functions or energy spectra, whereas the directly observable fluctuating fields are considered to be realizations of random functions.

Stochastic forcing models are by no means new to oceanography. Since the original work of Phillips, 1957, they have been used frequently in surface and internal gravity wave studies. More recently, this approach was applied to processes with much larger space and time scales (Frankignoul and Hasselmann, 1977; Lemke, 1977; Frankignoul and Müller, 1978). In the following, after a brief discussion of atmospheric forcing spectra, we consider a very idealized model for the generation of fluctuating currents which may help to relate observed oceanic current spectra to the meteorological fields.

SPECTRAL CHARACTERISTICS OF ATMOSPHERIC FLUCTUATIONS

The relevant atmospheric variables at the sea surface are wind

stress $\underset{\sim}{\tau}$, atmospheric pressure p^a, and buoyancy flux b^f. Instead
of the buoyancy flux which is poorly known from observations, we will
use the atmospheric temperature as variable, thereby implicitly assuming that the buoyancy flux is essentially proportional to the flux
of sensible heat, and that the air temperature fluctuates much faster
than the sea temperature. Within the framework of a correlation theory,
the statistical properties of these fields are completely determined
by the spectral tensor

$$F_{ij} (\underline{k}, \omega) = \int d\underline{r} \ d\tau < \zeta_i (\underline{x}, t) \zeta_j (\underline{x} + \underline{r}, t + \tau) > e^{i(\underline{k}\underline{r} - \omega\tau)} \tag{1}$$

Here $\underset{\sim}{\zeta} = (\zeta_1, \zeta_2, \zeta_3, \zeta_4) = (\tau_1, \tau_2, p^a, T^a)$ denotes the vector of atmosperic variables; otherwise the notation is standard.

We consider a typical mid-ocean, mid-latitude situation and disre-
gard complications due to inhomogeneities and instationarities in the
atmospheric fields, thereby excluding diurnal and annual signals.

As the network of weather stations over the oceans is rather sparse,
our empirical knowledge about $F_{ij}(\underline{k}, \omega)$ is not overly large. Recent
attempts to obtain information on F_{ij} from atmospheric data have been
made by Willebrand, 1978, and Emery et al, 1978. Frankignoul and
Müller, 1978, constructed a simple analytical expression for the
wavenumber structure of the stress-spectra at low frequencies, which
is consistent with several known characteristics of these fields. From
these investigations the most important properties of the atmospheric
spectra can be described as follows:

   i) The frequency autospectra, $F_{ii}(\omega) = \int F_{ii}(\underline{k}, \omega) d\underline{k}$ are w h i t e ,
      i.e. frequency independent, below a certain frequency ω_o. The
      value of ω_o corresponds to a period of 3 - 5 days for the stress
      spectra, and is somewhat longer for pressure and temperature
      spectra. At higher frequencies, all spectra fall off approxima-
      tely as $\omega^{-\alpha}$, with $\alpha \approx 1.5$ for the stress spectra, and $\alpha \approx 2.5$ for
      pressure and air temperature.

  ii) At periods larger than 10 days, the fluctuations have no pre-
      ferred direction and are s y m m e t r i c in wavenumber, i.e.
      $F_{ii}(\underline{k}, \omega) = F_{ii}(-\underline{k}, \omega)$. Accordingly, there is an equal amount of
      eastward and westward propagating energy. Only at higher fre-
      quencies the eastward propagating disturbances (cyclones) domi-
      nate. With regard to the north-south direction, the fluctuations
      are nearly symmetric at all frequencies.

 iii) For zonal wave numbers in the range 8-15, the pressure spectrum
      behaves as $k^{-\beta}$ with $\beta = 0(5)$. (Strictly, the conclusion from

atmospheric data is a $k_1^{-\beta}$-law). It is not well known how far this power law extends to higher wavenumbers. Also, the directional structure of the wavenumber spectra is not known. At low frequencies one may assume horizontal isotropy which is in accordance with the observed symmetry, but other non-isotropic spectral distributions would also be consistent with the data.

The various components of the spectral tensor are of course not independent, their relation depending on the kinematics of the atmospheric flow. The simplest model is to relate the wind stress linearly to the geostrophic velocity, leading to

$$F_{11}(\underline{k},\omega) = c^2 k_2^2 \, F_{33}(\underline{k},\omega)$$
$$F_{22}(\underline{k},\omega) = c^2 k_1^2 \, F_{33}(\underline{k},\omega)$$
$$F_{12}(\underline{k},\omega) = -c^2 k_1 k_2 F_{33}(\underline{k},\omega) \tag{2}$$
$$F_{13}(\underline{k},\omega) = -i c k_2 F_{33}(\underline{k},\omega) \qquad \text{etc.}$$

with a suitably chosen constant c.

This crude model ignores the differences between geostrophic and surface wind and hence describes a non-divergent stress field, a deficiency which can be removed easily on the cost of slightly more complex algebra than in (2). Furthermore, the nonlinear character of the velocity-stress relation is ignored in (2). The actual stress spectra are therefore more flat (white) than those given by (2), both in frequency and wavenumber, especially at high frequencies (1-10 d period). Nevertheless, the model is suitable for qualitative discussions.

With regard to the buoyancy flux resp. air temperature, a relation to the pressure field is less obvious. Frankignoul and Hasselmann, 1977, assumed that temperature changes are due to advection only. With a mean north-south temperature gradient, this leads to $T^a \sim u_2$ or

$$F_{44}(\underline{k},\omega) = c^{*2} k_1^2 \, F_{33}(\underline{k},\omega) \tag{3}$$

and analogous relations for $F_{i4}(\underline{k},\omega)$, with another constant c^*.

Fig. 1 shows the observed coherence between air temperature and north wind velocity at Ocean Weather Station D. At periods from 2 days to 200 days both variables are well (though not perfectly) correlated, their phase difference being very small. Here the approximation (3) works reasonably well, whereas it fails both at lower and higher frequencies.

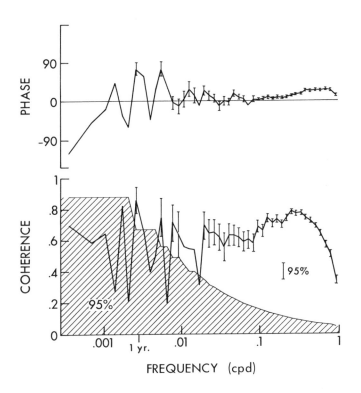

Fig. 1. Coherence and phase difference between north wind velocity and air temperature at Ocean Weather Station D (44°N,41°W), based on observations between January 1949 and April 1972. Phase positive if wind velocity leads air temperature.

A SIMPLE STOCHASTIC MODEL

The nature of the oceanic response to variable winds, air pressure or buoyancy flux strongly depends on the space and time scales of the fluctuations. Especially, it is crucial whether or not these scales coincide with those of free waves in the ocean, in which case resonance can occur. In the subinertial frequency range, the only free waves are Rossby waves which have a westward phase propagation. Because the atmosphere is dominated by eastward propagating cyclones, it has been argued that the resonance mechanism is rather unlikely to occur. However, this argument only applies to frequencies above 0.1 cpd, as the atmospheric asymmetry fades away at lower frequencies. The situation is visualized in the frequency-wavenumber diagram of Fig. 2. The shaded area indicates for each frequency where most of the atmospheric energy is concentrated. This distribution is based on an analysis of weather maps (Willebrand, 1978) and representative for the North Pacific from 30° - 50°N. It is clear that resonant generation of barotropic Rossby waves by fluctuating winds is by no means unlikely. (Baroclinic Rossby waves have a much longer time scale and are not considered here). The resonance mechanism should be most effective around a period of 10 days, and the lower (upper) frequency limit will depend mainly on the east-west (north-south) dimension of the oceanic basin.

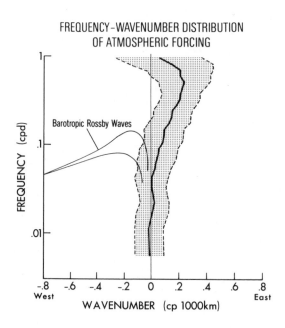

FREQUENCY-WAVENUMBER DISTRIBUTION
OF ATMOSPHERIC FORCING

Fig. 2. Frequency-wavenumber diagram indicating energy distribution in the atmospheric pressure field over the North Pacific. The heavy full line is the mean zonal wavenumber, defined as $\bar{k}_1(\omega) = \int d\underline{k}\ k_1\ F_{33}(\underline{k},\omega) / \int d\underline{k}\ F_{33}(\underline{k},\omega)$. The shaded area is limited by $\pm$ one standard deviation. The dispersion curves for barotropic Rossby waves are plotted for two different values of the meridional wavenumber

How will the ocean react if forced at off-resonant frequency-wave-number combinations? Philander, 1978, has demonstrated that the response can be trapped near the surface, the vertical trapping scale depending on ω and $\underline{k}$. It turns out that, for the time scales of interest here, a critical horizontal scale is O(100 km). For larger scales, the vertical trapping scale exceeds the depth of the ocean, and the forced motion is nearly depth-independent, i.e. barotropic, whereas for smaller scales the forced motion is trapped and baroclinic. The actual amount of atmospheric energy at these small scales is poorly known, and thus an estimate of the baroclinic response must be based on extrapolation of available atmospheric spectra (cf. Frankignoul and Müller, 1978).

The large scale response, however, can be deduced directly from atmospheric observations, as e.g. weather maps. The simplest model is a vertically integrated, linear model with quasigeostrophic dynamics. Neglecting variations in bottom topography, the vorticity balance in terms of the stream function $\psi(\underline{x},t)$ takes the familiar form

$$\frac{\partial}{\partial t}\, \nabla^2 \psi + \beta\, \frac{\partial \psi}{\partial x_1} = f_A \tag{4}$$

where f_A is the atmospheric forcing function. It is sufficient to consider the case of wind stress forcing, $f_A = (\nabla \times \underline{\tau})_3$. The corresponding expressions for pressure and buoyancy flux forcing are given by Magaard, 1977. Pressure forcing is generally negligible in the time scale range considered here (except for its influence on sea level). The importance of buoyancy flux forcing depends on the diffusivity of the upper layer of the ocean and has yet to be established.

Eq. (4) has two limiting cases, depending on the ratio of the time scale T to the characteristic time scale T_R of Rossby waves. For $T \ll T_R$, the first term on the left hand side of (4) dominates, and planetary effects are unimportant. For $T \gg T_R$, the planetary term is dominating, and the solution is in form of a Sverdrup balance which adiabatically adjusts to changes in the wind field.

In either case, the frequency spectra of oceanic variables can be expressed in terms of the spectrum of curl τ, $F_c(\underline{k},\omega)$. In the high-frequency limit, that relation is

$$F_\psi(\omega) = \{\ \int d\underline{k}\, F_c\, (\underline{k},\omega)\ /k^4\}/\omega^2 \tag{5}$$
$$= \overline{F_c(\omega)\ k^{-4}}\omega^{-2}$$

with the definition $\overline{(...)} = \int d\underline{k}(...)\, F_c(\underline{k},\omega)/\int d\underline{k}\, F_c(\underline{k},\omega)$.

The quantity $\overline{k^{-4}}$ in (5) depends only weakly on ω. From the discussion of the wind spectra in the previous section, we therefore expect a power law $F_\psi(\omega) \sim \omega^{-q}$ with $q \approx 3.5$ at periods between 1 and 3 days and $q \approx 2$ between 3 and 10 days. Analogous conclusions can be derived for the spectra of current components.

In the low-frequency limit, the corresponding relations are in terms of the current spectra

$$F_{u_1}(\omega) = \beta^{-2} F_c(\omega) \overline{k_2^2/k_1^2}$$

$$F_{u_2}(\omega) = \beta^{-2} F_c(\omega) \tag{6}$$

The magnitude of $\overline{k_2^2/k_1^2}$ depends on details of the directional distribution of the wind spectrum. Typically, one finds $k_2^2/k_1^2 \approx k_1^2/k_2^2 >> 1$ almost independent of frequency, and hence the magnitude of zonal current fluctuations will dominate that of meridional currents, even if there is no preferred direction in the atmosphere. Furthermore, both current spectra will be white, as the wind stress curl spectrum is also white in that frequency range.

Fig. 3 shows current spectra taken from a numerical model which calculated the oceanic response to actually observed wind stress fluctuations of the North Pacific (Willebrand et al., 1979). That model relaxes several of the constraints which were used to derive (5) and (6), e.g. linearity, quasi-geostrophy, idealized wind field, absence of lateral boundaries. Nevertheless, the results support the above conclusions. The spectral peaks around periods of 10 -20 days can be identified as resonant basin modes, their amplitude and location is determined by basin size and frictional effects.

It is remarkable that the rather steep slopes of current spectra immediately below the inertial frequency, which are generally observed in deep-water current meter records (cf. Thompson, 1971), can be deduced from a simple atmospheric forcing model. Thus, the model constitutes an alternative to nonlinear cascade arguments which over the well-known k^{-3}-law for geostrophic turbulence and a Taylor-hypothesis also lead to a ω^{-3} spectral law.

However, the shape of energy spectra of oceanographic variables is notoriously insensitive to different theories regarding their origin, and more specific consequences of the atmospheric forcing mechanism must be considered. From (4) one might expect that atmosperic and oceanographic fluctuations are correlated. A detailed analysis of (4) shows, however, that the local correlation is generally low,

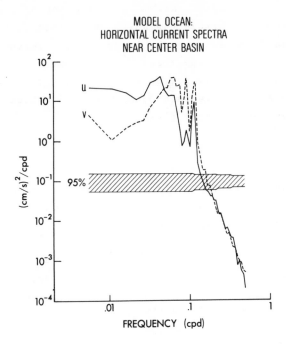

MODEL OCEAN:
HORIZONTAL CURRENT SPECTRA
NEAR CENTER BASIN

Fig. 3. Frequency spectra of east (u) and north (v) components of
ocean currents, computed from a numerical model of an idealized ocean basin resembling the North Pacific which was
driven by observed wind stress fluctuations.

partly due to the broad-band forcing spectrum, and partly due to the
wavelike nature of oceanic response (Willebrand et al, 1979). Only
at the highest frequencies, the simplified model (4) predicts coherence between certain atmospheric and oceanographic variables. That
coherence is reduced further by inhomogeneities in bottom topography
and the forcing fields which are ignored in (4). This picture agrees
with observational experience: no local correlation has been observed
at low frequencies, whereas at high frequencies occasionally significant (if marginal) correlation has been found (e.g. Brown et al, 1975;
Baker et al, 1977; Meincke and Kvinge, 1978).

In conclusion, we state that a stochastic model for the atmospheric
generation of oceanic variability can, in contrast to a deterministic
one, account for some observed features of oceanic current fluctuations,
namely the longer time scales in the ocean (oceanic spectra are much
more red than atmospheric spectra) and the lack of strong local co-

herence between atmosphere and ocean. However, only barotropic motions can be explained in this way, because the horizontal scale of wind fluctuations as inferred from weather maps is too large to generate baroclinic motion, except at very low frequencies. In order to explain baroclinic eddies, which frequently dominate upper ocean variability, in terms of atmospheric forcing, we need much more information on the small-scale structure of the meteorological fields.

ACKNOWLEDGEMENTS

We thank Mr. R.C. Pacanowsky for assistance in the numerical computations. This work has been supported through the Geophysical Fluid Dynamics Laboratory - NOAA Grant No. 04 - 7 - 022 - 44017.

REFERENCES

Baker, D.J.Jr., W.D. Nowlin,Jr., R.D. Pillsbury and H.L. Bryden, 1977. Space and time fluctuations in the Drake passage. Nature, 268:696-699

Brown, W., W. Munk, F. Snodgrass, H. Mofjeld and B. Zetler, 1975. MODE bottom experiment. J. Phys. Oceanogr. 5:75-85

Emery, W.J., A. Gallagos and L. Magaard, 1978. Frequency-wavenumber spectra of wind stress and sea surface temperature in the eastern North Pacific. J. Phys. Oceanogr. (submitted)

Frankignoul, C., 1979. Large scale air sea interactions and climate predictability. In: J.C.J. Nihoul (Editor), Marine Forecasting, Elsevier (this issue)

Frankignoul, C. and K. Hasselmann, 1977. Stochastic climate models, part II. Tellus, 29:284-305

Frankignoul, C. and P. Müller, 1978. Quasi-geostrophic response of an infinite β-plane ocean to stochastic forcing by the atmosphere. J. Phys. Oceanogr. (in press)

Lemke, P., 1977. Stochastic climate models, part III, application to zonally averaged energy models. Tellus, 29:385-392

Magaard, L., 1977. On the generation of baroclinic Rossby waves in the ocean by meteorological forces. J. Phys. Oceanogr., 7:359-364

Meincke, J. and T. Kvinge, 1978. On the atmospheric forcing of overflow events. ICES, C.M. 1978/C:9, Hydrographic committee

Philander, S.G.H., 1978. Forced oceanic waves. R. Geophys. Space Phys. 16:15-46

Phillips, O.M., 1957. On the generation of waves by turbulent wind. J. Fluid Mech. 2:417-445

Thompson, R., 1971. Topographic Rossby waves at a site north of the Gulf stream. Deep-Sea Res. 18:1-19

Willebrand, J., 1978. Temporal and spatial scales of the wind field over North Pacific and North Atlantic. J. Phys. Oceanogr. (in press)

Willebrand, J., S.G.H. Philander and R.C. Pacanowsky, 1979. The oceanic response to large-scale atmospheric disturbances. In preparation

A DISCUSSION OF WAVE PREDICTION IN THE NORTHWEST ATLANTIC OCEAN

C. L. VINCENT[1] and D. T. RESIO[1]
[1]Wave Dynamics Division, Hydraulics Laboratory
[2]U. S. Army Engineer Waterways Experiment Station, Vicksburg, Miss. (USA)

ABSTRACT

   Simulation of a wave climate for the Atlantic Ocean through hindcasting is
discussed in terms of methods for obtaining historical wind fields and available
numerical models for hindcasting directional spectra. Available, gridded pressure
and wind data produced recently by the U. S. Navy Fleet Numerical Weather Central
are shown to be distorted near major storms leading to the necessity of redigitizing
major storm areas from synoptic weather charts. The optimal method for using
most of the available oceanographic data to produce the wind fields is one in
which the surface wind field is estimated from the pressure field, and temperature
(air and sea) fields through a planetary boundary layer model. Afterwards,
the available ships wind-field observations are blended into the estimate. A
root-mean-square error of less than 3 mps on speed appears obtainable.
   Several numerical models for directional spectral wave estimation are reviewed.
Each model is examined in terms of source mechanisms and propagation schemes.
Models with wave-wave interaction source terms appear to perform better in tests
of wave growth with fetch as well as in field estimation. Models with a fourth-
order or ray propagation schemes appear adequate for oceanic hindcasts while first
order propagation schemes appear to disperse.

INTRODUCTION

   The prediction of sea state through the use of numerical models has become an

important method for estimating wave climates. Although the simple wave prediction

schemes based on the research of Sverdrup and Munk (1947) as modified are still

widely used, several recent studies have used a variety of numerical models that

calculate directional spectra for hindcasting wave climates. Included are major

studies on the Great Lakes (Resio and Vincent, 1976), the Atlantic and Pacific

Oceans (Lazanoff and Stevenson, 1977), and the North Sea (NORSWAM, 1977). The

rise in application of numerical methods is in part due to cost-effective computer

technology, and in part due to the short time frame required to compute a wave climate compared with the time and cost of performing an extensive wave-measuring program. Recent studies indicated that sea-state prediction with numerical models can have random error below 1 metre, which leads to the expectation that wave climates constructed numerically will have no more error than those measured, particularly when the difficulties inherent in an observational program are considered. An additional benefit of the hindcast studies is the estimation of the wave direction since this parameter is generally not measured in observational programs.

The U. S. Army Corps of Engineers has major responsibilities in the United States in the areas of navigation in coastal waters and shore protection, and, consequently, is one of the primary wave-data consumers in the U. S. As part of its Field Data Collection Program, the Corps of Engineers at the U. S. Army Engineer Waterways Experiment Station (WES) is sponsoring a study to hindcast wave climates for the Atlantic and Pacific Ocean and Gulf of Mexico coastal areas based on the concepts developed in the Corps-sponsored studies of the Great Lakes (Resio and Vincent, 1976). The study provides for hindcasts in the open ocean that serve as a boundary for hindcasts, at a finer scale, on the continental shelf and in nearshore waters. The shelf and nearshore studies will include refraction and shoaling effects. Also included will be a hindcast of storm tides so that the joint probabilities of water level and wave height in shallow waters can be estimated. A final part of the project will be the development of a computer-based information system to contain the data base and allow site-specific calculations based on a combination of detailed refraction studies and the data base by users at the many Corps field offices around the U. S. Further, it is the aim of this project to thoroughly document the procedures used and to provide an estimate of the error involved in each method.

This paper presents a discussion of some practical problems encountered in designing and operationalizing a system of numerical routines coupled with a data base to produce a deep-ocean wave climate from historical meteorological information. The sources of meteorological data and models available to compute ocean wind fields at a hindcast wind level and the errors involved in such an effort will be discussed. Several available methods for predicting sea state will be discussed in terms of their capability to reproduce theoretical results as well as their ability to predict observed sets of wave data. Discussion will center upon tests in the Northwest Atlantic Ocean off the Canadian Maritime provinces and some results obtained in wave tests in the Great Lakes. Later phases of the project will treat verification of the wave model in diverse locations.

WIND DATA

Sources of Meteorological Data

Our primary source of historical meteorological data over the ocean is the millions of observations taken aboard ship and archived on magnetic tape at the National Climatic Center in the U. S. To augment these data are land stations in the world meteorologic network. The U. S. Navy Fleet Numerical Weather Central (FNWC) has recently contracted with Meteorology International Incorporated, Monterey, California, to rework and edit these data into a 63×63 point regular, square grid over the Northern Hemisphere of both winds and surface pressures on synoptic time levels for FNWC's own hindcast efforts. The WES study initially accepted these data as a basis for WES wind estimates; but on the basis of a published review of the wind estimates by Lazanoff and Stevenson (1977) and concurrent examination of the data at WES, WES has decided to apply a modified method for obtaining winds. An additional effort jointly funded by WES and FNWC is the addition of several million ship observations available but not in the marine deck to this data base.

Review of the 63×63 Grid Pressures

Lazanoff and Stevenson (1977) reported that the hindcast winds obtained by Holl's (1976) method had an approximate root-mean-square (RMS) error in wind speed of 7 mps and a bias of 3 mps compared with selected observations. It was further reported that the winds were consistently too low above speeds of 15 mps. Clearly some improvement in the wind speed estimates is desired for accurate predictions of extreme wave conditions.

WES review of the 63×63 data began by a comparison (Fig. 1) of contoured 63×63 pressures to correspond to U. S. National Weather Service (NWS) surface weather charts, North American and Northern Hemisphere series, which served as an initial basis of the FNWC charts. In the 63×63 grid, the averaging or blending procedures to mix ships observations into the meteorologic data tend to smooth the sharp pressure gradients in storms, particularly the smaller ones. This is likely due to the inclusion of Laplacian-type operators in the blending process of Holl (1976) which requires evaluation of derivations over several of the grid steps in the 63×63 grid and may result in averaging of values over 1000 miles apart. For a selection of large storms, WES calculated the maximum geostrophic winds in each quadrant of each storm for both the NWS and FNWC charts (Fig. 2). The results show an under estimation of the NWS values. A plot of central pressures suggests also that the central pressures in the NWS charts tend to be lower than FNWC-sponsored estimates (Fig. 3).

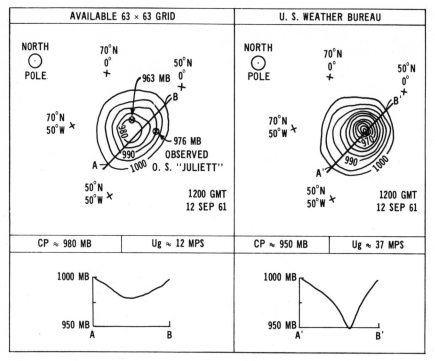

Fig. 1.   Comparison of low pressure areas on the 63×63 grid and the NWS charts. The difference in central pressure is 30 mb and in geostrophic wind is 25 mps. The transects AB and A'B' show the difference in the storm cross sections.

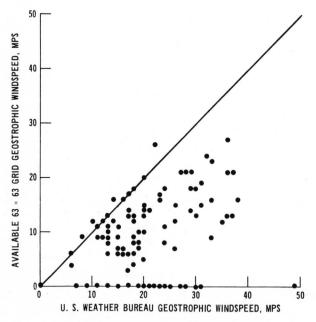

Fig. 2.   Comparison of geostrophic winds analyzed on the 63×63 grid and NWS charts.

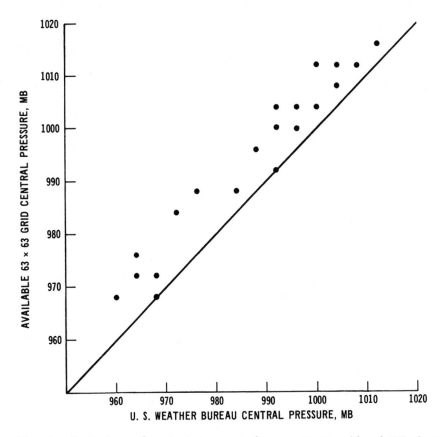

Fig. 3. Comparison of central pressures between 63×63 grid and NWS charts.

These comparisons were primarily drawn for large storms exiting the North American land mass and are, therefore, of primary interest for prediction of waves on the U. S. Atlantic coast. This area has many meteorologic stations and the coastal areas are major shipping lanes. In this area, the observational data are quite good; and the storms have been intensively monitored on their track across the North American continent. Consequently, there is little reason to expect a consistent error in these charts immediately after the storm has departed the mainland. Our results tend to reinforce the conclusion of Lazanoff and Stevenson (1977).

A Method for Improved Pressure Estimates

WES initially attempted to find a simple empirical corrective function to readjust the 63×63 grid data. No function however appeared satisfactory. As a result, the central areas (approximately 720 nautical miles square) of every major storm along the U. S. Atlantic coast for the 25-year period 1952-1977 is

being redigitized from the NWS charts and will be inserted into the corresponding
63×63 grid to produce a new pressure field. From the old 63×63 grid, pressures
will be interpolated on an approximately 50-mile grid. In the storm center
areas where new digitized data are available, the pressure data will be replaced
by a new value p

$$p = \alpha(x,y)p_1 + \beta(x,y)p_2 \tag{1}$$

where

p_1 is the old pressure estimate,

p_2 is the new value, digitized from the NWS charts,

$\alpha(x,y)$ and $\beta(x,y)$ are blending coefficients chosen so that $\alpha + \beta = 1$ with $\beta = 1$
$\alpha = 0$ in the 200-mile square around the storm center and $\alpha = 1$, $\beta = 0$ at the
edge of the digitized square.

This preserves the NWS data at the storm center but blends the data smoothly
into the FNWC pressure values away from the storm center.

Derived Wind Estimates

Since the FNWC pressures are being reconstructed, it is feasible to use a
planetary boundary layer (PBL) model to derive wind velocities at the 19.5-
metre level required in the wave method rather than the empirical method of Holl
(1976). A PBL is desirable because it is a physical model of the processes
relating geostrophic and lower level winds and provides an opportunity to incorporate
both the stability and baroclinicity of the lower atmosphere into the wind
estimates. The PBL model chosen is a recent upgrade of that of Cardone (1969).
This model has been shown to produce results with an RMS error of less than
2 mps (Overland and Street, 1977) for cases of prediction from geostrophic
conditions. Air-sea temperature differences will be derived from the 5-day mean
sea surface temperatures and ships observations of atmospheric temperature.

The construction of the air temperature fields from ships data is made difficult
by the erratic locations of the ships. A value is constructed at sites where
there are no data by an algorithm which accounts for both spatial and temporal
gradients.

Blending of Ships Wind Data

The ships observations of wind speed will be blended into the wind field derived
through the PBL model from the pressure estimates because they are observations
made independent of pressure. The quality of the wind observations is varied.
There are inconsistencies in observation level and method. However, comparison
of ships wind speed and direction to instrumented observations at Sable Island
indicates reasonable agreement between the two data sets. Accordingly, the
ships wind data are a valuable addition to the data base, and it is expected
that these data will tend to correct wind fields that are somewhat smoothed
because of their pressure grid origin. The method used will be of a restricted
type. Ships observations of winds will be allowed to influence only the three

nearest grid values and will result in a smoothing only on the order of 100-200 miles. The blending algorithm is of the form

$$q'_i = q_i + \sum_{j=1}^{n} \alpha_j \Delta q_j \tag{2}$$

where

q'_i is the blended value at grid location i,

q_i is the value at grid i derived from the pressure field,

α_j is a weighting based on the position of the ships observations,

Δq is the difference between the value of q at the ship and the grid location,

n is the number of ships within the triad of grid points.

The value q may be wind speed or a wind-speed component. The blending is restricted to the nearby grid points to prevent oversmoothing of the wind field.

Sources of Error in Wind Estimates

The wind velocities that are input to a wave model for a series of hindcasts represent an amalgamation of atmospheric and oceanographic data from different sources, observed at different levels by differing methods, that have been used to estimate spatial-field values from which wind velocities are eventually derived through a series of numerical models and approximations. In this process, there are many sources of error; some can be minimzed, but the rest are inherent to the data sets and cannot be reduced.

Since the precise level, placement, and manner in which the millions of ships observations of wind were made is not known, it is impossible to remove these sources of error. Locations at which the data were taken are fixed in history, and it is obviously impossible to ascertain historical data for places and times where it does not exist. Attempts to extrapolate data in time and space into regions of sparse data are subject to a high degree of uncertainty which cannot be removed by objective analysis. It also is impossible to account for observer error in any precise sense, although an editing scheme may catch the larger inconsistencies.

The errors that generally can be minimized result from the processing of the basic data into derived quantities usually through the use of numerical models. The interpolation functions used to form gridded data and the numerical difference schemes to compute gradients are based on subjective concepts of what is reasonable. Their accuracy generally increases as the grid system on which they are applied is more closely scaled to the size and magnitude of variations which they approximate. Thus, error is reduced by proper selection of grid size and appropriate order of derivative approximations. The transformation of the wind velocities from a geostrophic level to a level suitable for input to the wave model is through a numerical model that solves for the wind profile near the water surface. It is appropriate to select a model that is both unbiased and has minimal random error.

78

The objective of this exercise is to provide methods that take the basic data available, massage it, and transform it in such manner that the information is not degraded by the analyses and only a minimal amount of error is introduced by subsequent extension of the data. The ultimate evaluation of the methods is through comparison of diagnosed and observed winds. A small set of comparisons has been completed for Sable Island anemometer (Fig. 4) and shows reasonable agreement. The RMS error in these wind-speed estimates is about 2.5 mps. More evaluations involving a wide range of sites are under way. Ultimately, the error comparisons will involve all major NOAA data buoys in the Atlantic, Pacific, and Gulf of Mexico and should provide one of the more extensive tests of wind estimates over the open ocean.

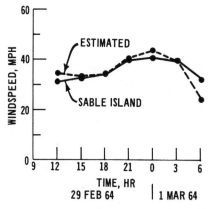

Fig. 4. Comparison of wind speed observed at Sable Island, by ships nearby, and prediction from the geostrophic wind. The geostrophic winds were derived from NWS charts. The Cardone (1969) boundary layer model was used to predict the wind speed, neglecting baroclinicity.

WAVE MODELS

Directional spectral wave prediction models generally contain two primary sets of computational algorithms excluding input-output and peripheral information handling: source term calculations and energy propagation. Source term calculations are the numerical mechanisms which simulate energy (a) transfer from the atmosphere to the spectrum, (b) transfer within the spectrum, and (c) dissipation through breaking. The energy propagation algorithms are the numerical mechanisms which simulate the propagation of energy across the water body. For purposes of WES evaluation of wave models, the source and propagation algorithms of a number of published wave models were considered separately under the presumption that the best wave model would be a combination of the best source terms with the best propagation scheme. It would be difficult to conceive of a propagation scheme whose errors counterbalance the errors of a set of source terms over a diverse range of wave-generation conditions.

Source Terms

The calculation of the energy input and exchange on the spectral calculations can be considered of two types. Parametric source term models are those in which some property of the spectrum such as wave height or period is estimated in terms of wind speed, duration, and fetch. Discrete spectral source term models are models that treat the spectrum in a discrete number of frequency and direction bands. The following discussion will treat only those source terms considered in the WES study and were selected on the basis of widespread use. A wider discussion of wave source terms is given in Resio et al. (1978).

Discrete Spectral Methods

In these methods, the energy balance equation is solved for a number of discrete frequency-direction (f, θ) bands and the source mechanisms directly contribute to each band. Integral properties of the spectrum are obtained through direct integration over the frequency direction space. In a simple expression, the energy transfer in and out of a spectral component (f, θ) is given by

$$\frac{\partial F}{\partial t} (f, \theta) = S_1 + S_2 + S_3 \tag{3}$$

where

S_1 is the energy exchange with the atmosphere,

S_2 is the energy exchange within the spectrum by conservative, nonlinear wave-wave interactions,

S_3 is the irreversible loss of energy due to turbulent interactions and wave breaking.

The discrete source term models treated in this study are those of Barnett (1968), Ewing (1971), and Salfi (1974) and a modification of Barnett and Ewing source terms involving a relaxation on the equilibrium range "constant" as described by Resio and Vincent (1977).

The source terms of Barnett and Ewing can be represented as in the energy balance equation by

$$\frac{\partial F}{\partial t} (f, \theta) = [a + b \, F(f, \theta)] \, [1 - \mu] + \Gamma - \tau F(t, t) \tag{4}$$

In both Ewing and Barnett

a parameterizes the Phillips (1957) resonance mechanism,

b parameterizes the Miles (1957) mechanism,

Γ, τ parameterize the nonlinear wave-wave interactions of Hasselmann et al. (1973),

μ parameterizes the wave-growth limiting term.

The details of the formulae of a, b, Γ and τ are not given here, but the differences between the models are relatively minor. In both models, a fully directional spectrum is treated.

Salfi (1974) source terms (used in the U. S. Navy Fleet Numerical Weather

Central model) are quite different. Salfi only solves the one-dimensional energy balance equation

$$\frac{\partial E}{\partial t} (f) = A[1 + \lambda]^{1/2} + BE'[1 - \lambda] \tag{5}$$

where

E is the one-dimensional (frequency) energy density,

A is a linear mechanism and a function of f, and wind speed,

B is a nonlinear mechanism,

E' is the energy density within 90° of the wind direction,

λ is a ratio E/E_∞,

E_∞ is a fully developed energy density.

It is clear that his model does not treat the details of the directional spread of energy and will clearly only be appropriate where there is not much energy propagating transverse to the wind or where there are not sudden shifts in wind direction.

In comparing, in a one-dimensional sense, these source terms with those of Ewing and Barnett, it is evident that they can be equivalent only if A incorporates the effects of a + Γ and B accounts for b - τ . Resio et al. (1978) point out that on the forward face of the spectrum wave growth in the Ewing-Barnett type of terms versus those of Salfi can only be calibrated to give the same source for only a narrow range of energy. It can also be seen that if Γ is proportional to E^n (Barnett gives it as E^3) then A must be a function of E to be equivalent to a + Γ ; however A is not a function of E .

Parametric Models

The parametric model considered in the study is the model proposed by Hasselmann et al. (1976). In this model a pair of differential equations are written to solve for the nondimensional peak frequency of the spectrum $\hat{f}$ and the nondimensional Phillips equilibrium coefficient $\hat{\alpha}$ as a function of wind stress. This simplification is achieved by assuming constancy of spectral shape and assuming that the spectrum changes sufficiently fast so that the wave direction rapidly adjusts to the wind direction. This model only treats active wave growth and decay and must be interfaced with a swell propagation routine.

Tests of the Source Terms

In the field it is very difficult to measure the contribution of the individual source term mechanisms. The detailed field studies of Mitsuyasu (1968) and Hasselmann et al. (1973), however, provide excellent evidence for the growth of wave height with fetch and are in close agreement. Unfortunately, no such well-defined curves are available for growth with time. It was decided to examine the growth of wave height

$$\hat{H} = \frac{g\sqrt{E}}{u_*^2} \qquad (6)$$

with dimensionless fetch.

$$\hat{F} = g\frac{F}{u_*^2} \qquad (7)$$

where

  g is gravitational,

  u_* is friction velocity of air for the wind condition,

  F is fetch,

  E is total energy in the wave field

predicted by the differing sets of source terms. Two wind velocities were input
15 and 30 mps.

  Fig. 5 presents the results of the tests. Barnett and Ewing are closer to
the curves of Mitsuyasu and Hasselmann than the curves generated from the FNWC
model (Salfi, 1974). The Barnett and Ewing curves do deviate considerably at
30 mps. The parametric model of Hasselmann et al. (1973) is not shown because
it was derived to fit his field data.

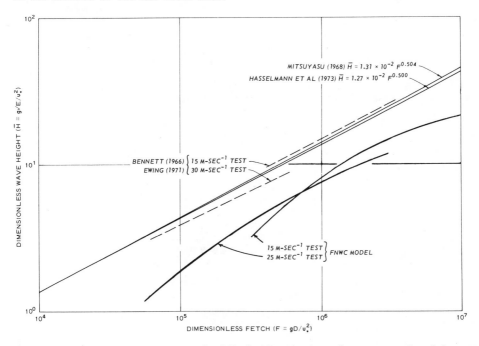

Fig. 5. Comparison between growth-with-fetch relations from spectral models
and empirical studies.

  Resio and Vincent (1977) show that the difficulties with Barnett's parameterizat
are due to the assumption of a constant Phillips equilibrium value α of 0.0081.
By parameterizing α as a function of dimensionless wave height, the Barnett

curves are brought into agreement with the field data (Fig. 6). This result is confirmed by a range of field data (Fig. 7).

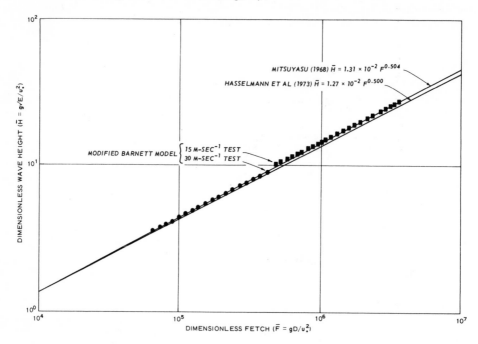

Fig. 6. Comparison between results from the model of Resio and Vincent (1977 and empirical formulae for relations between nondimensional fetch and wave height.

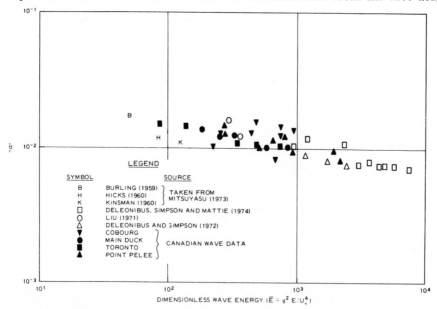

Fig. 7. Variation of Phillips equilibrium coefficient as a function of dimensionless wave energy.

Propagation Schemes

In studies which seek only to describe local sea, the swell propagation problem is not important. However, calculation of a wave climate that will be used for sediment transport requires solution of this problem. The general equation in one dimension for propagation is

$$\frac{\partial E}{\partial t} = C_g \frac{\partial E}{\partial X} \tag{8}$$

and can be solved by finite differences or through ray methods (sometimes called phonon or method of characteristics (techniques).

Finite Differences

Finite difference solutions are based on the approximation of the propagation equation based on a Taylor series expansion in time and space. Three types of finite difference solutions are treated here: a first order scheme (used in the FNWC model), a fourth order scheme (used by Ewing, 1971) and a Lax-Wendroff (Lax and Wendroff, 1960) scheme modified by Gadd (Golding, 1977) which involves second order expansions in space and time. The equations of each scheme are presented in Table 1.

These propagation schemes were evaluated by examining their abilities to propagate a one-dimensional wave envelope. In a practical sense, the most desirable propagation algorithms are those which conserve wave energy and do not change its spatial distribution. Fig. 8 provides the results of tests to examine

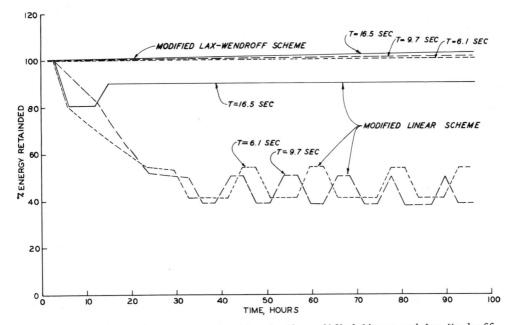

Fig. 8. Comparison of energy conservation in the modified linear and Lax-Wendroff propagation schemes. (Note that the analytical, ray, and fourth order methods all essentially retain 100 percent of the energy and are not shown).

TABLE 1

Propagation Schemes

A. FNWC (Salfi, 1974)

$$E_i^{n+1} = E_i^n + \mu(E_{i-1}^n - E_i^n)$$

Unconditionally stable upwind differencing scheme modified by computational logic to reduce diffusion of swell.

B. Ewing (1971)

$$E_i^{n+1} = E_i^n + [\frac{1}{12} \mu/(1 - \frac{1}{12} \mu)] \ [E_{i+2}^n + 8(E_{i-1}^{n+1/2} - E_{i+1}^{n+1/2}) - E_{i+2}^{n+1}]$$

Stable for $\mu \leq 2$ requires alternate grid/staggered mesh

C. Lax-Wendroff (Golding, 1977)

Step 1

$$E_{i+1/2}^{n+1/2} = \frac{1}{2} (E_i^n + E_{i+1}^n) - \frac{1}{2} \mu(E_{i+1} - E_i)$$

Step 2

$$E_i^{n+1} = E_i^n - \mu[(1+a) \ (E_{i+1/2}^{n+1/2} - E_{i-1/2}^{n+1/2}) - \frac{a}{3} (E_{i+3/2}^{n+1/2} - E_{i-3/2}^{n+1/2})]$$

Stable for $\mu \leq 2$, requires two-step calculation.

i   grid point

n   time level

$\mu = C_g \ \Delta t/\Delta x$ with C_g wave group velocity

$a = 3/4 \ (1-\mu^2)$

energy conservation as a function of wave period for the modified Lax-Wendroff and the FNWC schemes. The Lax-Wendroff and fourth order schemes had very similar results and have very slight amplification of energy (2 percent). To reduce confusion in the figure, the fourth order results were deleted. A comparison of the modified Lax-Wendroff and fourth order schemes is given in Golding (1977). The FNWC propagation scheme showed a marked dependence upon wave period, with, a significant loss of energy. For a propagation time of 24 hours, the loss was 10 percent for a period, in all cases, of 16.5 sec, and over 50 percent for both 6.1- and 9.7-sec waves. Fig. 9 provides examples of the diffusive properties of the scheme. Again the Lax-Wendroff and fourth order schemes characteristics are only weakly frequency-dependent and show excellent preservation of wave envelope shape. The FNWC scheme is markedly frequency-dependent and has the undesirable property of severely distorting the wave envelope.

Ray Methods

Propagation of wave energy through so-called ray or phonon methods is analagous to rays developed through geometrical optics for wave refraction. The water body is covered with a dense network of rays for each propagation direction. Along these rays are a series of storage locations spaced as a function of wave frequency through which the wave energy is jumped in time. Such methods have propagation properties equivalent to the analytical solution for energy features with space scale larger than the spacing of points along a ray. This method tends to be computationally swift, but requires considerable computer storage. Preliminary analysis of this technique of the Atlantic Ocean indicates that 1,800,000 storage locations requiring approximately 230,000 60-bit words on a Cyber 176 computer are needed.

Combination of Source Term and Propagation

One major reason for splitting the analysis of wave models on the basis of source terms and propagation was to obtain comparison of the wave models against common standards. Table 2 presents a compendium of the major models that embody the various operational combinations of the source terms and propagation schemes and provides estimates of the model errors in published verification studies. Unfortunately, it is virtually impossible to put into operation all these models in the same grid and test them because of the diverse natures of the techniques and in some instances the operationalization of the models in highly site-specific manners.

Based on the theoretical tests discussed in this paper, the following comments can be made concerning the models:

(1) The source terms used by Hasselmann et al. (1976) and Barnett (1968) as modified by Resio and Vincent (1977) and Ewing (1971) provide a more accurate growth with fetch than the FNWC source terms of Salfi (1974).

86

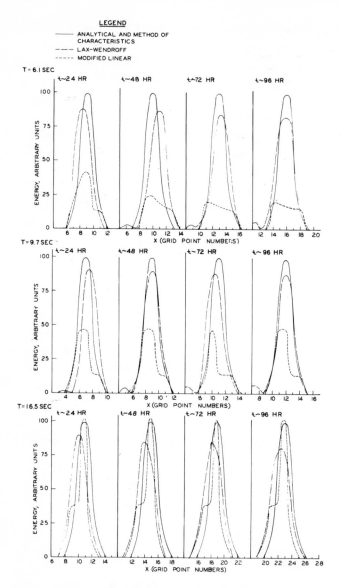

Fig. 9. Comparison of energy diffusion characteristics of the finite difference propagation schemes.

(2) The fourth order, modified Lax-Wendroff and ray methods provided suitable propagation of wave energy over the entire range of the spectrum of interest.

(3) The first order, modified scheme of Salfi (1974) should not be used for propagation of intermediate period (6-12 sec) wave energy for periods longer than a few hours because it does not conserve energy and greatly distorts the wave energy envelope.

Examination of Table 2 would indicate that the performance of these models in verification studies parallels the theoretical tests. The Hasselmann, Barnett,

TABLE 2

Model Comparisons

Model	Characteristics	Source Terms	Growth with Fetch	Model Comparisons
Salfi (1974)	Modified First Order Scheme	L/E; WB	Low by 25%-50%	Lazanoff & Stevenson (1975) N=11; σ=1.1m Bias=+0.7m Lazanoff & Stevenson (1977) Ocean Data Systems, Inc. (1975) N=76; σ=1.5m Bias=+0.1m A. NOAA-EB-01 N=59; σ=1.3m Bias=-5% B. NOAA-EB-03 N=48; σ=2.5m Bias=+20%
Ewing (1971)	Fourth Order Scheme	L/E; WW; WB	30% Low to 20% High	Ewing (1971) N=32; σ=1.0m Bias=-0.7m
Barnett (1968)	Method of Characteristics	L/E; WW; WB	30% Low to 20% High	Barnett (1968) N=12; σ=0.6m Bias=0.7m
Resio, Vincent (1976)	Hybrid Linear-Analytical	L/E; WW; WB	Unbiased	Resio & Vincent (1978) N=123; σ=0.5m Bias=0.1m
Hasselmann et al. (1976)	Mixed Finite Difference and Method of Characteristics	P	Unbiased	NORSWAM Project σ=1.0m Bias=-5%

[1] L/E - linear and experimental wind input; WW - wave-wave interactions; WB - wave breaking; P - parametric.
[2] N is number of comparisons; σ is rms error in significant wave height.

and Ewing type models generally perform better than that of Salfi. Part of the increased error in the Salfi model studies may be the difficulties of wind field over the ocean; however, a significant part would appear due to deficiences in the source terms and propagation scheme. The FNWC version of this model is currently having its propagation algorithms changed.

Parametric Versus Discrete Models

Parametric models such as Hasselmann et al. (1976) are the most recent advance in wave modeling. Because of their computational simplicity, they require less computer time. However, for the general open-ocean problem they must be coupled to a swell propagation routine. Two assumptions of the single parameter model which may not be strictly valid in an ocean model are that spectral shape is invariant and that the wave angle so rapidly adjusts to the wind angle that the wind angle and the wave angle are equal. The invariance of spectral shape for duration limited waves has been questioned by Mitsuyasu (1968). The relaxation time for wave angle adjustment may also be sufficiently long that considerable variation in directionality and angular dispersion of an active sea away from the wind angle occurs. Forristall et al. (1978) show a hurricane spectrum in which the direction for forward face of the spectral peak is approximately 90° different from the higher frequency portion of the spectrum. Parametric models using more parameters are being formulated and will likely resolve these difficultie The discrete spectral models can be adjusted to represent the effects of variable wind angles as demonstrated by Forristall et al. (1978) and do not require assumptions regarding spectral shape, other than in the parametric version of the wave-wave interaction source terms. At present they appear more flexible in terms of handling the many ambiguous wave generation cases that occur.

Summary

The hindcast of wave conditions requires an accurate resolution of the wind field and an equally accurate solution of the wave generation and propagation equations. Results of this study indicate that great care must be taken to obtain wind data on a grid that does not smooth out the pressure gradients responsible for large wave conditions. These data can be obtained from a detailed analyses of the synoptic weather charts and ships observations with an RMS error of about 3 mps. The prediction of sea state can apparently be made with those source terms incorporating conservative nonlinear wave-wave interactions to a higher accuracy than source terms without them. Wave propagation through a fourth order or Lax-Wendroff finite difference schemes or through a ray propagation technique appears more accurate than linear schemes.

ACKNOWLEDGEMENT

The authors wish to acknowledge the permission of the U. S. Army Engineer Waterways Experiment Station to publish this paper.

REFERENCES

Barnett, T. P., "On the Generation, Dissipation, and Prediction of Windwaves," Journal of Geophysical Research, Vol. 73, No. 2, 1968, pp. 513-529.

Cardone, V. J., "Specification of the Wind Distribution in the Marine Boundary Layer for Wave Forecasting," Tech. Rept. 69-1, Geophysical Science Laboratory, New York University, 1969, 99 pp.

Ewing, J. A., "A Numerical Wave Prediction Model for the Atlantic Ocean," Deutsche Hydrographische Zeitschrift, Vol. 24, 1971, pp. 241-261.

Forristall, G. Z., E. G. Ward, L. E. Borgman, and V. J. Cardone, "Storm Wave Kinematics," to be presented at the 1978 Offshore Technology Conference, Houston, Texas, 8-11 May 1978.

Golding, Brian, W., "A Depth Dependent Wave Model for Operational Forecasting," Meteorological Office, Berkshire, England, 1977, unpublished manuscript.

Hasselmann, K., T. P. Barnett, E. Bonws, H. Carlson, D. C. Cartwright, K. Enke, J. Ewing, H. Gienapp, D. E. Hasselmann, P. Kruseman, A. Meerburg, P. Muller, D. J. Olbers, K. Richter, W. Sell, H. Walden, "Measurements of Wind-Wave Growth and Swell Decay During the Joint North Sea Wave Project (JONSWAP), Deutshes Hydrographisches Institut, Hamburg, 1973, 95 pp.

Hasselmann, K., D. B. Ross, P. Muller, and W. Sell, "A Parametric Wave Prediction Model," Journal Physical Oceanography, Vol. 6, 1976, pp. 200-228.

Holl, M. M. "The Upper Air Analysis Capabilities, FIB/UA: Introducing Weighted Spreading, Project M-213, Final Report, Contract No. N-000228-75-C2374, for Fleet Numerical Weather Central, Monterey, California, 1976.

Hydraulic Research Station, NORSWAM Technical Advisory Group Report, Hydraulic Research Station, Wallingford, England, 1977

Lax, P. D. and Wendroff, B., "Systems of Conservation Laws," Communications on Pure and Applied Mathematics, Vol. 13, 1960, pp. 217-237.

Lazanoff, S. M. and Stevenson, N. M., "An Evaluation of a Hemispheric Operational Wave Spectral Model," Technical Note No. 75-3, Fleet Numerical Weather Central, 1975.

Lazanoff, S. M. and Stevenson, N. M., "A Northern Hemisphere Twenty Year Spectral Climatology," Preprint of Paper presented at NATO Symposium on Turbulent Fluxes through the Sea Surface, Wave Dynamics and Prediction, Ile de Bendon, France, 12-16 September 1977.

Miles, J. W., "On the Generation of Surface Waves by Shear Flows," Journal Fluid Mechanics, Vol. 3, 1957, pp. 185-204.

Mitsuyasu, H., "On the Growth of Wind Generated Waves (I)," Rept. Res. Inst. for Appl. Mech., Kyushu University, Fukuoker, Japan, Vol. 16, 1968, pp. 459-482.

Overland, J. and R. Street, "Winds in the New York Bight," Journal of Physical Oceanography, 1977, V. 7, pp. 200-228.

Phillips, O. M., "On the Generation of Waves by Turbulent Wind," Journal Fluid Mechanics, Vol. 2, 1957, pp. 417-445.

Resio, D. T. and C. L. Vincent, "Design Wave Information for the Great Lakes, Report 1, Lake Erie," U. S. Army Engineer Waterways Experiment Station, CE, TR H-76-1, Vicksburg, Miss. 1976, 54 pp.

Resio, D. T. and Vincent, C. L., "A Numerical Hindcast Model for Wave Spectra on Water Bodies with Irregular Shoreline Geometry, Report. 1: Test of Nondimensional Growth Rates," U. S. Army Engineer Waterways Experiment Station, CE, MP-H-77-9, August 1977, 53 pp.

Resio, D. T. Vincent, C. L., "A Numerical Hindcast Model for Wave Spectra on Water Bodies with Irregular Shoreline Geometry, Report 2: Model Verification with Observed Wave Data," U. S. Army Engineer Waterways Experiment Station, CE, MP H-77-9, to be published in 1978.

Resio, D. T., A. W. Garcia, and C. L. Vincent, Preliminary Investigation of Numerical Wave Models, Coastal Zone 78, ASCE, March 1978, p. 2085-2104.

Salfi, Robert E., "Operational Computer Based Spectral Wave Specification and Forecasting Models," City University of New York, University Institute of Oceanograp 1974, 130 pp.

Sverdrup, H. U. and W. H. Munk, Wind, Sea, and Swell: Theory of Relations for Forecasting," H. B. Pub. No. 601, U. S. Navy Hydrographic Office, Washington, D. C., 1947, 44p.

WAVE HEIGHT PREDICTION IN COASTAL WATER OF SOUTHERN NORTH SEA

S. ARANUVACHAPUN [1]

[1]Mekong Secretariat, ESCAP, United Nations, Bangkok (Thailand)

ABSTRACT

The paper discusses local effects such as wave refraction due to the irregular bottom topography in the nearshore region around the East Anglian Coast. It demonstrates how wave refraction and shoaling effects could be introduced into a simple wind-wave relation in order to form a basic model to estimate wave height in the area. Results from the study suggest that locality is rather important to the predictibility of the sea surface waves in the region being investigated.

INTRODUCTION

Although there are numerous wave prediction models in the literature, very few of these have taken regional effects into account. The models that allow for local effects can be very important because they may be more reliable than general models, where locality is a dominant effect. To demonstrate local affects, wave predictions in the nearshore region around the East Anglian coast are presented. The wave characteristics in this part of the Southern North Sea are very complex, due to the sand bank system (see Figure 1) which makes the bottom topography highly irregular and causes wave refraction. It has been shown that wave refraction in such an area is rather pronounced and that wave height can be affected by the refraction (Aranuvachapun, 1977a). General wave prediction models neglecting the refraction effect on wave height for example, by assuming a flat sea floor (as in the early model introduced by Sverdrup and Munk (1947) and improved by Bretschneider (1958), were applied to the area. The results from these models are no more satisfactory than the results from a simpler model which allows for the effect of wave refraction.

The refraction can be simply introduced into any prediction model by using the refraction coefficient (K_R) and the shoaling coefficient (K_S). The product K_R and K_S is given by

$$H/H_o = [(b_o/b) \cdot (C_{Go}/C_G)]^{\frac{1}{2}} = K_R \cdot K_S \tag{1}$$

where H_o is the wave height in the deep water with distance between two orthogonals b_o, and group velocity C_{Go} similarly, H, b and C_G are for the shallow water.

Fig. I

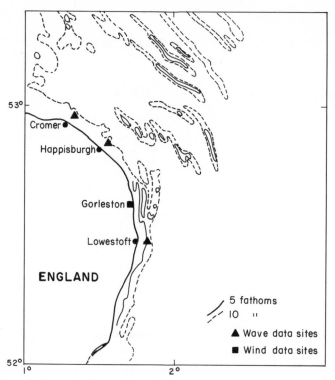

Figure 1. Map of the area investigated.

Fig. 2

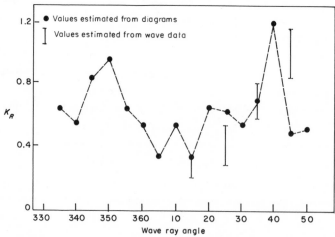

Figure 2. A graph of K_R values, from wave refraction diagrams against ray angles
compared with the wave data K_R values.

This derivation (1) may not be reliable when b = 0, such as at the crossing of orthogonals, H/H_o tends to infinity. Pierson (1951) suggested from his experimental work that at the crossing of orthogonals or at the caustic point, there might be phase shift in waves. In his experiment, the phase velocity of the waves seemed to vanish at the caustic point and reappeared again after. This phenomenon may not be observed in the real environment due to the randomness of the actual sea surface (Chao, 1974).

Refraction Coefficient (K_R)

To investigate how applicable the derivation (1) is to the estimation of wave height in areas of pronounced refraction, K_R values (in (1)) were determined from the field data (H and H_o) in the southern North Sea and the shoaling coefficient from

$$K_S = \left[\frac{L_o}{2\pi d} \right]^{\frac{1}{4}} \left[1 + \frac{4\pi d/L}{\sinh 4\pi d/L} \right]^{-\frac{1}{2}} \tag{2}$$

For five second waves in water, depth $d_o \simeq 32.0$ m, the associated wavelength is $L_o = 39.62$ m. Similarly in shallow water ($d \simeq 5.0$ m), the wavelength is $L = 29.87$ m. Substituting these values in (2) gives K_S equal to 0.86 and the derivation (1) becomes

$$K_R = \frac{1}{0.86} \cdot \frac{H}{H_o} \tag{3}$$

H and H_o at different wind directions were obtained (for details see Aranuvachapun, 1977a), and K_R values corresponding to each wind sector are tabulated in the following table. K_R was also evaluated from wave refraction diagram (Wilson, 1966) constructed by using topographic data of the area. A series of these refraction diagrams at various angles of wave rays gives values of K_R associated with wave ray directions as shown in Figure 2. (dotted line). Where wave direction is highly coherent with the wind direction, K_R values in Table 1 are compared to the values on the graph of Figure 2. The error bars represent standard deviation of the calculated K_R (Table 1). It may be seen that there is some agreement between the two sets of K_R suggesting the simple idea of wave refraction (1) can be applied to this area of southern North Sea.

TABLE I

Refraction coefficients

Wind direction	Mean values of K_R	S.D. of K_R
10° – 20° NE	0.2350	0.0902
20° – 30° NE	0.2996	0.2589
30° – 40° NE	0.5952	0.1933
40° – 50°		

Wind and Wave Relationship

Kraus (1972, section 4.4) shows by dimensional arguments that the mean square of the sea-surface displacement $\bar{\xi}$ is related to the shipboard - level wind speed U by

$$\bar{\xi}^2 = C_1 U^4 / g^2 \tag{4}$$

where C_1 is a constant and g is the acceleration due to gravity. Longuet-Higgins (1952) derived a theoretical expression for the expected maximum wave height in the form

$$H_{max} = C_2 \ (\bar{\xi}^{\ 2})^{\frac{1}{2}} \tag{5}$$

where $C_2 = (\log N)^{\frac{1}{2}}$ and N is the number of wave crests in the record. For this study the sample size of N is approximately 10,000, so $C_2 \simeq 3.04$. From (4) and (5), the expected wave height in terms of wind speed should be

$$H_{max} = C_2 C_1^{\frac{1}{2}} U^2 / g \tag{6}$$

Table 2 summarises the correlations between wind speed at Gorleston (see Figure 1) and wave heights at Cromer, Happisburgh and Lowestoft (details on data collection can be found in Aranuvachapun, 1977b). The highest correlation coefficient is found at Lowestoft, for which the distribution of wind speed and wave height is replotted onto a log-log scale as shown in Figure 3. The slope of the best fit line from linear regression is equal to 1.85. Hence relationship (6) between H_{max} and U agrees with field measurements in this coastal area of southern North Sea, and is a reasonable fit in the region being investigated.

TABLE II

The correlation coefficients of wind speed at Gorleston and wave height at three different stations for various wind directions.

Stations	Wind directions	Correlation coefficients
Cromer	all directions	−0.034
	onshore direction	0.480
	offshore direction	−0.139
	270° − 360° NW, 360° − 100° NE	0.534
Happisburgh	all directions	0.277
	onshore direction	0.657
	offshore direction	0.130
	330° − 360° NW, 360° − 100° NE	0.702
Lowestoft	all directions	0.669
	onshore direction	0.690
	offshore direction	0.648
	360° − 70° NE, 150° − 200° SE	0.710

Fig. 3

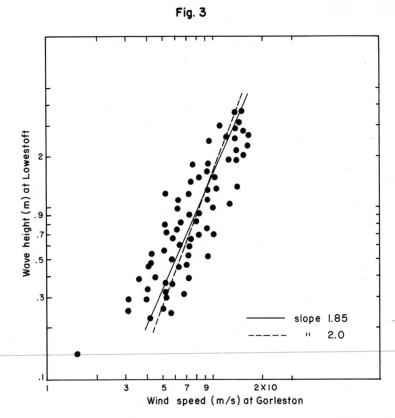

Figure 3. A graph of wave height (H_{max}) at Lowestoft against wind speed at Gorleston on a log-log scale.

Wave Prediction Model

To incorporate some effects of wave refraction in the area, a modified form
of (6) that includes refraction and shoaling coefficients, is proposed as

$$H_{max} = CK_S K_R U^2/g \qquad (7)$$

where C is a constant, K_S = 0.86 and K_R values from Figure 2. Predictions of
wave height at the three stations are carried out using (7). This work concen-
trates on wind sector from 330°NW to 50°NE as it is the regime of onshore winds
for this coastline.

Figure 4 shows results from the predictions compared with the actual measured
data. Other models are also used to predict wave height at the same three
stations in order to compare their accuracy. The predictions made using the
model of Darbyshire and Draper (1963), and the Sverdrup-Munk-Bretschneider
model, are presented in Figures 5 and 6 respectively (in the similar manner as
in Figure 4). The correlation coefficient between predicted values and the
measured values for each model is tabulated in Table 3.

TABLE III

The correlation between predicted wave height and the actual measured values at
the three stations.

Station	Model of predictions	Correlation coefficient	Figure associated
Cromer	Darbyshire and Draper	0.69	5a
	Sverdrup-Munk-Bretschneider	0.42	6a
	Simple relation (7)	0.62	4a
Happisburgh	Darbyshire and Draper	0.63	5b
	Sverdrup-Munk-Bretschneider	0.61	6b
	Simple relation (7)	0.72	4b
Lowestoft	Darbyshire and Draper	0.66	5c
	Sverdrup-Munk-Bretschneider	0.71	6c
	Simple relation (7)	0.78	4c

It is found that the proposed formula (7) gives, in general, a better
correlation than the Sverdrup-Munk-Bretschneider model in the area under investigation here. This result suggests that for the area like the nearshore region
around the East Anglian coast where the wave refraction is important, the
shoaling and refraction effects represented by K_S and K_R should be included in
the model. It also demonstrates that despite its simplicity, the model (7)
gives more satisfactory results than the complex model of Sverdrup-Munk-
Bretschneider for the area studied. The increase in accuracy may be due to the
fact that model (7) allows for local effect (i.e., refraction and shoaling
effects) while the other does not by assuming the flat sea floor, implying

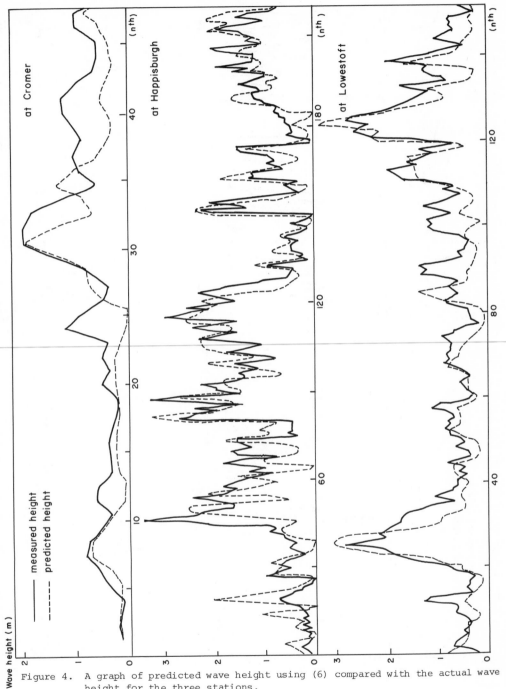

Figure 4. A graph of predicted wave height using (6) compared with the actual wave height for the three stations.

98

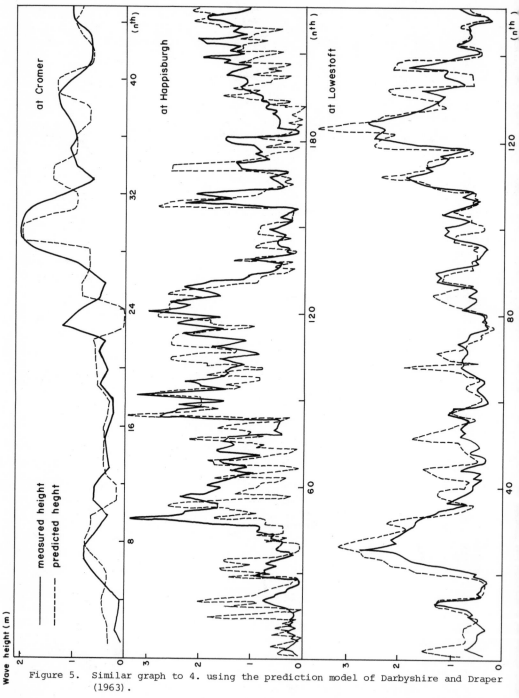

Figure 5. Similar graph to 4. using the prediction model of Darbyshire and Draper (1963).

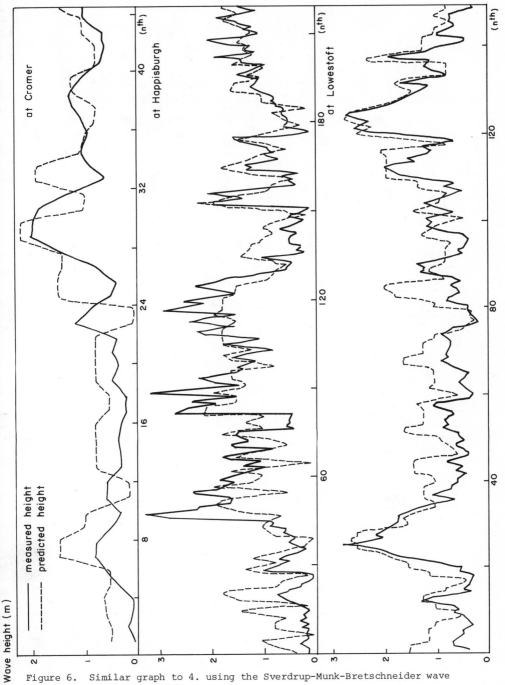

Wave height (m)

at Cromer

at Happisburgh

at Lowestoft

measured height
predicted height

Figure 6.   Similar graph to 4. using the Sverdrup-Munk-Bretschneider wave
            prediction model.

the locality can be very important to the predictibility of the sea in this region.

ACKNOWLEDGEMENTS

The author wishes to thank the Mekong Secretariat for aiding the presentation of this paper at the conference, Dr. Phadej Savasdibutr and Dr. P. Brimblecombe for their helpful discussions. The ideas expressed here are of my own and not necessarily those of the Secretariat.

REFERENCES

Aranuvachapun, Sasithorn, 1977a. Wave refraction in the southern North Sea. Ocean Engineering, 4:91-99.
Aranuvachapun, Sasithorn, 1977b. Wave Climate in the southern North Sea and Sediment Transport on the East Anglian Coast. PhD Thesis, University of East Anglia, Norwich, U.K.
Bretschneider, C.L., 1958. Revisions in wave forecasting: deep and shallow water. Proceedings of the Sixth Conference on Coastal Engineering, ASCE, Council of Wave Research.
Chao, Yung-Yao, 1974. Wave Refraction Phenomena Over the Continental Shelf Near the Chesapeake Bay Entrance. U.S. Army Corps. of Engineering, Coastal Engineering Research Centre, Tech. Memo. No. 47.
Darbyshire, M. and Draper, L., 1963. Forecasting wind - generated sea waves. Engineering, 195:482-484.
Kraus, E.B., 1972. Atmosphere-Ocean Interaction. Oxford University Press, pp. 268.
Longuet-Higgins, M.S., 1952. On the statistical distribution of the heights of sea waves. Journal of Marine Research, 11:245-266.
Pierson, W.J., Jr., 1951. The Interpretation of Crossed Orthogonals in Wave Refraction Phenomena. U.S. Army Corps. of Engineering, Coastal Engineering Research Centre Training, Memo. 21.
Sverdrup, H.U. and Munk, W.H., 1947. Wind, sea and swell; theory of relationships for forecasting. U.S. Navy Hydrographic Office, Washington, Publication No. 601.
Wilson, W.S., 1966. A Method for Calculating and Plotting Surface Wave Rays. U.S. Army Corps. of Engineering, Coastal Research Centre, Tech. Memo. No. 17.

CORRELATION BETWEEN WAVE SLOPES AND NEAR-SURFACE OCEAN CURRENTS

S. SETHURAMAN

Department of Energy and Environment, Brookhaven National Laboratory, Upton, NY, USA.

ABSTRACT

The development of wind generated currents in the ocean was studied with simultaneous observations of mean wind speed, wind direction, surface wave parameters
and near-surface ocean current. The measurements were carried out during February
23 - March 14, 1976 as part of a coastal ocean boundary layer and diffusion study
off Long Island, New York in the Atlantic Ocean.
The results show a high correlation between wave slope and near-surface current
indicating the possibility of wave age playing a significant role in the generation
of current. Wave age is known to cause variations in momentum transfer (Kraus, 1972;
SethuRaman, 1978). The wind generated current was found to have a broad spectral
peak as compared with tidal currents. This peak was found to occur at approximately
the same frequency as wind speed spectral peak. Integral time scales associated
with wind and near-surface current were about the same, indicating the dominance
of wind forcing near the ocean surface for this period of observations.

INTRODUCTION

As wind blows over water, wind-generated currents are produced in the water due

to the transfer of momentum from air to water at the interface and by friction be-

tween adjacent layers within the water. The downward, horizontal momentum flux

from the atmosphere is partly spent on the generation of waves and the rest on

drift currents or wind generated currents. The mechanism of momentum transfer is

not yet fully understood, but the magnitude seems to depend on the aerodynamic

roughness of the sea surface (SethuRaman and Raynor, 1975) which is a function of

sea state conditions (Neumann, 1968; Kitaigorodskii, 1973; SethuRaman, 1977).

Variations in wave age caused by the changes in mean wind direction, duration and

fetch appear to influence the momentum transfer significantly (SethuRaman, 1978).

Partially developed waves have steeper slopes and move at a lower speed than the

low-level winds contributing to higher frictional and form drags. On the other

hand, fully developed waves have flatter slopes and move at significantly higher

speeds relative to near-surface winds. The relationship between wind speed and

drift current has been investigated in the past by several investigators in the field and in the laboratory (Hughes, 1956; Carruthers, 1957; Shemdin, 1972; Wu, 1975). There seems to be a general agreement that the ratio between the surface drift velocity and the surface wind speed assumes an asymptotic value of about 3 per cent at long fetches.

The purpose of this study is to investigate the possible mechanism by which the wind generated currents are produced and maintained. Wave height and wave period measurements and near-surface current and wind observations were used to study the variations in wind-induced drift. Spectral analysis of various parameters were performed to determine the dependence of one on the other. Wave slope, wind speed and surface current were some of the variables considered important to determine as to whether the pattern of variability of atmospheric momentum is followed in the process of current generation.

MEASUREMENTS

The oceanographic measurements consisted of a moored instrument array 5 km off shore in the Atlantic Ocean near Long Island (Fig. 1). Observations of currents, salinity and temperature at different depths were recorded with this spar buoy. A description of the development of this telemetered, moored instrument array is given by Dimmler, et al (1975). Wave heights and wave periods were observed with a "waverider" which is essentially a buoy that follows the movements of the water surface and measures waves by measuring the vertical acceleration of the buoy. The spherical buoy was 0.7 m in diameter and was provided with an antenna for the tranmission of data to the shore. Mean wind speed and direction were measured at a height of 24 m at the coastal meteorological station at Tiana Beach (TB). The analyses reported here are based on measurements made for a period of three weeks from February 23 to March 14, 1976.

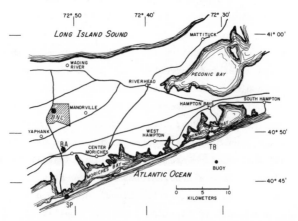

Fig. 1. Map of eastern Long Island showing the location of the oceanographic spar buoy, and the meteorological tower at Tiana Beach (TB). Wave rider was deployed close to the buoy.

ANALYSIS

Passage of synoptic meteorological systems over Long Island and the vicinity
causes variations in near-shore wind speeds and wind directions. A typical time
period of mean wind speed measured at Tiana Beach (TB in Fig. 1) at a height of
24 m is shown in Fig. 2, for the duration of this study. Wind speeds varied from
1 to 18 m sec^{-1}. Observations at the beach are approximately representative of
over-water winds (SethuRaman and Raynor, 1978). Time histories of the wave height
and wave periods are given in Fig. 3. Increase in wind speeds and wave heights
with the approach of storms can be seen in Figs. 2 and 3, respectively.

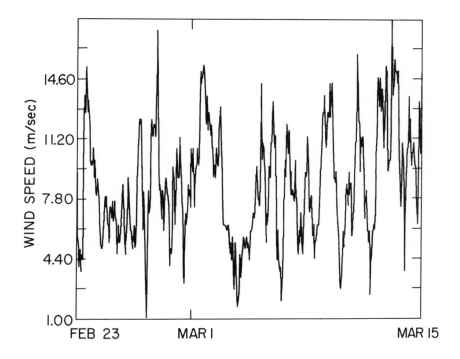

Fig. 2. Time history of one-hour mean wind speeds at Tiana Beach at a height of
24 m for the duration of the experiments.

One of the objectives of the analysis was to separate the wind generated current
from the total current and study its variation. Separation of the tidal component
is a difficult procedure due to its dependence on several factors. The tidal ellipse
seems to have an along-shore component of about 17 cm/sec with the ellipse inclined
to the shore (Scott and Csanady, 1976). Analysis of near-surface currents during
low wind periods indicated tidal amplitudes of comparable magnitude. A tidal

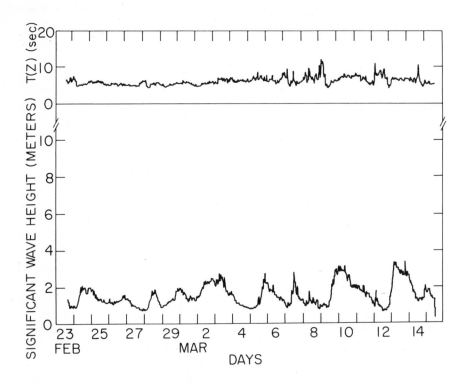

Fig. 3. Time history of 20-minute wave heights and wave periods near the buoy
(see Fig. 1).

amplitude of 15 cm sec^{-1} was used to help estimate the wind generated currents
from along-shore current observations. Observations of tides at Shinnecock Inlet
were used to get tidal cycles. This inlet is in the vicinity of the measurements
site. Any possible nonlinear interactions between the tidal and wind generated
currents were neglected in the present study. Some of the analyses were also
performed without separating the tidal current to provide an alternate interpretation. The measurements used here to study the wind generated current were made
at an average depth of 3 m below the water surface.

Wave slopes

The significant wave height, H, obtained from the waverider is the average
height of the highest 1/3 of the waves. Time periods, T, of the waves averaged
over 20 min. duration are used here. Mean wind speeds also corresponded to the
same 20 min. Wave length, L, was obtained from the relationship,

$$L = \frac{g\,T^2}{2\pi} \qquad\qquad (1)$$

where g is the gravitational acceleration. Mean slopes of the waves were then
estimated from

$$s = \frac{H}{L/2} \qquad\qquad (2)$$

A case study

A typical high wind period has been chosen to study the simultaneous variation
of wind direction, wave height, wave slope and wind generated current. High winds
with a long fetch over the water occurred on February 24 after a high pressure
system moved over the ocean. This caused a change in wind direction from off
shore to along shore. Wind speeds increased from about 3 m/sec to 15 m/sec.
Maximum wind speeds corresponded with maximum wave height observations shown in
Fig. 4. A time history of mean wind directions and mean wave slopes computed
from Eq. 2 are given in Fig. 5. The slopes are the steepest immediately after a
significant change in wind direction. The wave slope then reaches an asymptotically
constant lower value as the wind direction becomes more persistent. This phenomenon is believed to be due to different stages of development of waves or in
other words due to wave age. An increase in surface drag was observed immediately
following significant changes in wind direction in previous studies (Neumann, 1968;
SethuRaman, 1978). Mean wind speeds and the estimated wind generated currents
are shown in Fig. 6. The maximum wave slope and the highest current lag maximum
wind speed by a few hours. As the wind speed, wind direction and the wave slope
reach approximately constant values, wind generated current also tends to approach
an asymptotic value. This average equilibrium value can be estimated to be about
13 cm/sec for current and about 6 m/sec for wind for this case. Assuming fully
rough conditions, friction velocity u_* for the air can be estimated as 24 cm/sec
(SethuRaman and Raynor, 1975) yielding an average ratio of wind generated current,
V, to friction velocity u_* as 0.54. Values close to 0.53 have been found by Wu
(1973) and Phillips and Banner (1974).

Spectral analysis

Frequencies associated with wind generated currents will be readily apparent
in a spectral analysis of the time series data since the tidal frequencies are

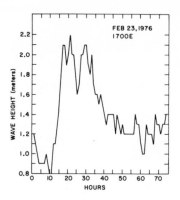

Fig. 4. Wave height variations for February 23-26. Starting date and time are
also indicated. Increase in wave height due to increase in mean wind
speed is seen.

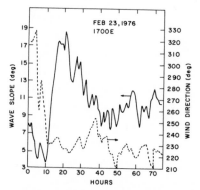

Fig. 5. Variance of wave slopes and mean wind directions for February 23-26.
Starting date and time are as indicated. Solid lines represent the
wave slope and dashed line the wind direction. Close correlation
between wave slope and wind direction seems to exist.

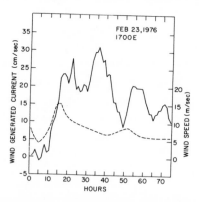

Fig. 6. Wind generated current (estimated) and wind speed for February 23-26.

diurnal and semi-diurnal. One advantage of this analysis is that there is no need
to separate the tidal currents. The variance spectrum of the one-hour along-shore
mean wind speeds for the duration of the study is shown in Fig. 7. The spectrum
has a pronounced peak around .014 cycles per hour which corresponds to a time
period of about 3 days. This time period represents the average time elapsed between two successive high wind episodes caused by the movement of synoptic systems
and is in agreement with a similar analysis made with observations collected continuously over one year (SethuRaman and Brown, 1977). A small diurnal peak can
also be seen in Fig. 7. Variance spectra of along-shore currents at depths of
3.1 m, 14.3 m, and 24.6 m are shown in Fig. 8. A pronounced, but narrow peak for
all depths with constant amplitude was found at a frequency corresponding to semi-
diurnal tidal period. An estimate of the semi-diurnal along-shore tidal current
from Fig. 8 gives about 16 cm/sec. A value of 15 cm/sec was assumed here and a
value of 17 cm/sec was reported by Scott andCsanady (1976). Diurnal tidal currents
did not produce a pronounced peak but was found to be present at all depths with
decreasing amplitudes. Decrease in spectral amplitude between the depths of 14.3
and 24.6 m was more than that between 3.1 and 14.6 m. Bottom friction might be
the reason for this difference. Depth of water at the site of the buoy was about
30 m. The frequency associated with wind-generated current is also seen in Fig. 8
which corresponds to the dominant peak of wind speed spectra in Fig. 7. A comparison of the spectral amplitudes at this frequency would yield a ratio of 3
per cent between the wind-generated current and wind speed which has been found
to be the equilibrium value by several investigators (Wu, 1973). Spectral densities for wind speed and along-shore current at 3 m have been plotted as a function of frequency in Figs. 9 and 10, respectively. The current spectra (Fig. 10)
seems to follow Kolmogorov's inertial subrange relationship at frequencies more
than 0.1 cycle per hour. With a mean current of 18 cm sec^{-1} this corresponded
to a wave length of about 2 m which was approximately equal to the depth of
measurement. Atmospheric turbulence was found to obey Kolmogorov's relation at
frequencies above 0.1 Hz (SethuRaman, et al., 1974).

Time scales

Autocorrelation function for wind speed $R_u(t)$ defined as

$$R_u(t) = \frac{\overline{u(t_1)\,u(t_2)}}{\overline{u}^2} \tag{3}$$

where $\overline{u(t_1)\,u(t_2)}$ is the autocovariance of wind speed and $\overline{u}^2$ is the variance,
is a function of only the time difference $t_2 - t_1$, and describes the memory of
$u(t)$. A similar function, $R_c(t)$ can be defined for the water current. Variation of $R_u(t)$ and $R_c(t)$ with time lags are shown in Figs. 11 and 12, respectively.

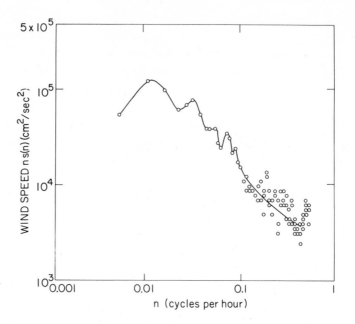

Fig. 7. One-dimensional variance spectrum for one-hour mean wind speeds at Tiana Beach.

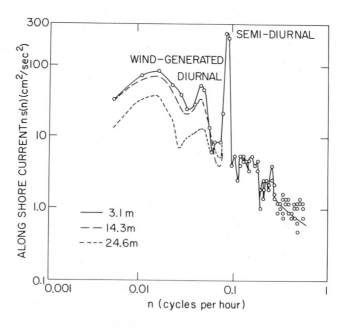

Fig. 8. One-dimensional variance spectrum for one-hour mean along-shore currents at different depths.

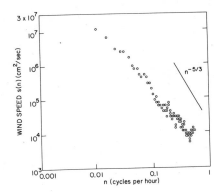

Fig. 9. Variation of spectral densities as a function of frequencies for mean wind speeds at Tiana Beach.

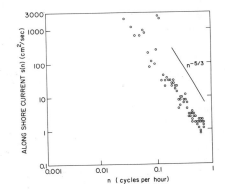

Fig. 10. Variation of spectral densities as a function of frequencies for one-hour mean along-shore near-surface currents (depth $\simeq$ 3 m).

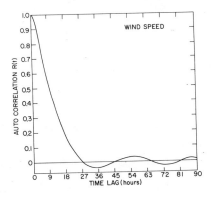

Fig. 11. Autocorrelogram for wind at Tiana Beach.

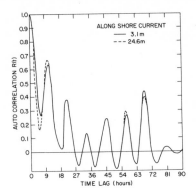

Fig. 12. Autocorrelogram for along-shore currents at 3.1 m and 24.6 m depths.

The atmospheric autocorrelation function falls off rather rapidly, but the autocorrelation for current has several peaks and falls off slowly indicating longer
memories and different forcing functions. Semi-diurnal and diurnal peaks can
also be seen in Fig. 12. An integral time scale,

$$\tau_u = \int_0^\infty R_u(t) \; dt \tag{4}$$

can be defined for wind speed and a similar one for current. This time scale
was estimated to be about 10.5 hours for wind from Fig. 11 and about 11.5 hours
for current from Fig. 12. The closeness of these two values suggests the dominance of atmospheric forcing on the ocean.

Coherence

A measure of the correlation between wave slope, s, and along-shore current,
c, as a function of frequency can be obtained by computing the coherence, Coh_{sc},
given by

$$Coh_{sc} = \frac{Co^2_{s,c}(n) + Q^2_{s,c}(n)}{S_s(n) \; S_c(n)} \tag{5}$$

where Co(n) and Q(n) are the cospectra and quadrature spectra, respectively,
and S is the individual spectrum at different frequencies, n. Values of Coh
are shown in Fig. 13 as a function of frequencies. A maximum coherence of about
0.55 occurs at a frequency of 0.16 cycles per hour corresponding to a time period of about 6 hours. This indicates that there is a good correlation between
wave slope and near-surface current and the maximum currents lag maximum slope
by about six hours.

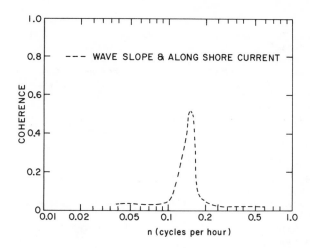

Fig. 13.   Coherence between wave slope and along-shore current as a function
of frequency.

CONCLUSIONS

Analysis of simultaneous observations of surface wave parameters, wind speed
and near-surface current indicates the possibility of wave age playing an important role in the generation of wind drift currents. Maximum currents appear
to lag maximum wave slopes by about 6 hours.

ACKNOWLEDGEMENTS

Many members of the Division of Atmospheric Sciences and the Division of
Oceanographic Sciences participated in the experiments. Assistance in computer
programming was provided by C. Henderson and J. Tichler and in data analysis by
J. Glasmann and K. Tiotis. The author wishes to thank T. S. Hopkins and G. S.
Raynor for studying the manuscript and offering some valuable suggestions.

The submitted manuscript has been authored under contract EY-76-C-02-0016
with the U. S. Department of Energy. Accordingly, the U. S. Government retains
a nonexclusive, royalty-free license to publish or reproduce the published form
of this contribution, or allow others to do so, for U. S. Government purposes.

REFERENCES

Carruthers, J. N., 1957. A discussion of "A determination of the relation between wind and surface drift." Quar. J. Roy. Met. Soc., 83: 276-277.

Dimmler, D. G., Greenhouse, N. and Rankowitz, S., 1975. A controllable auto-
mated environmental data acquisition and monitoring system. Proc. 1975
Nuclear Science Symposium, San Francisco, California, November 1975.

Hughes, P., 1956. A determination of the relation between wind and sea surface
drift. Quart. J. Roy. Meteor. Soc. 82: 494-502.

Kitaigorodskii, S. A., 1973. The physics of air-sea interaction. Translated
from Russian by A. Baruch, Israel Program for Scientific Translations,
Jerusalem, pp. 237.

Kraus, E. B., 1972. Atmosphere-Ocean Interaction. Clarendon Press, Oxford,
England, pp. 275.

Neumann, G., 1968. Ocean Currents. Elsevier Scientific Publishing Company,
New York, N. Y., pp. 352.

Phillips, O. M. and Banner, M. L., 1974. Wave breaking in the presence of wind
drift and swell. J. Fluid Mech., 66: 625-640.

Scott, J. T. and Csanady, G. T., 1976. Nearshore currents off Long Island.
J. Geophys. Res., 81: 5401.

Shemdin, O. H., 1972. Wind-generated current and phase speed of wind waves.
J. Phys. Ocean. 2: 411-419.

SethuRaman, S., 1977. The effect of characteristic height of sea surface on
drag coefficient. BNL Report 21668, pp. 33.

SethuRaman, S., 1978. Influence of mean wind direction on sea surface wave
development. J. Phys. Ocean., 8: (in press).

SethuRaman, S. and Brown, R. M., 1977. Temporal variation of suspended parti-
culates at Upton, L. I., N. Y. AMS Conference on Applications on Air Pollution Meteorology, November 28-December 2, 1977. Preprint Volume: 16-18.

SethuRaman, S. and Raynor, G. S., 1975. Surface drag coefficient dependence on
the aerodynamic roughness of the sea. J. Geophys. Res., 80: 4983-4988.

SethuRaman, S. and Raynor, G. S., 1978. Effect of changes in upwind surface
characteristics on mean wind speed and turbulence near a coastline. Am.
Meteor. Soc. Fourth Symposium on Turbulence, Diffusion, and Air Pollution,
Reno, Nevada, January 15-18, 1977. Preprint Volume (in press).

SethuRaman, S., Brown, R. N. and Tichler, J., 1974. Spectra of atmospheric
turbulence over the sea during stably stratified conditions. Am. Meteor.
Soc. Symposium on Atmospheric Diffusion and Air Pollution, Santa Barbara,
California, September 9-13, 1974. Preprint Volume: 71-76.

Wu, J., 1973. Prediction of near-surface drift currents from wind velocity.
J. Hyd. Division, ASCE, 99: 1291-1302.

Wu, J., 1975. Wind-induced drift currents. J. Fluid Mech., 68: 49-70.

THE TOW-OUT OF A LARGE PLATFORM

B. MacMahon

Civil Engineering Department, Imperial College, London

ABSTRACT

The time mean resistance experienced by a floating body being towed against
waves is found to depend on its motions.   If these are small compared to the
wave amplitude the appropriate frame of reference for calculations is Eulerian.
The magnitude of the wave force in this system agrees with Lagrangian determina-
tions which however point to a different line of action.   From a consideration
of the impulsive generation of the motion the line of action of the Eulerian wave
force is shown to be at a distance $ka^2/2$ below mean water level.   This leads to
some qualitative conclusions on frequency changes and induced currents at plane
barriers and some observations on the transformation of the "high level" Eulerian
momentum flux when the waves are attenuated by internal and bottom friction.
   The study of the stability of the vertical motion of large platforms when
sinking on to a prepared position on the sea bed leads to a confirmation of the
usual engineering experience that stability is dearly bought.   The possibility
of replacing an expensive static stability by a relatively cheaply acquired
inertial stability is seen to be impractical.

INTRODUCTION

The studies which follow arose out of a series of calculations to determine

the most adverse conditions of waves and currents compatible with towing and

sinking operations on a large concrete oil production platform.

Typically this comprises a cylindrical base about 100-150 metres in diameter

and 50 metres high from which arises a number of towers 150-200 metres long to

support the production deck which may be about 60-70 metres square.   Depending

on available water depth, the sequence of construction operations and questions

of stability and towing resistance,platforms may be towed out at various draughts

with the water line passing through either the base or the towers.   The floating

phase of the life of a platform is a vulnerable one when a very valuable but

unshipshape structure is being towed to its final emplacement at great expense by a fleet of tugs.

Unlike the first order oscillatory force which varies linearly with wave height but against which no nett work needs to be done in towing,the time mean wave force depends on the square of the wave amplitude. The rapid rise of towing resistance with increasing wave height which is a consequence of this makes the prediction of a suitable weather window vital as beyond a certain critical wave height a platform is no longer towable.

The location on the vertical of the line of action of the unidirectional wave force is also a matter of great practical importance as it may govern the draught at which a platform should be towed in given wave conditions. If the water is deep enough it may be possible to lower the part of the structure having a large area of cross section below the line of action of the wave force. Although the drag due to the translational velocity through the water may increase substantially at the lower draught,it will be more than counterbalanced by the diminution in the wave resistance when conditions are severe.

As regards the sinking operation it is important to determine the relative order of magnitude of the forces controlling the stability of the descent. There have been at least two incidents in the North Sea of tangential gliding ("skidding off") of platforms approaching touchdown on the sea bed and some anxiety expressed over the possibility of a "falling leaf" oscillation developing. In several designs large additions in the form of buoyancy chambers have been necessary to secure a satisfactory righting arm at all draughts through which the platform may pass. It is a curious fact that these chambers,which constitute a substantial part of the structure,function during only a few minutes of the platform's lifetime. It was of interest therefore to see if the necessary stability could be achieved by other means.

THE TOWING RESISTANCE

Lagrangian analysis

It has been known since Stokes' 1847 paper that a second order time mean current is associated with the propagation of irrotational gravity waves of finite amplitude. Stokes obtained this result by a change from a Eulerian to a Lagrangian specification of the motions and taking account of the finite dimensions of the particle paths. This showed the current to be of magnitude $U_L = ck^2a^2e^{2kz}$ in the direction of propagation of the waves when the water is deep in relation to the wavelength (c = phase velocity, k = wave number, a = amplitude and z is the vertical co-ordinate of the orbit centres). The corresponding time mean momentum flux is given by $\int_0^\infty U_L V_{group} dz$ as wave properties such as energy and momentum are

propagated at the group velocity (V group) equal to half the phase velocity on
deep water. The result, $\frac{1}{4}\rho g a^2$, is in agreement with the Eulerian analysis of
Longuet-Higgins (1964). In finite depth the Lagrangian drift velocity is given
by $U_L = \frac{ck^2a^2}{2\sinh^2 kh}\cosh 2k(h+z)$. When this is substituted in the previous integral, taking account also of the new value of the group velocity, $\frac{c}{2}\left(1 + \frac{2kh}{\sinh 2kh}\right)$,
the momentum flux turns out to be $\frac{\rho g a^2}{4}\left(1 + \frac{2kh}{\sinh 2kh}\right)$, again in agreement with the
"radiation stress" approach of Longuet-Higgins (1977).

The location of the mean momentum may be obtained by taking moments about $z = 0$:

$$\int_{-\infty}^{0} U_L z\,dz = \int_{-\infty}^{0} U_L dz\ \bar{z} \quad \text{whence} \quad \bar{z} = \frac{\lambda}{4\pi} \quad (z \leqslant 0 \text{ is the "paramètre de repos")}.$$

The impulsive generation of the motion

Any irrotational motion may be considered to have been generated from rest by
a system of impulsive pressures $\rho\phi$ applied to the fluid boundaries. Following
Lord Kelvin (1887) a rigid corrugated sheet is imagined to cover the water surface.
It is struck an impulsive blow in the horizontal direction and immediately with-
drawn (in Kelvin's example the sheet was gradually accelerated up to the phase
velocity when the fluid pressures on it vanished). By a well known theorem the
impulse is equal to the total momentum of the fluid plus the impulsive reactions
on the boundaries at infinity. These reactions are finite and may be ignored* if
the sheet is envisaged as being many wavelengths long. This would appear to be
an example of the uncertainty principle where the uncertainty in the momentum may
be reduced by spreading the wave motion over a great length.

Fig. 1. Generation of surface waves by an impulse

The resolved part of the impulse in the horizontal direction on an element of
the profile is given approximately by $\rho\phi dx \frac{d\eta}{dx}$ where $\eta = a\sin kx$ is the surface
ordinate and $ds \simeq dx$ if the wave slope is not too large. As $\phi = \frac{ga}{\sigma} e^{kz}\cos(kx - \sigma t)$

*In the case of 2 dimensional or 3 dimensional bodies moving in an unbounded
fluid there is an interesting indeterminacy in the fluid momentum due to the
finite reactions at infinity.

on deep water where σ is the frequency, the impulse per wavelength is given by:

$$\int_o^\lambda \frac{\rho g a}{\sigma} \cos kx \, ka \, \cos kx \, dx$$ where z has been taken as zero over the profile.

The result, $\pi \rho a^2 c$, is the horizontal momentum per wavelength of a gravity wave train on deep water.   The nett vertical momentum per wavelength is of course zero.   The momentum flux is $\frac{\pi \rho a^2 c}{2T}$ where T is the period, the factor 2 being due to the group velocity as before.   On substituting from the dispersion relation we again obtain the expression $\frac{\rho g a^2}{4}$.

In finite depth where $\phi = \frac{ga}{\sigma} \frac{\cosh k(h+z)}{\cosh kh} \cos(kx - \sigma t)$ the mean momentum per wavelength is given by the same integral as before since an impulsive motion of the plane bottom can make no contribution to the momentum if the fluid is inviscid. When the result $\frac{\pi \rho g a^2}{\sigma}$ is multiplied by the appropriate group velocity the momentum flux is given as $\frac{\rho g a^2}{4}\left(1 + \frac{2kh}{\sin 2kh}\right)$.

The line of action of the wave force

In the absence of viscosity there is no mechanism by which the momentum generated by the impulse can be communicated to other fluid regions outside the "layer of action" of the hypothetical corrugated sheet.   In both the infinite and finite depth cases the wave momentum is entirely contained in the region of space between the bottoms of the troughs and the tops of the crests (Phillips, 1977).

The line of action of the force is at still water level as is evident from the symmetry of the sheet.   To the second order the wave profile is given by $\zeta = a \sin kx + \frac{ka^2}{2} \sin 2kx$ relative to the still water line on infinite depth.   Both components are symmetrical about this level but the mean of the resulting profile is raised by $\frac{ka^2}{2}$ due to the fact that the first harmonic lifts both the minima of the troughs and the maxima of the crests in the fundamental.   The line of action of the Eulerian wave force is therefore to the second order of approximation $\frac{ka^2}{2}$ below the mean water level.   This location of the wave force was first obtained by Longuet Higgins (personal communication, 1978).

The contrast in the location of the wave force in the two systems is perhaps only to be expected.   It is well known that the fundamental Eulerian and Lagrangian definitions of velocity are quite different.   In practice however the force on the towrope is unambiguous and the resolution of the paradox must lie in the oscillations of the platform, as hinted at by Havelock (1940) in the context of ship resistance. If the heaving motions of the structure are small relative to the wave amplitude, the appropriate frame of reference is, like the body itself, fixed in space, i.e. Eulerian.   On the other hand if the heaving motions are significant and of the order of the wave amplitude the body will experience a force even on the part of it which lies below the wave troughs.

It is an interesting thought that the force could be opposite to the direction

of the waves were the heaving period adjusted to be somewhat greater than the
wave period.   Here, were the damping small enough,the body's motion would be in
antiphase to that of the wave,so coming under the influence of a particle velocity
opposite to that of the direction of wave travel at the top limit of its vertical
excursion and a smaller forward velocity when its downward displacement is a maximum.
This would be true for shapes like the platform in relatively long waves with slender
towers and the base submerged.   It is well known that ships may drift opposite to
the direction of wave propagation but due to the large waterline section the direction
of the effects would be reversed.

Wave transmission and reflection at plane vertical barriers

Some consequences of the location of the mean Eulerian momentum in relation to
the effect of fixed barriers on wave trains may be of oceanographic interest.   If a
vertical plate, normal to the wave direction, be imagined to extend from the sea
bed in infinite depth to the bottom of the troughs of the resultant motion, the
entire mean momentum of the wave should be transmitted past this "Eulerian" obstacle.
It is well known however from first order theory (Dean,1945) that such a barrier
would reflect nearly all the wave energy ($a \ll \lambda$).   Taking into account the expres-
sions for energy momentum and volume flux on deep water ($\dot{E} = \frac{\rho g^2 a^2}{4\sigma}$, $\dot{M} = \frac{1}{4}\rho g a^2$,
$Q = \sigma a^2/2$), it appears that a frequency change is necessary at the barrier in order

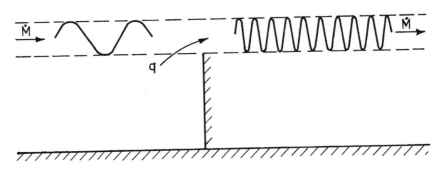

Fig. 2. Frequency doubling at a plane vertical barrier - schematic
        (additional to 1st order upstream reflected wave and reaction effects)

to convey the momentum of the incident wave and provide the reaction to that in the
reflected wave without transmitting more energy than first order theory permits.
The downstream increase in Q due to the frequency change would have to be balanced
by an induced current q from the upstream side.   If the downstream flowfield were
limited this current q could generate a set-up (Longuet-Higgins, 1967).

Were the motion everywhere irrotational, the barrier would experience no time
mean reaction as it is below the level of the wave force.   In this case, assuming

118

no set up, the mean momentum flux of the downstream fundamental and harmonics $\sum_{1}^{n} \tfrac{1}{2}\rho g a_n^2$ would be equal to that of the incident and reflected waves (approximately $\tfrac{1}{2}\rho g a^2$ if the reflection is nearly complete). In fact, the main departure from irrotationality is wave breaking, which if it did not lead to a significant force on the barrier, would increase the proportion of higher harmonics in the downstream motion, as less energy flux would be available for the same momentum flow rate (see Longuet-Higgins, 1977 for wave forces on extended submerged objects). The foregoing points are confirmed quite well by the experimental results of Jolas (1962 for wave transmission over a sill. An application of these effects is the "harmoni suppressor" (Hulsbergen, 1976) where a sill is positioned to act as an "anti-noise" device cancelling the emission of parasitic secondary waves by laboratory wavemakers

The opposite case of a barrier extending from above the surface down to the troughs, shows in the laboratory, as might be expected from the flux balances, the

Fig. 3. Frequency doubling at a fixed surface barrier - schematic
(additional to transmitted 1st order wave on downstream side)

double frequency harmonics on the upstream side, as in this case the wave momentum is reflected, but the energy is largely transmitted. The force on the barrier would be $\tfrac{1}{4}\rho g a^2$ per unit width, assuming almost complete transmission.

Low frequency surface waves can be seen from a physical viewpoint as a "device" to carry energy, momentum and a displacement of fluid across great distances while incurring very little penalty by way of rotationality. The "secret" is the locking of the mean Eulerian momentum into the surface deformation layer. The unlocking by fluid friction may be enhanced by frequency increases at shelves and submerged bars.

Wave attentuation

On deep water assuming laminar flow and zero surface stress the waves lose energy as a result of the viscous resistance to (at first) irrotational straining in the interior of the fluid (Stokes, 1845a). On the other hand the wave momentum cannot be destroyed by internal forces. The generation of a time mean second order vorticity at the water surface opposite in direction to the rotation of the

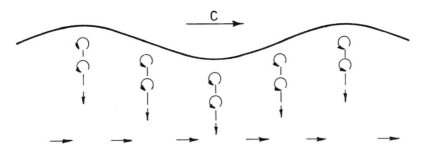

Fig. 4. Lowering of the wave momentum in deep water

particles in their orbits (Longuet Higgins, 1960) induces a velocity field in the
direction of propagation of the wave which affects the water at greater depths as
the vorticity slowly diffuses downwards.   In this way the wave momentum is removed
from the surface region.   When "separated" from the waves these second order flows
may come under the influence of the Coriolis force and manifest themselves as
inertia currents (Ursell, 1949).

In finite depth where there is greatly enhanced attenuation due to the no slip
condition at the bed the wave momentum must likewise be "brought down" from above
trough level so as it can be reduced by bottom friction.   Here the momentum is
transferred downwards,not by the vorticity generation at the surface which would
be altogether too weak (and the diffusion process too slow) but by pressure forces
consequent on a difference in phase between the motion in the boundary layer and
the irrotational interior.   This phase difference, first elucidated by Stokes
(1851), in the solution of his "second problem" results in a component of the
irrotational pressure field being in quadrature with the wave "profile" in the
boundary layer and a transfer of energy and momentum to the flow near the bed.
The convection velocities in the irrotational flow are such as to raise the vortex
lines where the spin is opposite in hand to the orbital motion and depress those
whose rotation is in the same sense.

The resulting vortex "street" creates a jet (Longuet Higgins, 1953) close to
the bed which has the function of reducing the wave momentum in step with the
energy loss.   The process can be followed visually when the motion is over a
sandy bed.   Sleath (1970) describes vortices being thrown out from behind the sand
grains at the end of each half cycle.   It appears from the diagrams in this
paper that those eddies which are opposite in hand to the orbital motion are on
a higher trajectory than their counterparts.

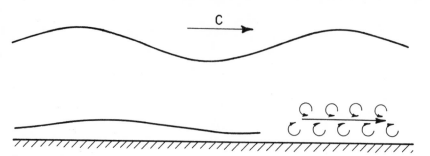

Fig. 5. Lowering and subsequent raising of the wave momentum in finite depth

STABILITY OF THE VERTICAL DESCENT

The stability of floating bodies against capsizing depends partly on the distribution of weight which provides a restoring or overturning moment depending on whether the centre of gravity is below or above the centre of buoyancy and partly on the volume and distance apart at the "wedges" that go into and come out of the water as the result of a heel. The stability at finite angles of heel is measured by the magnitude of the righting arm (GZ) where the restoring moment is given by the product of this term and the displacement. For infinitesimal displacements (initial stability) the slope of the curve of righting arm against angle of heel is taken as the criterion. The metacentric height (GM) is the slope of this curve at the origin and is measured in metres per radian.

For North Sea platforms the certification authorities specify a minimum metacentric height of about a metre for all draughts through which the platform may pass in its descent. This stipulation may require the construction of large buoyancy chambers in order to provide "wedges", particularly at the critical juncture when the water line passes through the roof of the base. It is not practical to provide this extra stability by adjustment of the weight distribution, as this would entail sacrificing completion of the deck and its equipment. In fact one of the main platform design objectives is to maximise the deck load, as this advances the date of commencement of oil production.

Stability, however, may also be acquired inertially if the fluid momentum associated with the translation of the body in the chosen orientation is greater than that for any other possible alignment. If the locus of the end points of the solid body translation velocity vector be plotted for constant associated fluid kinetic energy (T), the axes of the ellipsoid obtained give Kirchoff's three permanent directions of translation (Lamb, 1932).

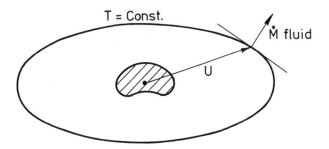

Fig. 6. The kinetic energy ellipsoid

The fluid momentum vector is normal to the energy ellipsoid as the latter is the vector rate of change of fluid kinetic energy with solid body velocity $\left(\frac{\partial T}{\partial U}\right)$. Unless the fluid momentum vector and the solid body momentum vector are parallel there will be a couple exerted on the body by the fluid. When the motion of the solid is in the direction of one of the axes of the ellipsoid, this couple is absent and the body once set translating, can, in theory, continue to do so without rotation. Of these three directions of permanent translation only one is considered to be stable although the complete validity of this rule in all cases has been disputed (Ursell, H.D., 1940).

For the most general motion of a body of arbitrary shape the kinetic energy of the fluid depends on 21 coefficients of inertia. In the restricted case of translation of solids of revolution (and of helicoidal symmetry), into which categories the descent of some platform designs fall, the number of coefficients reduces to two:

$$2T = Au^2 + B(v^2 + w^2),\quad \text{adopting Lamb's notation.}$$

In the particular case where the body moves with the axes always in one plane (that of x and y), the deflecting couple may be shown to be (A-B)u.v by a simple argument:

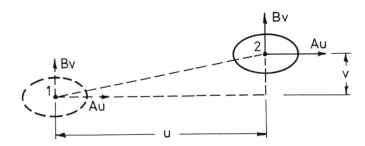

Fig. 7. The change of the fluid impulse in unit time

122

When the body translates from 1 to 2 in unit time the couple B.v.u - A.u.v. must be applied by the body to the fluid in order to effect the change in fluid velocity field. The reaction of the fluid on the body, the deflecting couple, is therefore (A-B)u.v about the z axis. In the case of a typical platform the couple (A-B)u.v may be established and compared to the static restoring moment.

The hydrodynamic restoring couple

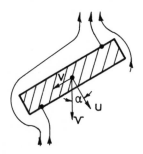

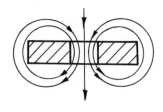

Fig. 8. Descent of base　　　　　Fig. 9. Annulus with circulation

In Fig. 8 , u = V cosα, v = V sinα where V is the vertical velocity of descent.
The fluid couple = $(A-B)u.v = \frac{A-B}{2} V^2 \sin 2\alpha$
Taking A-B to be of the order of the displacement M,
Maximum equivalent righting arm $(GZ) = \frac{V^2}{2g} \sin 2\alpha$
Metacentric height corresponding $(GM) = \frac{d(GZ)}{d\alpha}$ at α = 0

$$= \left| \frac{V^2}{g} \cos 2\alpha \right|_{\alpha=0}$$

Taking the maximum value of V to be 2 metres per minute:

$$GM_{equivalent} = \frac{(2/60)^2}{g} \simeq 0.1 \text{ mm}$$

This very small value compared to the design (hydrostatic) GM of 1 metre shows that fluid inertia plays an altogether negligible role in the stability of descent. The couple (A-B)u.v could be augmented in the case of a doubly connected (e.g. annular) body. If the impulse of the circulation is ξ the restoring moment will be augmented by a factor ξv. However the impulse of a frictionally generated circulation would be negligible at the velocities envisaged. More might be achieved by artificially creating a jet through the annulus but the practicalities seem doubtful. To develop a metacentric height of 1 metre by rapid descent would require a velocity of $\sqrt{\frac{Mmg}{A-B}}$ which could be in excess of 10 m sec^{-1}.

Falling leaf oscillation

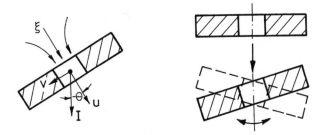

Fig. 10. Oscillations about the line of the impulse

The equation governing the oscillation of a body about a stable direction of translation was first given by Kelvin (1871).   Using the notation of Lamb (1932) and Gray (1950):

$Q\ddot{\theta}$ = (A-B)uv + ξ where Q is the combined solid body and hydrodynamic moment of inertia about an axis perpendicular to the plane of the figure.   If the generating impulse is I then:

$$I\cos\theta = Au + B\xi, \qquad I\sin\theta = -Bv$$

$$\text{The period } T = 2\pi\sqrt{\left(\frac{ABQ}{((A-B)I + B\xi)I}\right)}$$

In the absence of a static righting moment the period of the oscillation may be reduced and the stability of the descent enhanced by designing A to be greater than B, augmenting the equivalent I by increasing the rate of descent and if possible making use of the impulse ξ of the circulation.   In fact however the fear of this type of motion developing would appear to be groundless in view of the magnitude of the specified metacentric height.

It should perhaps be made clear that this inertial oscillation is only the initial part of the motions of a solid falling freely under gravity in a fluid.   Maxwell (1853) has shown in the case of a plane rectangular object (the effect is easily seen with his example of a slip of paper falling in air), that the descent after an initial "wavering" takes place at a constant angle to the vertical and is accompanied by a rapid rotation about a horizontal axis.   These last two effects are due to fluid friction,whereas the restoring force governing the initial oscillations depends as above on fluid inertia only.

Gyroscopic stability

If the platform, assumed a solid of revolution, is made to spin about the gener
ting axis, it will acquire stability in the manner of a "sleeping" top provided
the angular velocity is sufficiently great.   The point on the axis, about which
the motion would take place were precession to occur, corresponding to the point
of contact of a top with the ground, could in principle be obtained by minimising
the kinetic energy of the system as Wendel (1950) has done in locating the roll ax:
of ships.

If P is the solid body moment of inertia of the platform about its long axis and
Q the solid body and hydrodynamic moment of inertia about a perpendicular axis
through the fixed point, ignoring frictional couples, the conservation of angular
momentum gives:

Restoring torque = $P\omega\Omega\sin\theta - Q\Omega^2\sin\theta\cos\theta$ where ω is the spin about the
long axis, θ the inclination of this axis to the vertical and Ω the precession.
The left hand side of this equation may be written $Mgm\sin\theta$, where m is the metacent
height (GM).

The condition for the platform axis to return to the vertical after a small displacement, is, as in the case of the top, that the roots of the quadratic equation
in Ω be real, that is:

$$\omega^2 \;>\; \frac{4MgmQ*}{P^2}$$

The equality, $m = \dfrac{P^2\omega^2}{4MgQ}$, gives the equivalent metacentric height developed by the
spin.   Assuming the mass distribution about the axis of spin to correspond to that
of a solid cylinder of diameter 80 metres results in a value for P of 800 tonnes m^2
An estimate of the inertia distribution along the same axis relative to the fixed
point for a typical design gives $Q \simeq 1200$ tonnes m^2.

Substituting the values in the equation shows that a spin of 2.6 revolutions
per minute would develop a metacentric height of 1 metre.

CONCLUSIONS

The conviction that second order wave properties are sui generis and not mere
outgrowths from first order effects forces itself on one strongly.   The wave momen
when followed from its generation, through its conservation to its decay seems to
explain as many and as varied phenomena as the wave energy.

The basic Eulerian Lagrangian ambiguity manifests itself not in the magnitude
but in the line of action of the wave force.

* The problem of "Columbus'egg" remains to be solved here if there is water ballast
in the compartments of the base.

Wavering motions, typical of falling leaves, cannot occur on platforms descending at the usual speeds due to the smallness of the couples involved.   To acquire stability by sinking rapidly in a stable direction of translation is impractical. It is just conceivable that an addition of rotationally acquired stability might be used to pass through a section of low metacentric height in the descent.

ACKNOWLEDGEMENTS

I would like to thank the following for their help: Professor M.S. Longuet-Higgins of Cambridge University, Dr J.M.R. Graham of the Aeronautics Department, Imperial College, Mr J. Sioris and Dr C.D. Memos of the civil engineering hydrodynamics research group at Imperial College and Mr R.W.P. May of this group and the Hydraulics Research Station, Wallingford.   I am very indebted to Mr R.L. Jack of Noble Denton and Associates Ltd., London, marine consultants, for much information on platform behaviour.

REFERENCES

Dean, W.R., 1945. On the reflection of surface waves by a submerged plane barrier. Proc. Camb. Phil. Soc. 41:231-238.

Gray, A., 1918. A treatise on gyrostatics and rotational motion. Dover Editions, 1959, 530pp.

Havelock, T.H., 1940. The pressure of water waves upon a fixed obstacle. Proc. Roy. Soc. London. A175:409-421.

Hulsbergen, C.H., 1976. The origin effect and suppression of secondary waves. Publication No. 132, Delft Hyd. Laboratory.

Jolas, P., 1962. Contribution a l'étude des oscillations périodiques des liquides pesants avec surface libre. Houille Blanche, 758-769.

Kelvin, Lord, 1871. On the motion of free solids through a liquid. Phil. Mag. XLII: 362-377.

Kelvin, Lord, 1887. On ship waves. Popular Lectures and Addresses III:459, Macmillan, London (1891).

Lamb, H., 1932. Hydrodynamics. Art. 124, Cambridge University Press.

Longuet-Higgins, M.S., 1953. Mass transport in water waves. Phil. Trans. Roy. Soc. A245:535-581.

Longuet-Higgins, M.S., 1960. Mass transport in the boundary layer at a free oscillating surface. J. Fluid Mech. 8:293-306.

Longuet-Higgins, M.S., 1967. On the wave-induced differences in mean sea level between the two sides of a submerged breakwater. J. Mar. Res. 25:148-153.

Longuet-Higgins, M.S., 1977. The mean forces exerted by waves on floating or sub-merged bodies with applications to sand bars and wave power machines. Proc. Roy. Soc. London. A352:463-480.

Longuet-Higgins, M.S., and Stewart, R.W., 1964. Radiation stresses in water waves; a physical discussion, with applications. Deep Sea Res. 11:529-562.

Maxwell, J.C., 1853. On a particular case of the descent of a heavy body in a resisting medium. Camb. and Dublin Math. Jour. IX., Coll. Papers:115-118.

Navarro Pineda, J.M., 1972. Etude du passage de la houle sur un écran vertical mince immergé. Thesis. University of Grenoble.

Phillips, O.M., 1977. The dynamics of the upper ocean. 2nd Edn. Cambridge University Press, p. 40.

Sleath, J.F.A., 1970. Velocity measurements close to the bed in a wave tank. J. Fluid Mech. 42:111-123.

126

Stokes, G.G., 1845. On the theory of the internal friction of fluids in motion. Trans. Camb. Phil. Soc. 8:287, Papers 1:75.

Stokes, G.G., 1847. On the theory of oscillatory waves. Trans. Camb. Phil. Soc. 8:441-455, Papers 1:314-326.

Stokes, G.G. 1851. On the effect of the internal friction of fluids on the motion of pendulums. Trans. Camb. Phil. Soc. 9:8, Papers 111:1.

Ursell, F., 1947. The effect of a fixed vertical barrier on surface waves in deep water. Proc. Camb. Phil. Soc. 43:374-382.

Ursell, F., 1949. Wind and ocean currents. Nature 163:237-238.

Ursell, H.D., 1940. Motion of a solid through an infinite liquid. Proc. Camb. Phil Soc. 38:150-167.

Wendel, K. 1950. Hydrodynamische Massen und Hydrodynamische Massenträgheitsmomente Jahrb. d. STG 44.

A HYBRID PARAMETRICAL SURFACE WAVE MODEL APPLIED TO NORTH-SEA
SEA STATE PREDICTION

H. GÜNTHER and W. ROSENTHAL

Institut für Geophysik, Universität Hamburg, Hamburg (FRG)
and
Max-Planck-Institut für Meteorologie, Hamburg (FRG)

ABSTRACT

A hybrid parametrical wave model based on the ideas of Hasselmann et al. (1976)
is used to predict waves from surface wind fields on different sizes of prediction
area.

INTRODUCTION

We present here in a short version the performance of a new kind of surface wave
prediction model. Full details of the model and its applications can be found in
Günther et al. (1979a,b) and Ewing et al. (1979). These three papers give a
summary of an international effort to generate extreme value wave statistics for
the northern North-Sea. In the present contribution we describe qualitatively
the basic principles leading to our numerical model which can be found in more
detail in Hasselmann et al. (1973) and Hasselmann et al. (1976). As an example
of the applications of this model we discuss a hindcast on a small prediction area
using a dense computational grid, as well as a large hindcast area with a relatively
coarse grid.

DYNAMIC BEHAVIOUR OF GROWING WIND SEA AND SWELL SPECTRA

During the last twenty years there have been great advances in understanding
the behaviour of the energy distribution in surface gravity wave spectra. Nevertheless present day wave prediction models all use, along with well-established
research results, heuristic concepts which need further experimental and theoretical
investigations. Without going into details we give here an outline of our present
knowledge about the statistical behaviour of surface waves in deep water, which
is revealed by our model.

We describe the sea state at a fixed location for a certain time by the two
dimensional energy spectrum. For most applications the angular energy distribution

relative to the mean propagation direction in a wind-sea is known well enough to use the one dimensional energy spectrum following from an integration of the two dimensional spectrum over all wave directions. The one dimensional spectrum gives the energy density in frequency space and has a universal shape which can be characterized by a few parameters (see Fig. 1). Under atmospheric energy input the peak frequency of the spectrum shifts to lower values by nonlinear interaction

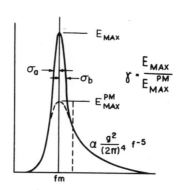

Fig. 1. Definition of the five JONSWAP parameters for a wind-sea spectrum.

processes between different wavenumber and frequency bands. The universal spectral shape is thereby retained also due to the nonlinear interaction. This concept may be compared to the establishment of a universal energy distribution in an ensemble of weakly interacting gas molecules resembling the Boltzmann-distribution, also during quasistationary external energy input. The spectral peak stops moving when the peak frequency is equivalent to a phase speed roughly equal to the surface wind speed

$$\frac{U_{10} \cdot f_m}{g} = 0.13, \tag{1}$$

where

U_{10} = wind speed at 10 m height

f_m = peak frequency

g = gravitational constant

Surface waves with larger phase speeds are not influenced by the atmosphere and can be treated as independent wave groups propagating with their group velocity. The described history of a growing wind-sea is clearly shown in Fig. 2 (taken from Hasselmann et al. (1973)), showing the energy spectra with increasing distance from shore for a homogeneous wind blowing perpendicular to the shore.

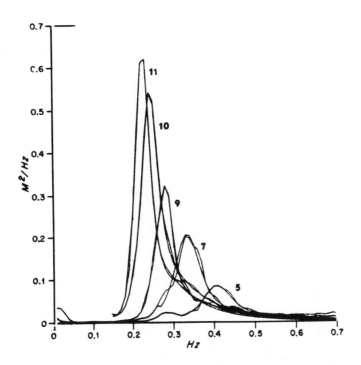

Fig. 2. Evolution of wave spectrum with fetch for offshore winds (11^h-12^h, Sept. 15, 1968). Numbers refer to stations with increasing distance to shore. The best-fit analytical spectra are also shown.

CLASSIFICATION OF NUMERICAL MODELS

Before showing some hindcast results of our model it may be appropriate to give a short description of the main features of wave models in use today.

Diagnostic models

These models can be used for special cases when the surface wave field is a function of the local wind field only, so that the prediction problem can be shifted completely to the prediction of the wind field. A model of this kind, for example, works remarkably well for prediction of hurricane sea states (Cardone et al., 1977).

130

Prognostic models

These models are based on an integration of the energy transport equation

$$\frac{dF}{dt} = S = S_{in} + S_{nl} + S_{dis} \qquad (2)$$

where

$\frac{d}{dt}$ = total time derivative

F = $F(r,f,\theta)$ = spectral energy density

S = source function

S_{in} = energy input from the atmosphere

S_{nl} = energy redistribution between different wavenumbers by nonlinear interacti

S_{dis} = dissipation of energy

The first models of this kind were spectral models developed by Pierson and his colleagues, for example Pierson et al., 1966. The energy input and dissipation is applied to each spectral component separately and so each component grows until its saturation state independent of the overall energy distribution. That means S_{nl}, the nonlinear interaction in the source function, is neglected.

Since the work of Phillips (1960) and Hasselmann (1963) on the nonlinear interactions between different spectral components and the confirmation of their importance in the first JONSWAP experiment (Hasselmann et al., 1973), wave predicti models were developed which used the nonlinear term S_{nl} in the source function in a rough parametrization (Barnett, 1968; Ewing, 1971).

Hasselmann et al. (1976) proposed a new technique to parametrize the complete prognostic equation instead of only the nonlinear source term, which is used in our numerical model for the wind-sea part. It does not use the energy content of the single frequency-direction bins of the spectrum as quantities to be predicte but the ensemble parameters characterizing the universal shape of the wind-sea spectrum. This results also in much lower demands on computer resources because of a smaller number of predicted parameters.

For swell frequencies characterized by

$$\frac{U_{10} \cdot f}{g} \leq 0.13 \qquad (3)$$

we use a technique, first applied by Barnett et al. (1969), in which the paths of individual wave trains are stored and at each time step the wave energy is propagated along these paths with the appropriate group velocity.

Hindcasting over a small area

This example compares hindcasted wave spectra with measurements taken during the JONSWAP 1973 experiment by DHI (Federal Republic of Germany), KNMI (Holland) and IOS (United Kingdom) and is given in more detail in Günther et al. (1979,b). The wind field is spatially homogeneous and depicted in the upper panel of Fig. 3.

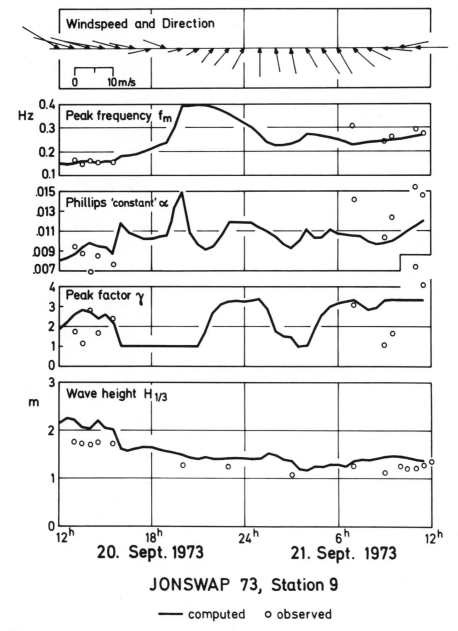

JONSWAP 73, Station 9

—— computed   o observed

Fig. 3. Time series for sea state parameters from Sept. 20th, 12 h until Sept. 21st, 12 h at Station 9 of the JONSWAP array. solid line: hindcast, circles: measurements. In the upper panel North is pointing vertically upward.

Incoming wave energy from the western, northern, or southern boundary is fed
into the model from waverider and pitch-roll buoys on the borders of the prediction
area. The eastern border is a coastline and boundary values, necessary for offshore
blowing wind, were modelled by the fetch laws of Hasselmann et al. (1973).

The computational grid of the hindcast had a spacing of 2 km and the time-
step was 5 min. For the swell part (transportation of wave energy along characteristics) we used 24 frequency and 8 direction bins. The hindcast area was a
square of 44 x 44 km^2 and the hindcast covered 24 hours. The computation was
done on a CDC Cyber 76 and required 3 sec CPU time per computational time step
and a total field length of 90 000 words.

Our hindcast example shows in Fig. 3 satisfactory agreement between measured and
hindcasted peak frequencies and wave heights. The spectral scale parameter α
and shape parameter γ do not agree as well.

The hindcasted spectra in Fig. 4,5 also show satisfactory agreement with the
measurements although differences can be recognized, which can be traced back
to the directional assumptions of the model. The directional assumptions will be
improved in future work. The Figures 3,4,5 only show the comparison at one
station near the center of the prediction area. From comparisons of the whole
set of measured spectra throughout the area and the appropriate hindcasted spectra
and from two similar hindcasts with rapidly varying wind fields in space and time
we conclude that the model is able to predict surface waves in an area of the
prescribed size as long as waves are short enough to be considered as deep water
waves. In rapidly turning wind fields the wind-sea is overestimated by the model.
The parametrization of this effect will be improved by numerical calculation of
the Boltzmann integrals of the nonlinear interaction in turning wind situations
and the comparison of the results with field data.

Hindcasting over a large area

Work described in this chapter was done at the Institute of Oceanographic
Science and the Harbour Research Station Wallingford. The same hybrid parametrical
model was used, differing only in numerical details. The goal was to generate an
extreme value statistic of surface waves in the northern North Sea. For that
purpose the model has to be checked on available measurements. The surface wind
field for 42 storms from the decade 1965-1975 and from the first half of 1976
have been analysed as described by Harding et al. (1978). The computational grid
and the points where surface winds are available are shown in Fig. 6. The spacing
of grid points is 100 km. The time step used for computation was 1 hour. For one
of the test storms Fig. 7,8 show a comparison at weather ship Famita between modelled
and measured significant wave height Hs and mean zero-up-crossing periods

$$T_z = (m_o/m_2)^{1/2} \tag{4}$$

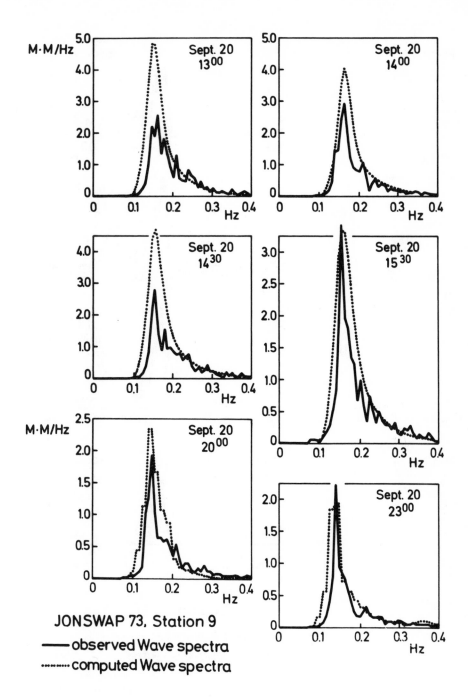

Figure 4. Wave spectra at Station 9 of the JONSWAP array. solid line: hindcast circles: measurements.

134

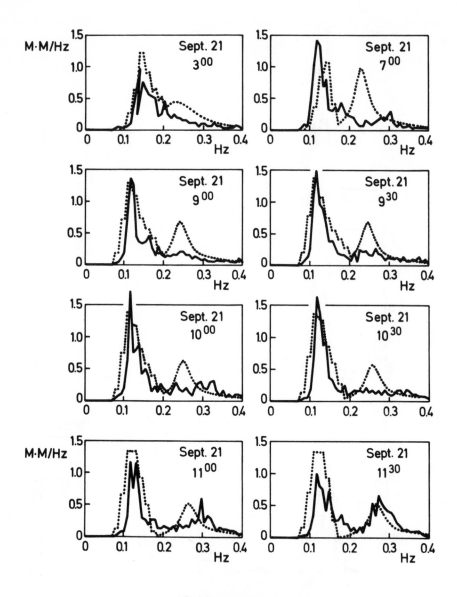

JONSWAP 73, Station 9

—— observed Wave spectra

·········· computed Wave spectra

Figure 5. Wave spectra at Station 9 of the JONSWAP array.   solid line: hindcast
circles: measurements.

where

$$m_n = \int_0^\infty E(f) \cdot f^n \, df$$

is the nth moment of $E(f)$, the energy spectrum, about the origin. Figure 9 shows the correlation between hindcasted and measured wave parameters for a series of test storms.

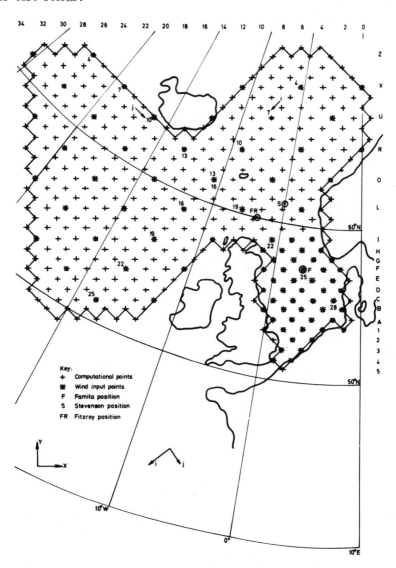

Figure 6   Model grid for NORSWAM hindcasts.

136

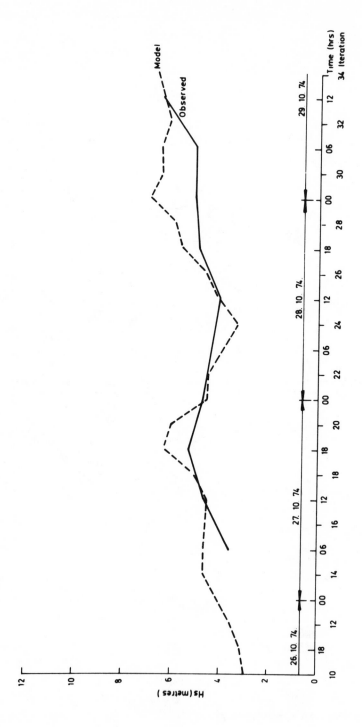

Figure 7. Modelled and observed significant wave height Hs at Famita position
(reproduced from HRS Report No. EX775, 1977).

137

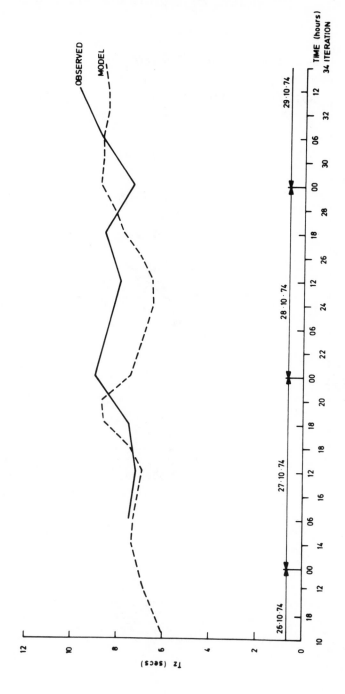

Figure 8. Modelled and observed mean zero-up-crossing periods T_z at Famita position (reproduced from HRS Report No. EX775, 1977).

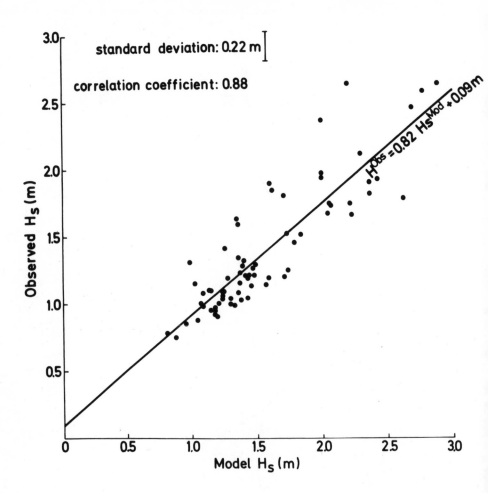

Figure 9. Correlation between measured and hindcasted wave heights at Stations 7 and 9 of the JONSWAP array.

CONCLUSION

From a number of test runs the correlations given for the small hindcast area in Fig. 9 show that the hybrid parametrical model in the present form is able to predict deep water surface waves. Although the overall performance confirms the underlying physical concept, there are some points which need further research. One is the energy transition from wind sea to swell, which will be tackled by clarifying the role of nonlinear wave-wave interaction in that case. Further investigation of nonlinear interaction and comparison with field data will also help to improve the model behaviour for rapidly turning wind situations. An extension of the model to shallow water wave prediction is under development.

REFERENCES

Barnett, T.P., 1968. On the generation, dissipation and prediction of ocean wind waves. J. geophys. Res., 73: 513-530.

Barnett, T.P., Holland, C.H. Jr., Yager, P., 1969. A general technique for wind wave prediction, with application to the South China Sea. Contract N62 306-68-C-0285, U.S. Naval Oceanic Office, Washington, D.C.

Cardone, V.J., Ross, D.B., Ahrens, M.R., 1977. An experiment in forecasting hurricane generated sea states. Proc. 11th Technical Conference on Hurricanes and Tropical Meteorology, Dec. 13-16, Miami, Florida.

Ewing, J.A., 1971. A numerical wave prediction method for the North Atlantic Ocean. Deut. Hydrogr. Z. 24: 241-261.

Ewing, J.A., Worthington, B.A., Weare, T.J., 1979. A hindcast study of extreme wave conditions in the North Sea. Accepted by J. geophys. Res.

Günther, H., Rosenthal, W., Weare, T.J., Worthington, B.A., Hasselmann, K., Ewing, J.A., 1979a. A hybrid parametrical wave prediction model. Accepted by J. geophys. Res.

Günther, H., Rosenthal, W., Richter, K., 1979b. Application of the parametrical wave prediction model to rapidly varying wind fields during JONSWAP 1973. Submitted to J. geophys. Res.

Harding, J., Binding, A.A., 1978. Wind fields during gales in the North Sea and the gales of 3 January 1976. Met. Magazine, 107: 164-181.

Hasselmann, K., 1962. On the nonlinear energy transfer in a gravity-wave spectrum, Part 1: General theory. J. Fluid Mech.,12:481-500.

Hasselmann, K., 1963. On the nonlinear energy transfer in a gravity-wave spectrum, Part 3: Evaluation of the energy flux and swell-sea interaction for a Neumann Spectrum. J. Fluid Mech., 15: 385-398.

Hasselmann, K., Barnett, T.P., Bouws, E., Carlson, H., Cartwright, D.E., Enke, K., Ewing, J.A., Gienapp, H., Hasselmann, D.E., Kruseman, P., Meerburg, A., Müller, P., Olbers, D.J., Richter, K., Sell, W., Walden, H., 1973. Measurements of wind-wave growth and swell decay during the Joint North Sea Wave Project (JONSWAP), Deut. Hydrogr. Z., Suppl. A., 8(12).

Hasselmann, K., Ross, D.B., Müller, P., Sell, W., 1976. A parametrical wave prediction model. J. phys. Oceanogr. 6: 201-228.

Hydraulics Research Station, Wallingford., 1977. Numerical wave climate study for the North Sea. Report No. Ex775.

Pierson, W.J., Tick, L.J., Baer, L., 1966. Computer-based procedures for preparing global wave forecasts and wind field analyses capable of using wave data obtained by a spacecraft. Proc. 6th Naval Hydrodynamics Symposium, Washington, D.C., 499.

Phillips, O.M., 1960. On the dynamics of unsteady gravity waves of small amplitude, Part I. J. Fluid Mech., 9: 193-217.

ON THE FRACTION OF WIND MOMENTUM RETAINED BY WAVES

M. DONELAN

Canada Centre for Inland Waters, Burlington, Ontario (Canada)

ABSTRACT

The fraction of the momentum or energy transfer across the air-water interface which remains in the wave field is a crucial parameter in many wave prediction models. In this paper, direct measurements of the air-water momentum transfer coupled with wave spectra measured at several fetches provide estimates of this parameter. Both laboratory and field measurements are used to provide a wide range of conditions. It is found that the fraction of momentum remaining in the wave field can be as much as ¼ of the total transferred locally across the air-water interface, and is largely dependent on the wave age.

INTRODUCTION

Given the surface wind over a body of water, the problem of wind wave prediction may, in the simplest instance of still deep water, be reduced to three distinct steps: estimation of the surface momentum flux, determination of the fraction of the momentum transferred which remains in the wave field, and finally, the solution of an appropriate momentum balance equation of the type:

$$\frac{\partial M_i}{\partial t} + \frac{\partial (v_j M_i)}{\partial x_j} = \gamma \tau_i \quad , \qquad i = 1, 2 \; ; \; j = 1, 2 \tag{1}$$

where the M_i, v_j, τ_i are the average wave momentum per unit area, average group velocity and surface wind stress components respectively; γ is the subject of this paper.

The first step, essentially that of the determination of the surface drag coefficient, has been the cause of intense empirical study over the past two or three decades. Some of this work is summarized by Stewart (1974) and will not further concern us here. The third step, the solution of an equation like (1) or its energy equivalent, is normally carried out through finite difference methods on a digital computer. Its solution, which clearly requires a link between the average group velocity and the momentum, can be approached in several ways, two examples of which are given in Hasselmann et al. (1976) and Donelan (1979).

It is to the second step that this paper is directed. We wish to establish the fraction Y of air-water momentum transfer which is retained by the wave field. Or, in other words, we seek the net momentum transfer to the wave field from the wind - the difference between wind input to the waves and local wave dissipation - in relation to the total momentum transfer across the air-water interface. The fraction Y might reasonably be expected to be a function of other dimensionless parameters which characterize the air-water interface, in particular the ratio of peak wave phase speed to wind speed, generally termed 'wave age', and some atmospheric stability index - here we use the bulk Richardson number. This paper describes field and laboratory experiments aimed at exploring the dependence of Y on inverse wave age U/c. The bulk Richardson number dependence of Y is investigated for a restricted range of U/c only.

EXPERIMENTAL DESIGN

The ideal experimental arrangement for the direct determination of Y obtains when mean wind and wave directions and the line joining measuring positions all coincide. In the light of figure 1, we may rewrite (1) in terms of the directional spectrum of wave energy $F(\omega, \theta)$ and the dispersion relation $\omega(k)$.

$$\frac{\partial}{\partial t}\left\{\iint \frac{Fk}{\omega} \cos\theta \, d\theta \, d\omega\right\} + \frac{\partial}{\partial x}\left\{\iint \frac{Fk}{\omega} \frac{d\omega}{dk} \cos^2\theta \, d\theta \, d\omega\right\} = -\frac{\gamma}{g}\frac{\rho_a}{\rho_w} \overline{u'w'} \tag{2}$$

where $F(\omega, \theta)$ and the corresponding one-dimensional frequency spectrum $E(\omega)$ are defined, in the usual manner, such that $\iint F \, d\theta \, d\omega = \int E \, d\omega = \overline{\zeta^2}$, the mean square surface displacement. We have, of course, made use of the equality of mean potential and kinetic energies applicable to small amplitude waves and the relation between momentum and energy valid for progressive waves of constant shape $\vec{M} = E\vec{k}/\omega$ (Phillips, 1977). ω and k are the radian frequency and radian wave number; ρ_a and ρ_w are the air and water densities; g, the acceleration due to gravity and $-\overline{u'w'}$ is the Reynolds stress in the air near the surface.

In order to precisely evaluate Y from (2), we need measurements of the directional distribution of wave energy $F(\omega, \theta)$, the dispersion relation $\omega(k)$ and the Reynolds stress as a function of x and t. The satisfaction of such stringent requirements being beyond our resources, we set out to estimate Y from the finite difference equivalent of (2) using two-point measurements of $E(\omega)$ with estimates of $F(\omega, \theta)$, $\omega(k)$ and $\overline{u'w'}$ at a single location. The estimates of $F(\omega, \theta)$ and $\omega(k)$ were non-dimensionalized with respect to the peak frequency and applied to the point measurements of $E(\omega)$. Time differences were rendered insignificant in the laboratory and were estimated in the field from consecutive 20 minute averages. It behooves us, at this point, to show the experimental arrangement before attempting to provide further details of the analysis procedure.

Figures 2 and 3 show the field and laboratory experimental arrangements. The field site consists of a tower, 1100 m offshore mounted on the bottom in 12 m of water, and a 'waverider' accelerometer buoy, moored an additional 2000 m offshore along the normal to

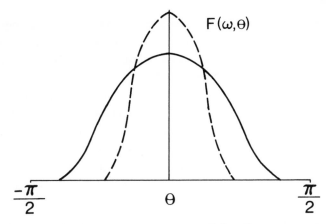

Fig. 1. Schematic of idealized directional spectra F (ω, θ) for two values of the radian frequency .

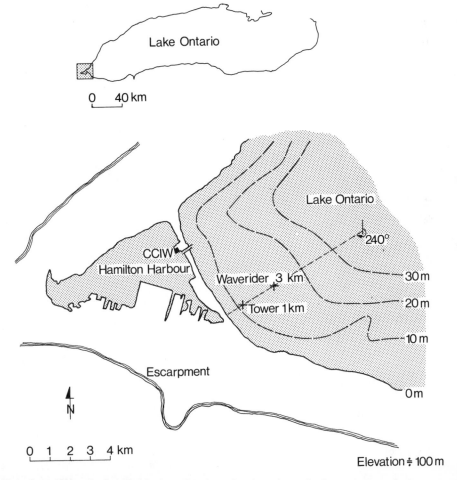

Fig. 2. Map of the field site showing the location of the tower and the waverider accelerometer buoy at the western end of Lake Ontario.

the shoreline through the tower. The tower supports an array of 14 capacitance wave staffs for estimating directional spectra and various micro-meteorological equipment for estimating the Reynolds stress and the bulk Richardson number. Further details of the tower and its measuring equipment are given by Birch et al. (1976) and Der and Watson (1977). Suffice it to say that the wind speed, direction and Reynolds stress measurements were made with a Gill anemometer bivane (Young, 1971) mounted at 11.5 m above the water and sampled five times a second per channel. The laboratory tests were carried out in a large closed wind-wave flume at the Canada Centre for Inland Waters (CCIW). Figure 3 provides information on the flume's dimensions and general characteristics sufficient for our purposes here. The directional spectra were estimated using a 1/28 scaled version of the tower array placed as indicated in figure 3. One-dimensional frequency spectra were measured using capacitance wire staffs at stations 4 and 5 (denoted in the figure by ζ_4 and ζ_5), and wind speed and Reynolds stress were obtained through x-film anemometry at the intermediate position shown, 26.2 cm above the mean water level.

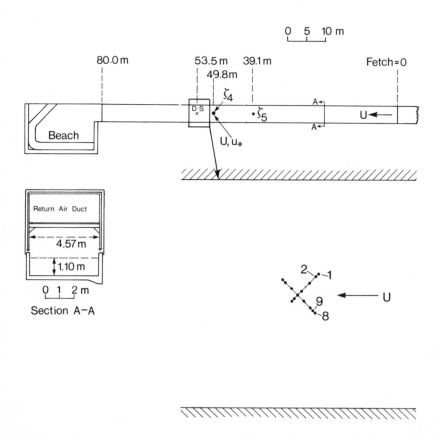

Fig. 3. Plan of the wind-wave flume at the Canada Centre for Inland Waters showing the disposition of wave ζ and wind U, u_* measurements. The location of the directional wave array is indicated by an X labelled "D.S.". The lower part of the figure shows an enlargement of the directional array.

DISPERSION RELATION AND SPREADING FACTORS

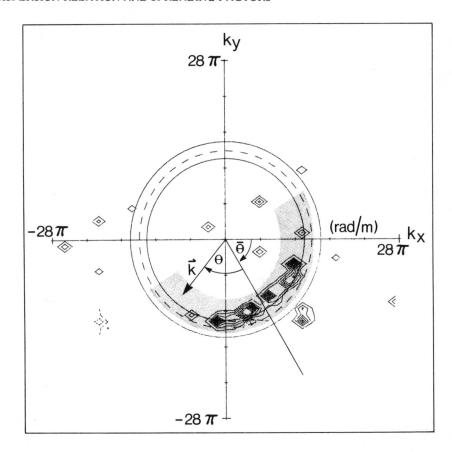

Fig. 4. The wave-number spectrum for a frequency band of width 1.97 rad/s centred on 20.7 rad/s, from laboratory run 3. The full circles represent the limits of the deep water dispersion relation corresponding to the edges of the frequency band analysed, while the dashed circle corresponds to the energy-weighted mean frequency of that band. The area of integration is shown shaded.

Although it is convenient to write (2) in terms of the directional spectrum F (ω, θ), the data are first analysed in terms of the frequency wave-number spectrum $\chi (\omega, \vec{k})$, which is, in fact, required to determine the dispersion relation ω (k). Figure 4 is a sample of the wave-number spectrum χ for a frequency band $\Delta\omega$. Details of the characteristics of the wave-number spectrum χ are given in Donelan et al. (1979) and only a few features need concern us here. The solid circles shown (figure 4) represent the limits of the deep water dispersion relation $(\omega = \sqrt{gk})$ corresponding to the limits of the frequency band analysed. The average wave-number $\bar{k}$ corresponding to the frequency band $\Delta\omega$ is evaluated as indicated in (3). It is assumed that k is a function of ω only and not of θ. In addition to the dispersion relation, we will need spreading factors associated with the two first terms of (2), which we will designate A(ω) and B(ω) and define thus:

$$A(\omega) = \frac{1}{\overline{k}(\omega) \cdot E(\omega)} \iint \chi(\omega, \vec{k}) \, k \cos\theta \, dk_x \, dk_y$$

$$B(\omega) = \frac{1}{\overline{k}(\omega) \cdot E(\omega)} \iint \chi(\omega, \vec{k}) \, k \cos^2\theta \, dk_x \, dk_y$$

$$\overline{k}(\omega) = \frac{1}{E(\omega)} \iint \chi(\omega, \vec{k}) \, k \, dk_x \, dk_y \tag{3}$$

$$E(\omega) = \iint \chi(\omega, \vec{k}) \, dk_x \, dk_y$$

Since noise, which is predominant at large wave-numbers, tends to bias the relations (3), the integration of (3) includes only a band of thickness $2\Delta\omega$ and π radians in angular width centred on the peak of the directional spectrum. Finally, the angle θ is referenced to the mean momentum direction $\overline{\theta}$ defined thus:

$$\tan\overline{\theta} = \frac{\iint \chi(\omega, \vec{k}) \, k \sin\theta \, dk_x \, dk_y}{\iint \chi(\omega, \vec{k}) \, k \cos\theta \, dk_x \, dk_y} \tag{4}$$

Noting that $\int Fk \cos^n\theta \, d\theta = \iint \chi \, k \cos^n\theta \, dk_x \, dk_y$ and that k and ω are independent of θ, it can be seen that the integrals of (2) can be reconstructed from the spreading factors $A(\omega)$ and $B(\omega)$ provided the dispersion relation and the frequency spectrum $E(\omega)$ are known.

Having determined $A(\omega)$, $B(\omega)$ and $\overline{k}(\omega)$ at one fetch, we are faced with the problem of applying the results to the one-dimensional frequency spectrum $E(\omega)$ at the other fetch. By analogy with the results of Hasselmann et al. (1973) for the one-dimensional frequency spectra, we would expect the spreading factors and the dispersion relation to scale with the peak frequency ω_p. We have not included any fetch dependence of the scaled spreading factors $A(\omega/\omega_p)$ and $B(\omega/\omega_p)$. Although Mitsuyasu et al. (1975) have indicated some fetch dependence of their spreading parameter, their results are not applicable to the high values of U/c encountered in our experiments. For example, at $U/c=3.5$ Mitsuyasu et al. (1975) would suggest an angular distribution of the form $\cos(\theta/2)$, which is certainly not in agreement with our observations (Donelan et al., 1979).

Figure 5 shows an example of the spreading factors $A(\omega)$ and $B(\omega)$. The frequency axis has been scaled by the peak frequency ω_p and a second order polynomial in $\{(\omega/\omega_p)^2 - 1\}$ fitted to the data points. A spreading factor of 1.0 corresponds to all waves at that frequency travelling in the same direction. It is seen that the spectrum broadens away from the peak. Spreading factors corresponding to the commonly assumed $\cos^2\theta$ or $\cos^4\theta$ distributions of $F(\omega, \theta)$ are indicated for comparison on figure 5.

The average wave-number $\overline{k}$ scaled by the theoretical value for deep water at the peak frequency ω_p^2/g is plotted in figure 6 against $(\omega/\omega_p)^2$ and a second order polynomial in $(\omega/\omega_p)^2$ fitted to the data points. In all cases, the polynomials were fitted so as to minimize $\Sigma(\Delta y_i)^2 \cdot E(\omega_i)$. This forces the fit to be closest near the peak of the spectrum. In the case of $k(\omega)$, the point (0,0) was added and given weight equal to that of the spectral peak. In these coordinates, the theoretical dispersion relation is the straight line through the origin shown (dashed) in figure 6.

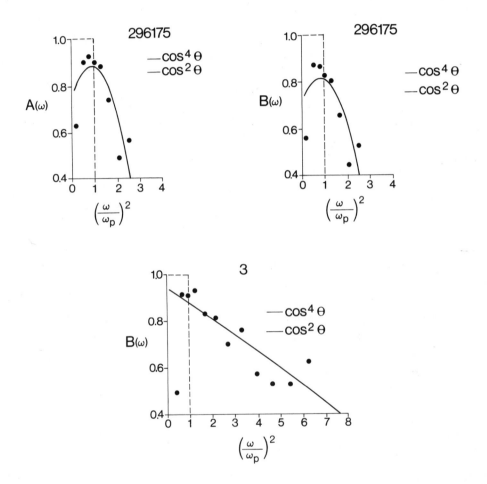

Fig. 5. Samples of directional spreading factors A (ω) and B (ω) for field (296175) and laboratory (3) measurements. The curves shown are fitted second order polynomials in $\{(\omega/\omega_p)^2 - 1\}$. The equivalent constant spreading factors for $\cos^2 \theta$ and $\cos^4 \theta$ directional distributions are also indicated by the horizontal lines. It can be seen that these yield reasonable approximations only near the peak.

The normalized one-dimensional frequency spectra are shown in figure 7 and, in figure 8, the calculated phase speed $\omega/\bar{k}$ is compared with its theoretical small amplitude, deep water value g/ω both normalized by the theoretical value at the peak g/ω_p. In all cases, the abscissa is the square of the frequency scaled by its value at the spectral peak. It is clear that the small amplitude, deep water dispersion relation is not exactly followed and, in fact,

148

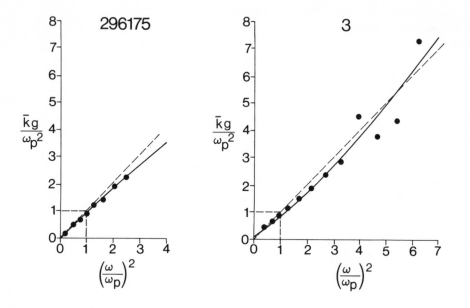

Fig. 6. Computed mean wave-numbers in each frequency band normalized by the theoretical value at the peak of the spectrum. The solid curves are fitted second order polynomials in $(\omega/\omega_p)^2$, while the dashed straight lines are the theoretical small amplitude, deep water dispersion relation. Only frequency bands whose wave-number spectra could be successfully resolved are included (Donelan et al., 1979).

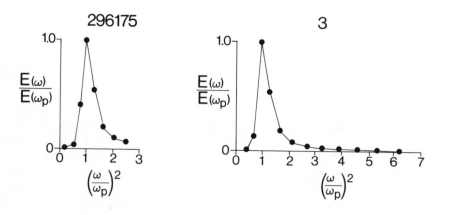

Fig. 7. Normalized one-dimensional frequency spectra corresponding to Figures 5, 6 and 8.

the disagreement becomes more pronounced at higher U/c values (Ramamonjiarisoa and Coantic, 1976 and Donelan et al, 1979).

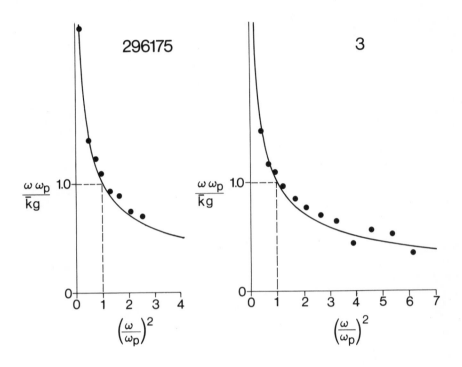

Fig. 8. Observed (dots) and theoretical (curve) phase speeds. Both are normalized by the theoretical peak value g/ω_p.

The polynomial $kg/\omega_p^2 = a_0 + a_1 (\omega/\omega_p)^2 + a_2 (\omega/\omega_p)^4$ may then be used to determine k and $d\omega/dk$ once the peak frequency ω_p is known. Similarly, the spreading factors $A(\omega)$ and $B(\omega)$ may be derived and applied, through (3), to the evaluation of the integrals of (2) for the fetch at which only $E(\omega)$ is known.

ANALYSIS OF FIELD DATA

Observations were obtained in runs of one hour under mini-computer control triggered by a change in the ten-minute average wind speed. During such runs, all wave staffs and relevant meteorological sensors were sampled at 5 Hz and, at the same time, the waverider recordings were made. Although there were a large number of these runs, only six satisfied the criteria that both wind and wave directions be within 25 degrees of the normal to the beach and that the peak frequency at the waverider be less than 3.14 rad/s - the upper limit of the waverider's reliable resolution. These six are summarized in table I. The tower runs,

150

TABLE I

Summary of field data (1976)

Symbol	V^+	$\bar{\theta}$	U	$-\overline{u'w'}$	Rb	ω_p	ω_p	$\overline{\zeta^2}$	$\overline{\zeta^2}$	U/c	γ
Units	m/s	Deg.	m/s	(m/s)²		rad/s	rad/s	cm²	cm²		
Height (m)	11.5	11.5	10	11.5	11.5	0	0	0	0	10	
Fetch (m)	1100	1100	1100	1100	1100	1100	3100	1100	3100	*	
Run No.											
29617	12.4	261	12.1	0.334	-0.005	3.3	2.4	84.4	178.9	3.4	0.060
30209	8.5	240	8.3	0.109	-0.034	4.3	3.1	20.5	84.1	3.1	0.071
30219	11.1	238	10.9	0.344	+0.004	3.7	2.7	40.4	115.2	3.5	0.028
30317	8.9	236	8.8	0.166	+0.019	4.1	3.1	22.3	39.2	3.3	0.034
31106	11.3	236	11.1	0.392	-0.008	3.7	2.8	42.4	121.0	3.6	0.060
31821	9.5	248	9.2	0.261	+0.003	4.0	3.0	28.0	96.0	3.2	0.052

+ V is the measured wind speed
* c is the average of the theoretical phase speeds of the peak frequencies at tower and waverider.

TABLE II

Summary of laboratory data

Symbol	V^+	U	$-\overline{u'w'}$	Rb	ω_p	ω_p	$\overline{\zeta^2}$	$\overline{\zeta^2}$	U/c	γ
Units	m/s	m/s	(m/s)²		rad/s	rad/s	cm²	cm²		
Height (m)	0.26	10	0.26	0.26	0	0	0	0	10	
Fetch (m)	49.8	49.8	49.8	49.8	39.1	$\frac{49.8}{53.5}$	39.1	$\frac{49.8}{53.5}$	*	
Phase Run										
I 12	6.54	9.70	0.121	+0.001	10.9	9.0	2.01	2.32	9.7	0.108
13	6.54	9.56	0.110	+0.001	10.9	9.9	1.78	2.32	10.1	0.205
14	6.55	9.71	0.120	+0.001	10.9	9.0	1.85	2.52	9.7	0.262
15	6.58	9.86	0.130	+0.001	10.4	9.4	2.12	2.71	9.9	0.191
19	6.45	9.50	0.112	+0.001	10.9	9.0	1.81	2.41	9.5	0.223
20	6.53	9.64	0.117	+0.001	10.9	9.0	1.91	2.18	9.7	0.096
23	11.69	20.94	1.034	0	7.2	6.6	16.16	20.89	14.7	0.183
24	11.71	20.99	1.038	0	7.9	6.6	15.09	18.80	15.4	0.143
25	10.63	18.22	0.696	0	7.9	7.2	10.86	15.29	14.0	0.265
26	10.67	18.54	0.749	0	7.9	7.2	10.31	12.82	14.3	0.140
28	4.06	5.49	0.025	+0.002	11.6	11.6	0.56	0.69	6.5	0.189
38	7.19	10.96	0.171	0	9.4	8.7	3.35	4.75	10.1	0.354
49	2.80	4.08	0.020	+0.003	15.3	13.9	0.22	0.24	6.1	0.043
51	10.31	17.67	0.651	0	7.9	7.2	9.99	14.67	13.6	0.304
52	9.30	14.85	0.371	0	8.9	7.5	7.84	10.50	12.4	0.311
II 2	3.29	4.58	0.020	+0.011	17.0	15.0	0.07	0.19	7.5	0.160
3	4.80	6.86	0.051	+0.004	13.0	11.5	0.51	0.55	8.5	0.019
5	7.79	11.94	0.208	+0.001	10.5	9.0	2.38	3.37	11.8	0.159
6	9.46	15.40	0.426	+0.001	9.5	8.4	3.93	5.54	14.0	0.121
7	10.87	18.83	0.766	+0.001	8.9	7.7	7.37	9.56	15.9	0.088
29	3.24	4.49	0.019	+0.007	16.7	14.9	0.16	0.19	7.2	0.057
30	12.00	21.88	1.179	+0.001	7.8	7.0	13.53	20.08	16.5	0.164

+ V is the measured wind speed
* c is the average of the theoretical phase speeds of the peak frequencies at both stations
In phase II - $\overline{u'w'}$ was not measured but inferred from figure 10.

identified by a composite number consisting of the Julian day and hour (GMT) of start, were divided into four consecutive sections of 13.64 minutes each. The waverider runs were divided into three consecutive 20 minute sections. Directional spectra and Reynolds stresses were available at the tower only. Thus non-dimensionalized characteristics of the directional spectra derived from the tower data were applied to the waverider data closest in time.

Evidently the reliability of this method, based on the difference between waverider and capacitance wave staff measurements, rests on accurate calibrations of both types of instruments. The capacitance wave staffs were calibrated in situ by immersion to marked intervals about the mean water level. The one-dimensional frequency spectrum discussed here is the average of the spectra from each of the 14 staffs. The original spectrum of 128 estimates is grouped into 15 equally spaced bands between 0.03 and 7.4 rad/s. The waveriders were calibrated on a rotating arm and their data analysed by the Marine Environmental Data Service, Ottawa (Wilson and Baird, 1972) who subsequently provided us with a tape of 62 equally spaced spectral density estimates from 0.32 to 3.14 rad/s.

These were then grouped into frequency bands corresponding to the tower spectra, and an ω^{-5} tail appended to fill the eight high frequency bands, which are beyond the frequency resolution of the waverider. The tail was computed from the peak spectral estimate and frequency in the following form:

$$E(\omega_i) = E(\omega_p) \cdot \omega_p^5 \cdot \omega_i^{-5} \qquad\qquad i = 8 \text{ to } 15 \qquad\qquad (5)$$

In order to compare the waverider and tower wave measurements, the waverider was moored 150 m to the northwest of the tower for a period of one month. Figure 9 shows the superimposed spectra of tower and waverider measurements for two cases in which conditions were very steady so that, although the tower measurements covered only the first 68% of the waverider observations, they could be expected to yield the same results. Evidently the comparison is very encouraging even when the peak frequency approaches 3.14 rad/s - the Nyquist frequency of the waverider analysis. The addition of a ω^{-5} tail is clearly necessary in such cases.

As mentioned before, the characteristics of the directional spectrum at the tower are applied to the waverider observations and the integrals of (2) are evaluated at both the tower and the waverider. Linear interpolation between pairs of tower results is applied to bring them into time coincidence with the waverider results. The finite differences of (2) are then evaluated using only the final two sections of each run, i.e. the last 40 minutes of each hour. In the evaluation of (2), the time difference was computed from the difference over these last two (20 minute) data sections of the space-average of the first integral of (2). Similarly, the space difference was computed from the difference between waverider and tower of the time-average of the second integral of (2). Since the wind speed change which triggered the recording was frequently accompanied by a wind direction change, the wave field was generally steadier during the latter part of the hour. In general, the second term of (2) was twice as large as the first.

152

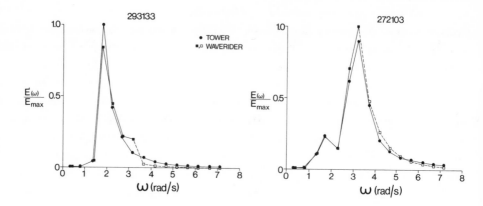

Fig. 9. Comparison of wave spectra measured by fixed capacitance wave staffs (TOWER) with those measured by the waverider accelerometer buoy (WAVERIDER). During the comparison the waverider was moored near the tower at the same fetch. The normalizer, E_{max} is simply the highest value of E (ω) from either tower or waverider. The dashed lines indicate the ω^{-5} tails appended to the waverider spectra which extend only to 3.14 rad/s.

The Reynolds stress and all other meteorological parameters were derived from averages of tower measurements over the last 40 minutes of the hour. The wind speed, measured at 11.5 m, was adjusted to 10 m using the measured friction velocity $u_* = (-\overline{u'w'})^{1/2}$ and a logarithmic velocity profile. The errors introduced by this procedure, through diabatic distortions to the logarithmic profile, are small, since the wind gradient is weak at these heights.

ANALYSIS OF LABORATORY DATA

The laboratory data were gathered in two phases. In phase I, there were 15 runs, during which recordings were made of surface elevation at stations 4 and 5 and of wind speed and Reynolds stress near station 4. The sampling frequency and duration were 60 hz and 4.5 minutes for nine of these runs, and were 30 Hz and 9.1 minutes for the other six. In phase II, the directional spectrum was measured 3.75 m downwind of station 4 (figure 3), the one-dimensional frequency spectrum at station 5 and the wind speed at station 4. There were seven of these runs of duration 13.7 minutes and sampling frequency 20 Hz. The still water depth was 125 cm in the first phase and 110 cm in the second.

In order to integrate the two phases, we will attempt to relate the vertical flux of wind momentum (phase I) and the horizontal flux of wave momentum (phase II) to the wind speed, which was monitored in both phases. The ratio of the measured friction velocity u_* to the measured wind speed as a function of wind speed is given in figure 10. The scatter is sufficiently small to insure a reasonably accurate estimate of the friction velocity from the measured wind speed in phase II and the two highest wind speed cases of phase I wherein the x-film anemometer was too frequently assaulted by water droplets to permit a reliable estimate of the friction velocity. A more appropriate parameter for the abscissa might be U/c, but this figure is meant only as an interpolation tool and the wind speed $U_{.26}$ is more accurately measured than U/c. Furthermore, in these experiments, c is determined almost entirely by the wind speed. For the same reasons, the dimensionless ratio of the downwind flux of wave momentum to the wave energy $(\overline{\zeta^2})^{-1} \iint Fk/\omega \ d\omega/dk \cos^2 \theta \ d\theta \ d\omega$ is given in figure 11 as a function of the measured wind speed. This is simply the ratio of the integral of the second term of (2) to the mean square surface dispacement. Other conditions being unchanged, this dimensionless ratio might be expected to be a function of U/c or $U_{.26}$ for these experiments. Here again the scatter is small enough to allow use of the sketched curve as an interpolation tool in the analysis of phase I data; i.e. to infer the integrals of the second term of (2) from the mean square surface displacement at the two fetches being considered.

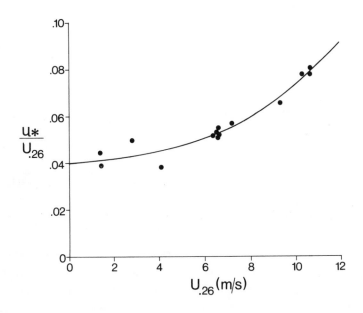

Fig. 10.   Ratio of friction velocity u_* to measured near surface (26 cm) mean wind $U_{.26}$ versus the mean wind. The smooth curve is fitted by "eye" to the laboratory (phase I) results.

For the sake of uniformity with the field data, the measured wind speed has been adjusted to 10 m assuming a neutral logarithmic profile. Both the 10 m and 26 cm wind speeds are reported in table II. It should be noted that the extrapolated 10 m wind speed is employed in this analysis only to construct the parameter U/c, which, at low values, is not sensitive to small errors in U. It is only at the lowest wind speeds that the stability index Rb indicate significant departure from neutrality. The bulk Richardson number Rb is defined in terms of the wind speed U and air temperature T_a evaluated at height Z and the water surface temperature T_w: Rb = $Zg(T_a-T_w)/(273+T_a) U^2$.

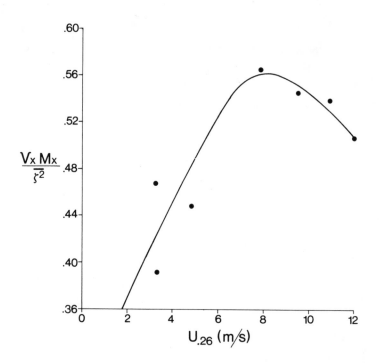

Fig. 11. The ratio of the downwind flux of wave momentum to wave energy versus the measured mean wind. The smooth curve is fitted by "eye" to the laboratory (phase II) results. This ratio would be 0.5 if all wave components were travelling in the same direction and were in strict obedience of the dispersion relation $\omega = \sqrt{gk}$. Lower values reflect significant directional spread, while higher values occur because the ratio of group to phase speeds approaches unity for trapped harmonics.

For the phase II data, (2) was solved by ignoring the first term (steady state) and applying (3) to the one-dimensional frequency spectrum at station 5 and the average of the spectra of the 14 staffs of the directional array. The relations (3) were determined separately for each of the seven phase II cases. Here, as in the field data, the original spectrum of 128 estimates was grouped into 15 equally spaced bands between 0.12 and 29.6 rad/s. In the case of the phase I data the flux of wave momentum at stations 4 and 5 was estimated using figure 11 and the mean square surface displacement $\overline{\zeta^2}$ measured at stations 4 and 5.

RESULTS AND DISCUSSION

The estimates of γ (figure 12), from both field and laboratory observations, show a definite dependence on U/c, although there is considerable scatter. The field observations

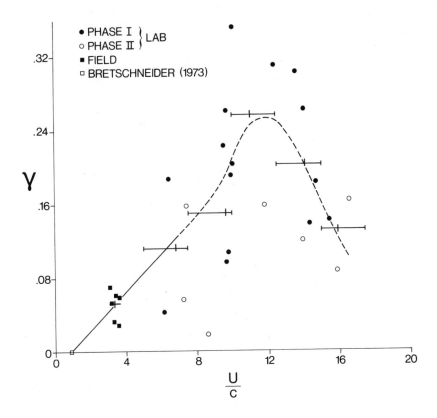

Fig. 12. The fraction of wind momentum retained by waves Υ versus the inverse wave age U/c. For the laboratory data the large crosses indicate the averages of the data assembled in groups of width 2.5 in U/c. The vertical bar indicates the average value of U/c in the group. The cross representing the field data has been placed at the value of Υ corresponding to neutral atmospheric stability (figure 13). The point of full development (U/c=0.83, Υ=0) according to Bretschneider (1973) provides another estimate of Υ. The straight line (solid) at lower values of U/c provides convenient access to Υ for wave prediction purposes. The extension of the line to higher values of U/c (dashed) serves only to illustrate the trend in Υ.

fall within a rather narrow band of U/c and the variability of the field estimates of Υ seem to be largely due to stability differences (figure 13). The smooth curve, fitted by 'eye' to figure 13, suggests that the appropriate value of Υ for neutral stability is 0.053. This value is indicated in figure 12 by a large cross centred on the mean value of U/c for these six field observations. The laboratory observations are rather badly scattered but, nonetheless, indicate a general increase in Υ as U/c increases from 6 to about 12, then, beyond that, a general decrease. The scatter in the laboratory values seems surprising at first sight in view of the steady and controlled conditions of these measurements. Stability effects can easily

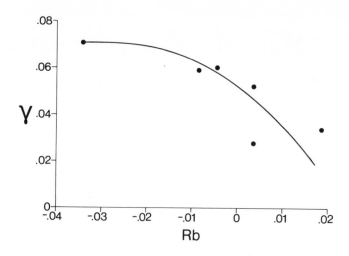

Fig. 13. γ versus bulk Richardson number Rb for the field data. The smooth curve has been fitted by "eye", and its intersection with Rb = 0 transposed to Figure 12.

be ruled out by examining the set of almost consecutive runs (12, 13, 14, 15, 19, 20) taken under nearly the same conditions. In these runs, γ is seen to vary from 0.10 to 0.26. The explanation appears to be that the mean value of $\overline{\zeta^2}$, averaged over several hundred wave periods, is a function not only of the x (down-tank) direction but also of the y (cross-tank) direction, and the y dependence is itself dependent on fetch x and the mean flow parameters: $\overline{\zeta^2} = f\left[x, y(x, U)\right]$. It is not our purpose here to explore the exact functional form f, but merely to demonstrate that the failure of a time average of ζ^2 along the centreline of the tank to represent the cross-tank average value at that fetch is much of the cause of scatter in figure 12. To do this we examine (table III) the values of $\overline{\zeta^2}$ from four of the staffs for the five runs (2, 3, 5, 6, 7) at the beginning of phase II. The four staffs are located in pairs (figure 3) at the ends of the long arms of the array. The staffs in each pair are separated by 3.6 cm and the pairs are separated by 73 cm symmetrically across the tank at the same fetch. Runs 29 and 30 are omitted from this comparison because the entire array was rotated by 15 degrees for these runs. Table III shows $\overline{\zeta^2}$ for staff number 1 and the ratio of $\overline{\zeta^2}$ from the other staffs to this. While there are differences of about 6% between the members of the pairs, these differences are nearly constant and reflect calibration differences of about 3%. Much larger and erratic differences occur between the pairs. Also shown in table III is the ratio of the average $\overline{\zeta^2}$ from the 14 staffs of the array to $\overline{\zeta^2}$ from station 5. The average of this ratio is 1.51. Evidently, the cross-tank variability of about 12% contributes significantly to the error in determining γ. These cross-tank differences do not appear to be due to reflected waves from the beach since these would be detected in the directional spectra and they were not. It seems probable that they arise as a consequence of reflection from the side walls. It would be interesting to know whether similar cross-tank variations have been detected in other tanks. In retrospect, it would have been preferable to have made several cross-tank measurements at each fetch, and thereby reduce the scatter of figure 12. Lacking these, we have averaged the estimates of γ in each of five equal bands of U/c. These are shown as large crosses on figure 12.

TABLE III

Cross-tank variance of $\overline{\zeta^2}$

Run No.	Right Pair		Left Pair		
	$\overline{\zeta_1^2}$	$\overline{\zeta_2^2}/\overline{\zeta_1^2}$	$\overline{\zeta_8^2}/\overline{\zeta_1^2}$	$\overline{\zeta_9^2}/\overline{\zeta_1^2}$	$\overline{\zeta_{D.S.}^2}/\overline{\zeta_5^2}$
	cm^2				
2	0.18	.994	1.14	1.10	2.62
3	0.58	1.06	0.85	0.81	1.09
5	2.93	1.07	1.25	1.17	1.42
6	5.37	1.05	1.05	1.03	1.41
7	9.17	1.06	1.09	1.02	1.30
Average		1.047	1.076	1.026	1.51
Std. Dev.		0.03	0.13	0.12	

* The subscripts 1, 2, 8, 9 refer to individual wave staffs in the directional array, D.S. refers to the average of all 14 staffs in the array and 5 denotes the upstream station 5.

It is commonly held (Bretschneider, 1973) that full development occurs when the wave age reaches 1.2 (U/c=0.83). That is, at this stage there is no net change in the momentum of the wave field (γ=0). This point of full development has been added to figure 12. Practical wave forecasting, on all but the smallest ponds, is concerned only with the region of U/c less than 6 or so. In this region, γ is approximated by the straight line shown:

$$\gamma = 0.02 \frac{U}{c} - 0.017 \quad , \quad 0.83 < \frac{U}{C} < 7 \tag{6}$$

At larger values of U/c, applicable to very short fetches or unusually sharp wind transients, γ is seen to increase to about 0.24 and then to decrease again as U/c exceeds 12. At these very high values of U/c, whitecapping becomes an important dissipative process. Evidently the non-linear wave-wave interactions are not sufficiently rapid to transfer all the absorbed momentum to longer waves. The waves at the peak of the spectrum have become quite steep, showing pronounced harmonic distortion (Donelan et al., 1979), and are unable to retain the absorbed wind momentum themselves, losing most of it through whitecapping. Thus, at these high values of U/c, the momentum retained by the wave field must come partly from non-linear transfer to frequencies below the peak and partly from direct wind input to these longer waves. An interesting discussion of the momentum balance may be found in Hasselmann et al. (1973).

The value of γ for over-developed waves (0 < U/c < 0.83) and for waves in an adverse wind (U/c<0) cannot be inferred from these measurements.

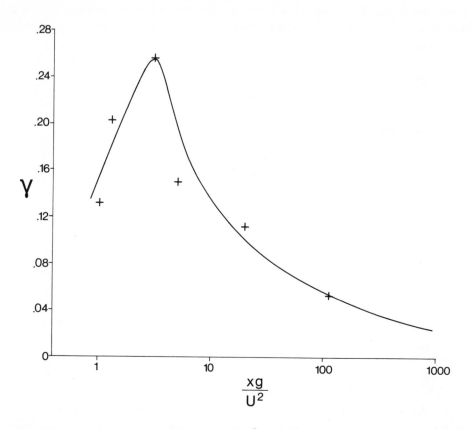

Fig. 14. γ versus non-dimensional fetch xg/U^2. The crosses correspond to those in figure 12. The freehand curve shown illustrates the general trend.

Finally (figure 14), we examine γ as a function of non-dimensional fetch xg/U^2 (Kitaigorodskii, 1962). It is interesting to note that Hasselmann et al. (1973), using an assumed $\cos^2 \theta$ spreading factor for $F(\omega, \theta)$, an assumed overall drag coefficient of 1.0×10^{-3} and the fetch dependence of $\overline{\zeta^2}$, obtained an estimate for γ of 0.05. The fetch dependence of $\overline{\zeta^2}$ was based on laboratory measurements in the non-dimensional fetch range of 3.0×10^{-2} to 1.0 and field measurements in the range of 60 to 10^4. Our measurements lie largely between these two ranges and demonstrate that considerably more than 5%, perhaps up to 25%, of the locally transferred wind momentum may be retained and advected away by the wave field.

ACKNOWLEDGEMENTS

Many members of the management and staff of the Canada Centre for Inland Waters have generously supported this work.

I am grateful to T. Nudds, D. Beesley and J. Carew for assistance in the collection of field and laboratory data, to the staff of the Marine Environmental Data Service, Ottawa for waverider field support and analysis, and to M. Dick, P. Hamblin, J. Hamilton, W. Hui and M. Skafel for valuable comments and criticisms.

REFERENCES

Birch, K.N., Harrison, E.J. and Beal, S., 1976. A computer based system for data acquisition and control of scientific experiments on remote platforms. Proc. Ocean'76 Conf., Wash., D.C.

Bretschneider, C.L., 1973. Prediction of waves and currents. Look Lab/Hawaii, 3.1:17.

Der, C.Y. and Watson, A.S., 1977. A high-resolution wave-sensor array for measuring wave directional-spectra in the nearshore zone. Proc. Ocean'77 Conf., Los. Angeles, Calif.

Donelan, M.A., 1979. A simple numerical model for wave and wind stress prediction. (In press).

Donelan, M.A., Hamilton, J. and Hui, W.H., 1979. Directional spectra of wind-generated waves. (In press).

Hasselmann, K., Barnett, T.P., Bouws, E., Carlson, H., Cartwright, D.E., Enke, K., Ewing, J.A., Gienapp, H., Hasselmann, D.E., Kruseman, P., Meerburg, A., Muller, P., Olbers, D.J., Richter, K., Sell, W. and Walden, H., 1973. Measurements of wind-wave growth and swell decay during the Joint North Sea Wave Project (JONSWAP). Deut. Hydrogr. Z., Suppl. A, 8, No. 12, 22 pp.

Hasselmann, K., Ross, D.B., Muller, P. and Sell, W., 1976. A parametric wave prediction model. J. Phys. Oceanog., 6:200-228.

Kitaigorodskii, S.A., 1962. Applications of the theory of similarity to the analysis of wind-generated wave motion as a stochastic process. Bull. Acad. Sci. USSR Geophys. Ser. No. 1:105-117.

Mitsuyasu, H., Tasai, F., Suhara, T., Mizuno, S., Ohkusu, M., Honda, T. and Rikiishi, K., 1975. Observations of the directional spectrum of ocean waves using a cloverleaf buoy. J. Phys. Oceanog., 5:750-760.

Phillips, O.M., 1977. The dynamics of the upper ocean. 2nd edition. Cambridge University Press, Cambridge, 336 pp.

Ramamonjiarisoa, A. and Coantic, M., 1976. Loi expérimentale de dispersion des vagues produites par le vent sur une faible longueur d'action. C.R. Acad. Sc. Paris, Série B. 282:111-114.

Stewart, R.W., 1974. The air-sea momentum exchange. Boundary-Layer Met., 6:151-167.

Wilson, J.R. and Baird, W.F., 1972. A discussion of some measured wave data. Proc. Thirteenth Conf. on Coastal Engineering, Vancouver, B.C., pp. 113-130.

Young, R.M., 1971. Precision meteorological instruments. Catalogue of the R.M. Young Company, Traverse City, Michigan.

THE NUSC WINDWAVE AND TURBULENCE OBSERVATION PROGRAM (WAVTOP); A STATUS REPORT

DAVID SHONTING[1,2] and PAUL TEMPLE[2]

[1]Naval Underwater Systems Center, Newport, R.I. 02840 (USA)

[2]Graduate School of Oceanography, University of Rhode Island, Kingston, R.I. 02881 (USA)

ABSTRACT

Wind generated white caps have been identified as providing the key mechanism by which a floating oil slick is fragmented and injected as droplets beneath the sea surface. Ambient motions of the wind waves and smaller scale turbulence then mix the oil further downward. After the initial breaking, the degree of downward oil mixing is dependent upon the kinetic energy and Reynolds stresses of the wind wave motions.

Our wave and turbulence observation program (WAVTOP) has developed instrumentation and techniques to gather field data of the wind generated motions in the upper 5-10 m. The sensor package (BLT) includes small fast-response impellors, and a capacitance wave staff system. The impellor systems, which were calibrated in ship wave-tow tank, are configured in arrays to record velocities at several depths simultaneously for estimates of kinetic energy content, Reynolds stresses, and mean shear of the motions. The system is battery powered and self recording, utilizing a microprocessor and digitizer.

Preliminary measurements were made in Narragansett Bay with the BLT mounted on a free drifting AESOP Stable Spar Buoy and from an extended boom on Gould Island. Measurements were made to relate the local wind with the growth of wind waves and their turbulent kinetic energy. Preliminary results portray the energy attenuating with depth following classical theory, where changes are highly correlated with the local wind speed. Also there appear phase lags of energy with wind changes and evidence of a downward propagating rate of energy at 0.5 cm/sec. Fluctuations in variance appear to occur in curious pulses over 3-5 minute periods and wave energy appears to "advect" downwind in packets. Vertically integrated energy content appears to correlate with the cube of the wind speed.

INTRODUCTION

A catastrophic tanker collision or grounding can quickly deposit large volumes of oil at the sea surface producing an immense slick. The prediction of the fate of this slick poses to some, an intractable problem. At best, the surface oil is advected by a mean current and dispersed laterally into the open sea. Less favorable conditions of wind and current can contaminate the coast.

At worst, oil churned up by high sea state is mixed downward, and in shallow regions, causes oil penetration into the bottom sediments as was seen in the AMOCO CADIZ Disaster (ENDECO, 1978). Horizontal advection can flush away surface and volume distributed oil, but, contamination in sediment may reside for months or years having disastrous effects on sea life near and at the bottom. It is thus important that we understand the mechanisms by which oil is mixed vertically; knowledge which is necessary to devise systems and techniques to combat the spread and vertical mixing of surface oil. It is further important to assess the energy and intensity of mixing as related to the sea state in order to judge at what point oil containment and recovery techniques become impractical.

The Wave and Turbulence Observation Program (Project WAVTOP) was initiated by the Naval Underwater Systems Center (NUSC) under sponsorship by the U.S. Coast Guard to define and assess the magnitudes of dynamic parameters which promote the vertical mixing of spilled oil on the open sea, and further, to develop a system and technique to directly measure wind wave and turbulent motions in the upper boundary layer.

The Project WAVTOP is designed to compliment several Coast Guard studies of oil spill dynamics which include: Wave tank modeling of breaking waves in an oil slick (Lin et al, 1978); Theoretical studies of the relation of sea state to the survival of oil spills (Raj, 1977); and Studies of oil mixing properties and breaking waves in a tank (Milgram et al, 1978). It is the necessary welding of theory, laboratory experiments, and actual field measurements from which we can forge workable predictor models of turbulent mixing of oil slicks. During the WAVTOP studies, we have developed a system to measure dynamic properties wind generated surface motions, and made several preliminary field observations. This paper serves as a status report of the WAVTOP effort.

PARAMETERS ASSOCIATED WITH OIL MIXING IN THE UPPER LAYER: WHAT TO MEASURE?

The calm ocean exhibits a minimal effect upon a deposited oil slick; a mean current may advect the oil, spreading it laterally. The only waves present per se, may be swells from a distant storm which produce essentially irrotational motions, transmitting no stress or oil vertically.

The onset of local wind, stressing the sea surface, produces a variety of surface motions. The initial response is the generation of capillary waves which in turn appear to feed momentum into the sea surface to form progressively larger wind waves which are accompanied by a mean vertical shear of horizontal wind-generated current. The waves increase in size with wind speed and form white caps when 6-7 m/sec is exceeded. It is these conditions in which we observe vertical mixing of an oil slick.

What specific parameters or mechanisms govern vertical transfer of oil through the sea surface and into the surface layer? Recent studies shed light on this problem:

1. Raj (1977), in summarizing oil mixing parameters (as a function of wind and sea states), specified the depth variations of the mean square velocity fluctuations, the integral scale of turbulence, and the energy spectrum.

2. Milgram et al (1978) determined from observations of breaking waves in a tank, that most significant influence on the dispersion of oil into submerged droplets is that of breaking waves. Once the oil film is transported by the high energy turbulence it is dispersed as droplets. The motion of these droplets is then determined by the balance of the buoyancy of the oil droplets (upward) and the Reynolds stresses tending to mix the oil (downward). The scales of motions which mix oil downward may be much larger than the breaking turbulence. In fact, it may be associated with the scales of the surface wave motions.

3. Lin et al (1978) observed from wind and wave flume measurements, that wave breaking initially introduces oil vertically as a "strong vertical burst motion like a plume plunging downwards below the water surface" with penetration about one wave height. Spectra of the motions show a dominant peak at the wave frequency with a "Kolmogorov" $(-5/3)$ fall-off for two decades of frequencies. Oil distributions (concentrations) resulting from the wave breaking were exponentially distributed with depth.

From the above results, it appears that the vertical transfer of oil through the sea surface is a two-step process. First, breaking white caps supply large "point" concentrations of high wave number turbulent stress which can overcome the surface tension by breaking the oil into globules and spewing them downward (Fig. 1A). Second, after the oil globules are scattered beneath the sea surface they will be mixed downward by those internal motions which are rotational and transfer stress. The motions existing in the upper layer are basically orbital motions of the wind waves themselves, having scales from a few centimeters to the orbital diameters of the largest wind waves present. The vertical extent of this mixing is directly related to the kinetic energy of the local orbital motions of the waves; found to decrease almost exponentially with depth (e.g. Shonting, 1967). Thus, the net effects of both white capping and the intensity of the subsurface wave motions will create a local steady state oil gradient distribution (Fig. 1B).

Oil mixing is associated with motions which transfer stress and kinetic

164

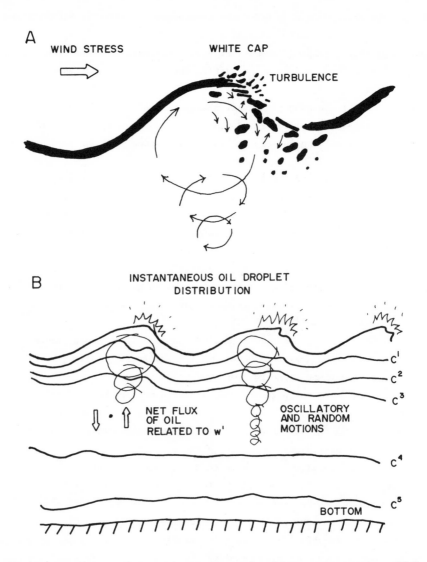

A. WIND STRESS   WHITE CAP   TURBULENCE

B. INSTANTANEOUS OIL DROPLET DISTRIBUTION

c^1
c^2
c^3

NET FLUX OF OIL RELATED TO w'   OSCILLATORY AND RANDOM MOTIONS

c^4

c^5

BOTTOM

Fig. 1 A. Oil slick broken into globules as breaking wave overcomes slick surface tension. B. Subsurface mixing distributes oil downward and produces an oil concentration gradient.

energies. Fig. 2 is a contrived spectrum suggesting the types of wind generated motions associated with stress and oil mixing. Calm conditions spectrum
(heavy line) has little energy in the scales or frequencies of our "mixing
motions" (orbital scales and smaller). The "windy" spectrum, up to the
breaking wave peak, delineates the energy bands producing stress and mixing

(beyond, the scales are so small and isotropic that little net stress and oil transfer should occur). From Fig. 2, it appears that our measurements should be of motions whose scales fall in the shaded regions.

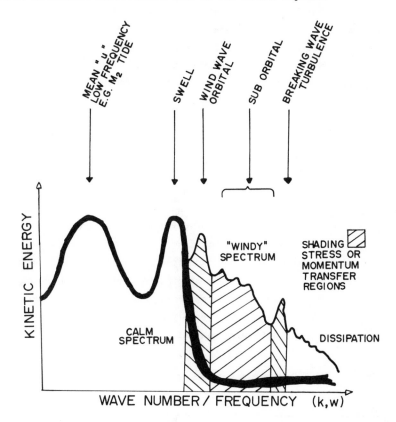

Fig. 2. Difference in auto-spectra with and without wind wave associated motions.

Very few measurements have been made in the open sea for verification of the effects of breaking waves and kinetic energy intensity upon oil slick mixing. Ideally we should make observations first in uncontaminated open sea conditions to obtain baseline data and then conduct similar observations in situ in an oil spill environment. This later study may be over ambitious since oil spills are not often provided specifically for such studies.

At the time of this writing, the Northern Cherbourg Peninsula of France is still recovering from a most disasterous oil spill resulting from the wreck of the 200,000 ton oil tanker, AMOCO CADIZ. After the spill, persistant high winds, and choppy seas served to mix and spread the light diesel oil over a wide stretch of the coastal zone.

Measurements were made by workers at ENDECO (1978) of the vertical concentration of oil at stations along the coast and in a brackish estuary.
Sections made from 7 km offshore to the shoal beaches, indicated high concentrations of oil dispersed 3-4 m downward offshore, but, near the beaches
where waves were breaking, the oil was totally distributed throughout the
water column from 5-12 m. These results indicated that the oil distribution
with the water was very dependent upon the presence of breaking and high energy
wind waves; a clear verification of the laboratory results sited above.

ON THE ENERGETICS OF THE WAVE-INDUCED MOTIONS

One objective of our study is to determine the energetics associated
with the wind wave motions and relate it to the character of the wind itself.
It is therefore useful to construct a framework to relate the variables
measured to physical concepts of momentum and energy. We formulate expressions for the time rate of change of energy and momentum related to wind waves
in terms of measurable quantities within the upper layer. The analysis follows
that of Starr (1968) who has utilized the Reynolds formulation of fluxes from
covariance relationships of fluctuating quantities to understand a wide variety
of transport phenomena in the atmosphere, oceans and even spiral galaxies.

Consider a surface layer of the ocean (Fig. 3) within which we wish to
measure motions associated with wind generated surface waves. The physical

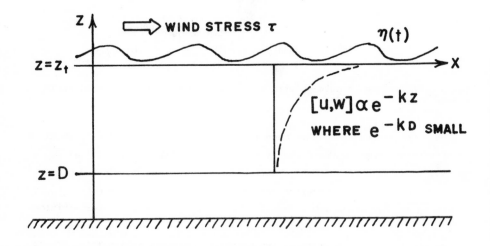

Fig. 3. A two-dimensional ocean from which we estimate the time variation of
mean and eddy kinetic energy. The layer above z_t is where the wind stress
is directly applied. Velocity measurements are made from $z = z_t$ to $z = -D$,
where the motions become negligible.

measurement of these motions, for practical reasons, must be made at or below the level Z_T which is just beneath the wave trough level (i.e. a region occupied by water at all times). The x **axis**, defining the Z_T level, points positive in the direction of the progressive wind waves and z points positive upward. For our formulations we make the following assumptions;

1. We consider motions only in the x and z directions, neglecting cross wave velocities i.e., parallel to wave crests.

2. Motions occurring at speeds producing particle displacements associated with gravity waves which are virtually unaffected by Coriolis forces.

3. The mean sea surface is essentially horizontal, and the local density is assumed constant. Therefore, no mean horizontal pressure gradients are present in the water other than those associated with a slowly varying baratropic tide.

4. The statistical properties of the velocity and pressure fluctuation are homogeneous in the horizontal over the local area as is the mean horizontal current. These properties may, however, undergo changes much slower than the periods of the wave motions (in fact, these changes such as those associated with wind variations are of principle interest).

5. Motions are associated only with wind generated waves, propagating in the +x direction and the mean tidal flow; no standing waves being present.

The momentum equations for the u and w motions in the x and z directions respectively, may be written as

$$\rho \frac{du}{dt} = -\frac{\delta p}{\delta x} - Fx \tag{1}$$

$$\text{and} \quad \rho \frac{dw}{dt} = -\frac{\delta p}{\delta z} - g\rho - Fz \tag{2}$$

where ρ is density, p is pressure, t is time, Fx and Fz are frictional retarding forces per unit volume in the x and z directions and g is the acceleration of gravity assumed constant. With the aid of the two dimensional continuity equation

$$\frac{\delta \rho}{\delta t} + \frac{\delta \rho u}{\delta x} + \frac{\delta \rho w}{\delta z} = 0 \tag{3}$$

and by multiplication of equations (1) and (2) by the respective velocities, expansion of the total derivatives, and invoking assumptions 1-5 we form the kinetic energy relation

$$\frac{\delta}{\delta t}\left(\rho \frac{u^2 + w^2}{2}\right) = -\left[\frac{\delta}{\delta x}\left(\rho \frac{u^2 + w^2}{2} u\right) + \frac{\delta}{\delta z}\left(\rho \frac{u^2 + w^2}{2} w\right)\right] - \left(\frac{\delta pu}{\delta x} + \frac{\delta pw}{\delta z}\right)$$
$$-g\rho w - uFx - wFz \tag{4}$$

Now we consider equation (4) as being averaged in time over a region

of ocean where again we may invoke statistical homogeneity assumption (4 above).
We then may define the properties of the measured variables in the form

$$u(t) = \bar{u}+u', \quad w(t) = \bar{w} + w'', \quad p(t) = \bar{p} + p' \tag{5}$$

$$\text{where } \overline{(\ \ )} = \frac{1}{T}\int_o^T (\quad) \, dt$$

with $\overline{(\ \ )}' = 0$, noting that the period of averaging T is much greater than
the periods of the surface wind waves and in spite of the slowly varying tide
level, we may safely assume $\bar{w} = o$.

We have thus, after the appropriate averaging

$$\frac{\delta}{\delta t}\left(\rho\frac{\bar{u}^2}{2}+\rho\,\overline{\frac{u'^2 + w'^2}{2}}\right) = \frac{\delta}{\delta z}\left(\rho\overline{u'w'}\,\bar{u}\right)_+ \frac{\delta}{\delta z}\left(\rho\overline{\frac{u'^2 +w'^2}{2}\,w'}\right) +\frac{\delta}{\delta z}\left(\overline{p'w'}\right)$$

$$-\bar{u}\,Fx - \overline{u'Fx} - \overline{w'Fz} \tag{6}$$

This relation is a balance of mean and wave (eddy) kinetic energy. To the
mathematician, the time derivative on the left hand term may be troublesome.
We note however, that the partial time derivative is meant to assess the
change over time intervals large compared to the wind wave periods. Thus,
the wave periods may range from 0.5-10 sec while we which to consider the
kinetic energy changes in response to wind stress changes which occur from
tens of minutes to several hours.

To obtain the energy balance of horizontal mean motion, we multiply
equation (1) by u and form

$$\frac{\delta}{\delta t}\left(\rho\frac{\bar{u}^2}{2}\right) = \frac{\delta}{\delta z}\left(\rho\overline{u'w'}\,\bar{u}\right) + \rho\overline{u'w'}\,\frac{\delta\bar{u}}{\delta z} - \bar{u}Fx \tag{7}$$

By subtracting equation (7) from equation (6) we obtain

$$\frac{\delta}{\delta t}\left(\rho\,\overline{\frac{u'^2 + w'^2}{2}}\right) = -\frac{\delta}{\delta z}\left(\rho\overline{\frac{u'^2 + w'^2}{2}\,w'}\right) -\rho\overline{u'w'}\,\frac{\delta\bar{u}}{\delta z} - \frac{\delta}{\delta z}\,\overline{p'w'}$$

$$- \overline{u'Fx} - \overline{w'Fz} \tag{8}$$

which is the balance of wave (eddy) kinetic energy.

We note that all terms on the right, in both equations (7) and (8)
(except the viscous dissipations) involve vertical spatial derivatives and
covariances, both indicative of transport properties of a fluid. The equations indicate the change occurring at a point. It is of interest also to
integrate the relation vertically since our concern is the energetics of a
finite water column and how surface wind stress imparts momentum and energy
through the level Z_T. We thus integrate equations (7) and (8) from the
trough level Z_T down to a depth D. All integrands, which are partial
derivatives with respect to z, need be evaluated only at $z = Z_T$ since at

Z = ·D, all fluctuations vanish (Shonting, 1967).

The integrated forms of equations (7) and (8) are

$$\frac{\delta}{\delta t}\int_{D}^{Z_T} \rho \frac{\bar{u}^2}{2}\, dz = \int_{D}^{Z_T} \rho\overline{u'w'}\ \frac{\delta \bar{u}}{\delta z}\, dz - \overline{\rho u'w'}\ \bar{u}\ \bigg|_{Z_T} - D_M \qquad (9)$$

and

$$\frac{\delta}{\delta t}\int_{D}^{Z_T} \rho\frac{\overline{u'}^2 + \overline{w'}^2}{2}\, dz = -\int_{D}^{Z_T} \rho\overline{u'w'}\ \frac{\delta \bar{u}}{\delta z}\, dz - \overline{\frac{u'^2 + w'^2}{2}\, w}\ \bigg|_{Z_T} - \overline{p'w'}\ \bigg|_{Z_T} - D_E \qquad (10)$$

The quantities D_M and D_e denote the frictional dissipation of the mean flow and of the eddy kinetic energies respectively for the unit column between Z_T and D.

The first integral on the rhs, in equations (9) and (10) represents the transformation term between the two types of kinetic energy or, in other words, energy transferred from the mean motion through the turbulent shear stresses or production of turbulent energy (Hinze, 1959) over the water column of unit cross-section and Z_T D meters high. The second term in equation (9) is the boundary transport of mean flow kinetic energy between the layer of direct forcing and the region below. This is what Hinze (1959) refers to as convective diffusion by turbulence of the kinetic energy. This effect is due to the Reynolds stress acting in the direction of the mean flow.

The second term of equation (10) is a boundary transport of eddy kinetic energy across the surface Z_T caused by the net advection associated with the covariance between w' and the eddy kinetic energy components (note that this is sum of two triple product correlations). Finally, the third term of equation (10) is the covariance of pressure and vertical velocity. It provides a second boundary flux of eddy kinetic energy across the surface which may be written as

$\overline{u'^2 w'}$ and $\overline{w'^2 w'}$

The sum of the vertical terms is the transport of eddy kinetic energy for the u' and w' velocity components.

From a practical point of view, we should note that wave observations have been made in Narragansett Bay where the tidal currents (i.e. values of u), depending upon the phase and location, vary in speed from 0-100 cm/sec. Clearly, for a vanishing mean current, equation (9) becomes identically zero. Equation (10) on the other hand, simplifies to

$$\frac{\delta}{\delta t}\int_{D}^{Z_T} \rho\frac{\overline{u'}^2 + \overline{w'}^2}{2}\, dz = -\rho\overline{\frac{u'^2 + w'^2}{2}\, w'}\ \bigg|_{Z_T} - \overline{p'w'}\ \bigg|_{Z_T} - D_E \qquad (11)$$

Now there is no mean eddy energy conversion term. However, we still have the advection terms of kinetic energy and pressure. This suggests the importance of assessing the ratio

$$\frac{\overline{u'}^2 + \overline{w'}^2}{\overline{u}^2} = f(t) \tag{12}$$

for a given set of observations.

The above derivations allow the observed parameters to be evaluated in the context of balanced energy equations. Specifically, the variances and Reynolds stresses can be used to calculate the order of magnitude of terms in equations (9) and (10). The Reynolds stress component $\tau_z = -\overline{\rho u'w'}$ defines a flux of horizontal momentum $\rho u'$ transported downward by the covariance between u' and w'. These same concepts can be applied to the vertical mixing of oil. Then, considering again Fig. 1, we could define a vertical oil transport term analagous to the Reynolds stress as an oil concentration fluctuation (oil)' correlating with w' i.e.

$$F_O = \overline{(oil)'w'} \tag{13}$$

Thus, in an actual oil spill, one could measure at a point in the ocean the time variation of oil concentration (possibly by use of a fluorometer system) and w' simultaneously to determine the flux F_O.

It is clear that our observational problem is first to measure and gather simple statistics of wind wave motions from the largest wave orbital diameters down to scales of breaking wave turbulence. Analysis should include;

1. The kinetic energy (variances) of the motions.
2. The dominant energy containing scales from the spectral density versus frequency and the spatial correlations.
3. Covariances between horizontal and vertical velocity components.

These statistical properties should be assessed in relation to depth, the local wind conditions, sea state (observed from free surface elevation statistics) and whitecap occurrance, since these data are shown to be relevant to oil mixing.

INSTRUMENTATION

Our theoretical discussion has defined a particular measurement problem such that instrumentation that we choose must contain certain specific attributes. Our sensing systems must register both the mean and fluctuating velocity components in the water column. The instruments should have a fast response to measure the smallest scale velocity fluctuations and long enough recording capacity to evaluate variability and mean values. Further, the systems must be arranged geometrically for simultaneous measurements at different depths.

Based upon the above considerations, an instrument package for the WAVTOP observations was developed consisting of three basic components; the sensors, electronics package and cabling connectors and pressure cases. The sensor system (Fig. 4A) is a ducted impellor device designed by Smith (1978) at the University of Washington. The plastic impellor is mounted with jewel

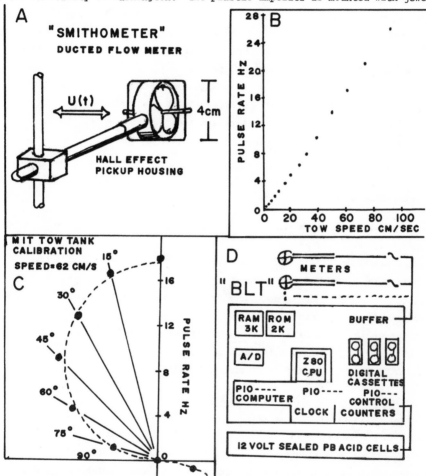

Fig. 4A. Ducted impeller meter. B. Calibration curve made in MIT towing tank. C. Off angle response calibration. Dots are actual data; broken line circle is the cosine relation. D. Solid state components in the electronics of the "Boundary Layer Thing" (BLT). Central processing unit (CPU) controls system, input from meters is sensed for ± sign, digitized and averaged and at intervals is stored onto digital cassetts. The read only memory (ROM) is pre-programmed for BLT to record data at prescribed intervals.

bearings. Two small magnets are mounted on a diameter of the impellor. As
the impellor spins in response to the fluid flow, the magnetic field of the
magnets interact with a "Hall effect" crystal mounted inside the tip of the
stainless steel support rod. With passage of the blade magnets, a charge
separation occurs in the Hall effect device forming a pulse which is amplified, shaped into a square wave by a Schmidt trigger, and coded by a sequence
of short/long, or long/short pulses to indicate the sense of rotation. The
Hall effect pickup offers almost no magnetic force field which could slow
the impellors through interaction with the magnets. Thus, the combination
of a neutrally buoyant impellor and the Hall effect pickup provides a velocity
threshold response of 3-4 mm/sec. The angular velocity of the impellors is
linear enough with flow speed (Fig.4B) so that the identical calibrations
curve may be used for all seven sensors with a maximum error of 2-3%. The
response of the impellors varied approximately as the cosine of the angle
subtended by the meter axis of rotation and the flow vector (Fig. 4C). A
summary of the flow meter characteristics is given in Table 1.

TABLE 1
Characteristic of Ducted Flow Meters

Diameter Impellor	4 cm
Responce Distance	3 cm
Frequency Response	5-10 H_z
Threshold Speed	3-5 mm/sec
Linearity	High (One curve can be used for all systems)
Cosine Response	Good 10-15% Max Error
Signal Output	Nominal 15 V- coded square wave pulses whose sign changes indicate a sense of rotation.
Max Depth Of Use	Tested to 20 atm (200 m)

Solid state electronics for signal preparation is incased in the stainless steel cylindrical support and tested for 20 atm or 200 m water depth.
The unit is powered by + 15 v dc via a water tight connector. The signals
from each of the sensor units are led to the BLT data processor which contains
a complex of solid state micro chips for complete data processing and storage.
Fig. 4D shows the types of I.C. chips used. The current meter pulses are
converted into signed digitized values of velocity and are averaged every
0.2 sec. Each of the six channels of time series data is recorded in three
Memodyne digital cassettes. The data is transferred to 7 channel Mag tape
for processing.

The lower limit of the scales of motions measured is limited by the dimensions of the impellor sensor. The 4 cm diameter impellor limits our detectable scale size of an eddy motion to about 10 cm. The response time of the Smithometer impellor sensor is about 5Hz (see Table 1). This is equivalent to our sampling frequency of 0.2 sec. This gives a Nyquist frequency of 2.5 Hz; the upper limit frequency from which we could expect any spectral information. The nominal continuous recording period is 3-4 hours of continuous data; the period being roughly the life of our battery pack.

THE VERTICAL DISTRIBUTION OF KINETIC ENERGY AND THE ENERGY INTEGRAL

We wish to measure the vertical distribution of kinetic energy from the sea surface down to a depth where the wave induced motions vanish. If we assume that the motions u' and w' defined by equations 1 and 2 are strictly wave orbital motions of deep water type (wavelength less than $\frac{1}{2}$ the depth) and can be represented as fluctuations at a fixed horizontal position by

$$u' = a \, \omega \, e^{\omega^2 z/g} \cos(-\omega t)$$

and $w' = a \, \omega \, e^{\omega^2 z/g} \sin(-\omega t)$

where a = amplitude

ω = frequency

g = gravity

then, by eliminating the trigonometric terms and multiplying by $\rho/2$ we obtain

$$KE(z) = (\overline{u'^2} + \overline{w'^2}) \frac{\rho}{2} = (a \, \omega \, e^{\omega^2 z/g}) \frac{\rho}{2} \tag{14}$$

which becomes the kinetic energy at depth z associated with a two dimensional progressive wave in the x-z plane.

Now if we integrate in the form

$$\int_{D(t)}^{\eta(t)} KE(z) \, dz \tag{15}$$

we obtain the total kinetic energy of the wave motions in the water column between the free surface $\eta(t)$ and the depth D where the u' and w' motions are negligible (say 1% of their surface value). A problem occurs in the interpretation of the energy integral at the surface. It is clear, since we desire the average kinetic energy, we shall integrate from the upper limit $\eta(t) = z = 0$ by definition.

One aim of our field measurements is to evaluate the energy density term $(\rho \overline{w'^2})$ and vertical integral by use of actual values of $\overline{u'^2}(t)$ and $\overline{w'^2}(z)$. These variances, however, are obtained only at discrete depths of instrument placement; usually, for the WAVTOP measurements, the uppermost

174

instrument was placed at 10-20 cm below the estimated mean trough level. The
uppermost region must be measured as accurately as possible since the energy
is an exponential function increasing to the surface.

OBSERVATIONS

Preliminary serial records of velocity components were taken at several
depths simultaneously allow comparison of the vertical distribution of kinetic
energy of the fluctuations with the character of the local wind. Two records
of wind wave observations are studied. Each was made over a period when
there occurred pronounced wind speed changes.

Series 002 (24/Mar/78): Six Smithometers were supported from the AESOP
Stable Spar Buoy (Shonting and Barrett, 1971) at depths of 10, 110, 210, 310,
410, and 610 cm beneath the mean trough level z_t (Fig. 5). The time series

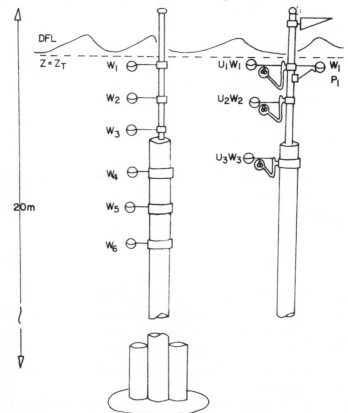

Fig. 5. Configurations of the impellor meters on AESOP to measure w(t) at
six depths simultaneously (left) and to measure u(t) and w(t) at three depths
and p(t) and w(t) at z_T (right).

of the vertical velocity w(t) were recorded at 0.2 sec intervals and running
means and variances were estimated for time intervals of 128 sec. Since the
wave periods ranged from 0.5-4 sec, this supplies over 30 cycles of the lowest
frequency waves - sufficient for a reasonable estimate.

Assuming we are measuring essentially two dimensional waves i.e. wind
waves whose orbits are in the x-z plane, the total kinetic energy of these
motions is

$$E_k = \frac{1}{2} \rho \left(2 \overline{w'^2} \right) = \overline{\rho w'^2} \tag{16}$$

The running variances for 002 are shown in Fig. 6A together with wind
speed values. The variance curves indicate the following:

1. The general trend of energy density follows the wind speed and is
seen to vary by over an order of magnitude at all depths for the wind speed
range of 2-9 m/sec.

2. The energy remains quite exponentially distributed with depth as
indicated by the similar vertical spacing with time.

3. A pronounced minimum appears in all records but appears time lagged
progressively with depth whereby the 610 cm minimum appears about 20 min
after the 10 cm value.

4. Pronounced fluctuations in energy occur at all depths. Sometimes
these pulses appear in phase with adjacent records above and bllow, but
other times appear randomly. The amplitudes with respect to the slow variation (i.e. the signal-to-noise ratio) appear independent of depth. There
was no visible indication that the AESOP buoy was fluctuating vertically (it
has a natural heave period of 26 seconds). It is tempting to speculate that
these fluctuations are wave train packets which are generated by irregular
wind stress pulses upwind in the bay.

Different wind conditions (and hence, sea states) will cause different
energy distribution to exist in the water column. An indication of the
different distribution may be seen by plotting variance vs depth on a logarithmic scale (Fig. 6B). For two different wind conditions, different
kinetic energy distributions were observed. With a 2 m/sec wind, the decrease
of kinetic energy with depth follows an exponential form. For the larger
wind speed, 8 m/sec, a larger amount of kinetic energy exists in the water
column. Also indicated is a larger amount of energy near the surface. The
fact that the energy - distributing process is not in a steady state is shown
by the non-exponential shape of the curve nearer the surface. The deeper
values tend to converge; perhaps an indication that surface energy has not
reached to the deepest sensors. Finally, extrapolation above the shallowest -

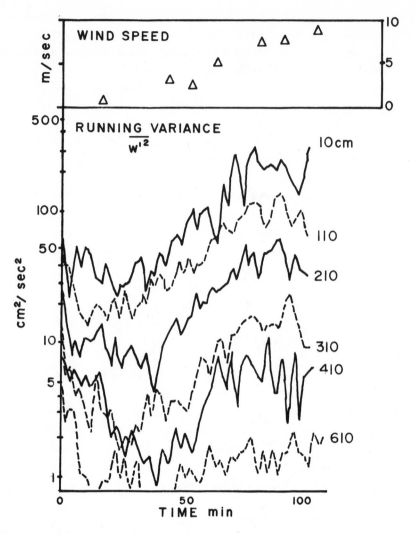

Fig. 6A. Running variance of the vertical velocity made from consecutive 128 sec records.

depth measurement may prove useful in investigating the energy input mechanism at the surface.

The total estimate of wind wave kinetic energy is obtained by numerical integration of the relation

$$E_k = \int_{D}^{Z_T} \overline{w'^2} dz \tag{17}$$

layer-by-layer over the depths of observation. This is shown in Fig. 6C (solid

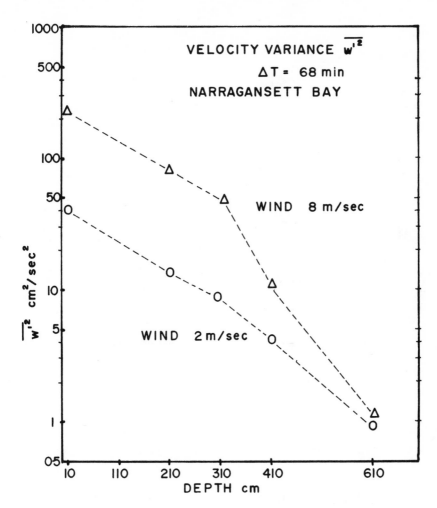

Fig. 6B. The vertical distribution of variance of the vertical velocity:
First record with wind at 2 m/sec showing an exponential decrease with depth;
second record (68 min later) with wind at 8 m/sec shows increase of energy
from 10 - 310 cm but a delay in energy propagation deeper.

curve) which represents four averages each over 1280 sec (21 min). Clearly,
this energy content is increasing throughout the record.

Denman (1973) specifies the turbulent energy input by wind stress as

$$\tau = \rho_A \, C_A \, U_{10}{}^2 \qquad\qquad (18)$$

ρ_A is air density, C_A is a drag coefficient and U_{10} is the wind speed observed
at 10 m height above sea level. The rate of working by the wind stress (or
power input) is given by

$$E_W = \tau U_{10} = \rho_A \, C_A \, U_{10}{}^3 \qquad\qquad (19)$$

178

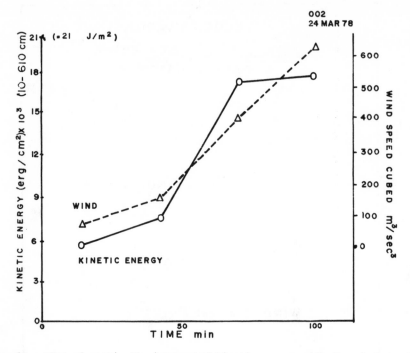

Fig. 6C. Plot of vertically integrated kinetic energy with the wind speed cubed.

The cube of the wind speed (dashed curve in Fig. 6C) shows a positive correlation with the increase in the integrated energy content.

Series 006 (9/Aug/78): This was made from the north end of Gould Is, Narragansett Bay where a 3-10 km fetch occurs to the northwest. The BLT system was positioned over 17 m water depth using a large stiffboom (Fig. 7). Five Smithometers were suspended from the boom support to register $w(t)$ at depths of 10, 50, 100, 150, and 200 cm below Z_T (Fig. 7). The wind was light and variable with calms occurring during and at the end of the record (Fig. 8A). The wind waves were small - lengths 2-3 m and heights 10-15 cm commensurate with the small wind stresses. The running variances (calculated for a sequence of 128 sec intervals) show energy perturbations which exponentially decay with depth (Fig. 8A lower curves).

The energy distribution also displays sensitivity to the fluctuating wind field. The drop in wind at approximately 15 min, is reflected in the variances (even down to 200 cm) within several minutes. The wind increase of wind peaking at 3 m/sec is followed, but at a 15-20 min time lag, by the

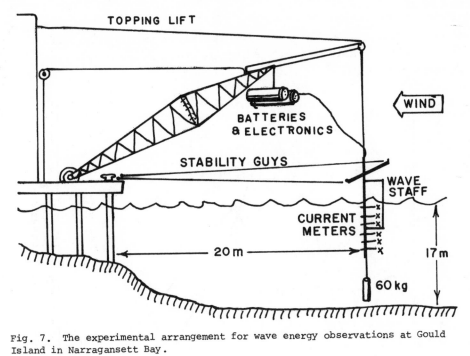

Fig. 7. The experimental arrangement for wave energy observations at Gould Island in Narragansett Bay.

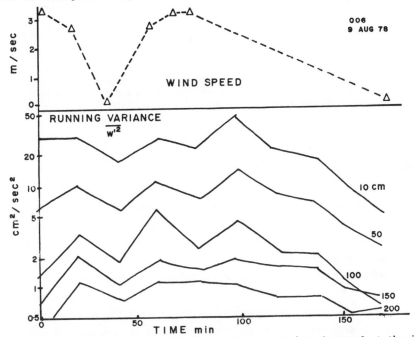

Fig. 8A. The running variances of the vertical velocity observed at the indicated depths plotted with the changing wind speeds (upper curve).

variances at 10, 50, and 100 cm depths. The plots of the wind speed cubed
and the vertically integrated energy demonstrates the time lag (Fig. 8B); the
kinetic energy certainly falls behind both the wind peak and the tail-off at
the end of the record.

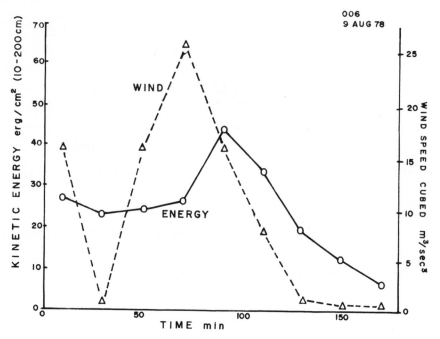

Fig. 8B. Vertically integrated kinetic energy and the wind speed cubed. Note
the apparent phase shift between the curves whereby the wind stress work leads
the kinetic energy of the wave motions by 15-20 minutes.

CONCLUSIONS AND PLANS FOR FURTHER STUDIES

From the present WAVTOP program, preliminary conclusions are made:

1. Review studies indicate oil slicks are mixed downward by breaking
waves coupled with the subsurface wind wave and turbulent motions; the degree
of mixing being dependent upon the kinetic energy of the fluctuating motions.

2. Equations have been derived which define the balance of the time
rate of change, mean and fluctuating kinetic energy, relative to dynamic
stress and pressure terms. These equations when integrated vertically, provide a physical framework from which observed values of wave motions and
dynamic pressure may be evaluated in terms of energy and momentum fluxes from
the wind.

3. The Smithometer ducted impellor systems developed at the University of Washington, used in conjunction with the NUSC developed microprocessor data logger, provide a tool to make a variety of observations of wave motion velocity, free surface and dynamic pressure.

4. The basic sensors are capable of registering wave motions from 4 mm/sec to 100 cm/sec and fluctuations up to a frequency of 5 H_z. The sensors can resolve orbital motion and turbulent components having scales down to a few centimeters; and can resolve orthogonal velocity components to estimate Reynolds stresses in the wave field.

5. Preliminary observations of wave motions show the relation of wind stress variation to the vertical distribution of wave kinetic energy and the total energy integral. Further, the actual phase shift of energy flow from the surface which indicates a downward speed of propagation of roughly 0.5 - 2 cm/sec. The work of the stress (varying as the wind speed cube) correlates with the vertically integrated energy.

These preliminary results of the study of the wind generated kinetic energy content in the upper layer, offer incouraging results. Further analyses and more observations will provide information on spectral composition, and Reynolds stresses as a function of the local wind field and the sea state. These measurements will explore the relationship of the wave orbital motions to the momentum and energy transfer of the wind downward through the water column. These observations will be similar to those made by Shonting (1971) and by Cavelari et al (1977). Added sensor systems will include wave staff (free surface) records and a dynamic pressure sensor. With this array of systems a step-by-step evaluation of the terms of energy equations will commence. Once the validity of the energy and stress relations is demonstrated these parameters can be used to verify oil mixing models and theory.

ACKNOWLEDGEMENTS

The design and construction of the microprocessor data logger was done by John Roklan and assisted by William Ryan of the Naval Underwater Systems Center. The field assistance and data analysis of Tony Petrillo, Dale Licata, Robin Robertson and Capt. Angelo Mazarrelli of the Department of Ocean Engineering of the University of Rhode Island, is appreciated. Mike Kinane, president of Shielco Corp. of Davisville, R.I. assisted in the designs and fabrication. Special thanks is owed to Ms. Eileen Domingos, our secretary for her typing efforts and helping us to organize our chaos. Guidance for the overall program was recieved from R. Griffiths of the U.S. Coast Guard R & D Office Washington, who sponsored this study (MIPR NO. Z-70099-7-71825-A).

REFERENCES

Cavaleri, L., Ewing, J., and Smith, N., 1977. Measurements of the pressure and velocity field below surface waves. NATO Symposium: Turbulent fluxes, through sea surface. Wave dynamics and prediction. Ile de Bendor, France 12-16/Sep/77.

Denman, K.L., 1973. A time-dependent model of the upper motion. J. Phys. Oceanogr., 3/3:173-184.

ENDECO, 1978. Measurement of dynamics of oil-in-water concentrations during the "AMOCO-CADIZ" oil spill. Data Report Environmental Devices Corporation, Marion, Massachusetts, 14 pp.

Hinze, J.O., 1959. Turbulence an introduction to its mechanism and theory. McGraw-Hill Book Co., New York, 586 pp.

Lin, J.T., Gad-el-Hak, M., and Ta-Liu, H., 1978. A study to conduct experiments concerning turbulent dispersion of oil slicks. Flow Research Corporation For the Department of Transportation (U.S. Coast Guard) No. 65 (DOT-CG-61688-A).

Milgram, J.H., Donnelly, R.G., Van Houten, R.G., and Camperman, J.M., 1978. Affects of oil slick properties on the dispersion of floating oil into the sea. Massachusetts Institue of Technology For U.S. Coast Guard Report No. CG-D-64-78, 374 pp.

Raj, P.K., 1977. Theoretical study to determine the sea state limit for the survival of oil slicks in the ocean. Arthur D. Little Corporation, Final Report No. CG-D-90-77, Task No. 4714.21, For the Department of Transportation (U.S. Coast Guard), 276 pp.

Shonting, D., 1967, Measurement of particle motions in ocean waves. Journal of Marine Research., 25/2:162-181.

Shonting, D., 1971. Observations of Reynolds stresses in wind waves. Pure and Applied Geophysics., 81/4:202.

Shonting, D. and Barrett, A.H., 1971. A stable spar-buoy platform for mounting instrumentation. Journal of Marine Research., 29/2:191-196.

Smith, J.D., 1978. Measurement of turbulence in ocean boundary layers. Proceedings of a working conference on current measures. University of Delaware, Newark, Delaware, pp. 95-128.

Starr, V.P., 1968. Physics of negative viscosity phenomena. McGraw-Hill Book Company, N.Y., 256 pp.

A NUMERICAL MODEL OF LONGSHORE CURRENTS

M. SABATON, A. HAUGUEL

E.D.F., Direction des Etudes et Recherches, Laboratoire National d'Hydraulique,
Chatou (France)

ABSTRACT

We introduce here a numerical model which permits to compute the longshore currents
for any shape of shoreline and any bathymetry.

The integration over the depth and the average over a wave period of the momentum
and continuity equations leads to a system of equations identic to those of Saint-
Venant in which supplementary stresses due to waves appear. The model first includes
the wave refraction computation which permits to estimate the driving forces and then
the computation of longshore currents themselves. This model has been calibrated in
the case of a rectilinear shoreline for which the influence of the various parameters
has been studied. At last, the model has been applied to a semi-circurlar bay.

INTRODUCTION

The longshore currents are important as they can be as strong as the tidal currents

and wind induced currents. Neverthelers, they are very different because from one

hand, they are generally restricted to the surf area and because on the other hand,

the velocity of the other currents is weak there. This let appear that the longshore

currents action is essential for littoral drift and dispersion of outfalls near the

coast. Consequently, they must be precisely defined. Thanks to some assumptions parti-

cularly concerning the wave breaking, it is possible to compute the current pattern

with a two dimensional numerical model.

THE MODEL

Assumptions and equations

The assumptions of calculation are as follows

H1 Incompressible fluid

H2 Irrotational Stokes waves

H3 The vertical component of the current is neglected

H4 For the calculation of wave currents, the varations of sea surface elevation are
   neglected compared to the total depth

H5 The mass current is assumed to be constant over the depth

H6 Wind action is neglected.

With these assumptions, the Navier-Stokes equations can be integrated over the total depth and time averaged over a wave period. Then the following equations are obtained

$$\frac{\partial \bar{\xi}}{\partial t} + \frac{\partial \bar{U}}{\partial x} + \frac{\partial \bar{V}}{\partial y} = 0$$

$$\frac{\partial \bar{U}}{\partial t} + \frac{\partial \overline{uU}}{\partial x} + \frac{\partial \overline{vU}}{\partial y} = - g(d + \bar{\xi}) \frac{\partial \bar{\xi}}{\partial x} + \tau x + \pi x(- d) + \nu \Delta \bar{U} + \lambda \bar{v}$$

$$\frac{\partial \bar{V}}{\partial t} + \frac{\partial \overline{uU}}{\partial x} + \frac{\partial \overline{vV}}{\partial y} = - g(d + \bar{\xi}) \frac{\partial \bar{\xi}}{\partial y} + \underbrace{\tau x}_{\substack{\text{wave} \\ \text{induced} \\ \text{driving} \\ \text{force}}} + \underbrace{\pi x(- d)}_{\substack{\text{bottom} \\ \text{friction}}} + \nu \Delta \bar{V} - \lambda \bar{U}$$

with $\bar{U} = (d + \bar{\xi}) \ \bar{u}$

$\qquad \bar{V} = (d + \bar{\xi}) \ \bar{v}$

$\qquad$ d depth

$\qquad \left.\begin{array}{c} \bar{u} \\ \bar{v} \end{array}\right\}$ mass current components

$\qquad \bar{\xi}$ mean sea surface elevation

$\qquad \lambda$ Coriolis parameter.

The parameter ν mainly takes into account the eddy viscosity of mass current. The one is certainly due to wave breaking mainly.

Wave driving forces

These stresses can be defined as follows

$$\tau_x = - \frac{1}{\rho} \left(\frac{\partial S_{xx}}{\partial x} + \frac{\partial S_{xy}}{\partial y} \right)$$

$$\tau_y = - \frac{1}{\rho} \left(\frac{\partial S_{xy}}{\partial x} + \frac{\partial S_{yy}}{\partial y} \right)$$

Longuet-Higgins has first obtained the tensor S called tensor of the flux of excess momentum in the waves. Under some assumptions (particularly H4), it can be written as follows.

$$S = E \left(\begin{array}{ll} (2n - \frac{1}{2}) \cos^2 \theta + (n - \frac{1}{2}) \sin^2 \theta, n \cos\theta \sin\theta \\ \\ n \cos\theta \sin\theta \qquad\qquad\qquad , (2n - \frac{1}{2}) \sin^2 \theta + (n - \frac{1}{2}) \cos^2 \theta \end{array} \right)$$

where $E = \frac{1}{8} \rho g H^2$ is the wave energy

$\qquad$ n $\qquad\qquad$ is the shoaling number

$\qquad \theta \qquad\qquad$ is the wave incidence.

Consequently the stresses τ_x and τ_y are functions of three quantities H (wave height), θ (wave direction) and n (ratio between group velocity and phase velocity). These three quantities depend on the mass current which modifies the wave propogation ; here is the difficult problem of the interaction between waves and current. It has been neglected here.

Thanks to this important assumption, the quantities H, θ and n only depend on the wave characteristics and bathymetry. They can be obtained from a numerical model of refraction.

Generally, this kind of model neglects the breaking of waves. The method we have used is to follow a wave orthogonal and to study in each point if there is or not breaking.

The Battjes criterion has been used. The waves break in a place of depth d if the wave height verifies

$$H > \gamma d$$

where γ is an experimental coefficient, function of bottom slope β and local wave steepness :

$$\gamma = f (\xi) \quad \text{with } \xi = tg\beta/\sqrt{H/L}$$

This coefficient determines the breaking shape. The wave height in the surf area is taken as

$$H = \gamma d$$

The relation obtained by Bowen, Inman and Simmons for a rectilinear shoreline and a constant bottom slope was in fact $H = \gamma(d + \xi)$.

The expression we have used involves two supplementary assumptions, from one hand the relation obtained by Bowen is exact even for a variable bottom slope, and on the other hand, the mean sea surface elevation can be neglected.

186

Bottom friction

The bottom friction has been assumed as proportional to the square velocity

$$\vec{\tau} = -\rho \frac{g}{c_s^2} \vec{u} \, || \, \vec{u} \, ||$$

In a first step, the orbital velocity due to waves has been neglected. Then, has been taken into account in the following form :

$$\pi_x = -\rho \frac{g}{c_s^2} u \sqrt{(u + \overline{u_{orb}})^2} + \sqrt{(v + \overline{v_{orb}})^2}$$

$$\pi_y = -\rho \frac{g}{c_s^2} v \sqrt{(u + \overline{U_{orb}})^2 + (\overline{v + v_{orb}})^2}$$

Scheme of model

The scheme of the model is as follow

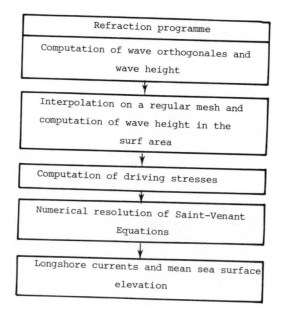

RECTILINEAR SHORELINE

Most of actual longshore currents theories have been applied to rectilinear and constant slope shore. In other respects, for this particular case, many measurements were made on scale models and few on full scale. In order to compare the results of the numerical model with the experimental data, it was of a great interest to cali-brate the model for a rectilinear shoreline.

Boundary conditions

Longshore currents along a rectilinear infinite shoreline must be parallel to the coast and in the direction of waves propagation. To simulate an infinite beach, it is possible to set as up stream boundary condition the downstream velocities computed in the previous time step : so the lateral boundaries are cancelled (see fig. 1)

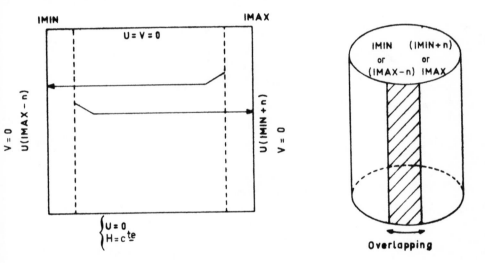

Figure 1.

Results and influence of parameters

The stresses computed in the refraction programm are applied to the fluid at rest.

188

When the fluid begins to move, the current pattern keeps parallel to the wave
stresses, but very soon, owing to the boundary conditions the current becames pa
to the shoreline. In this period, some water comes from deep sea through the bre
line, but as soon as the equilibrium is reached for the sea surface elevation, t
incoming flow disappears and the boundary condition is verified : no velocity ar
initial water depth.

See figures 2 and 3 for an example of results.

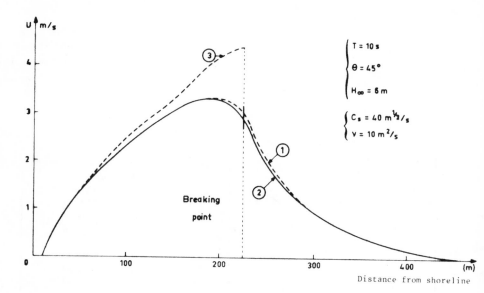

Fig. 2. Velocity profile : (1) computed
                              (2) computed with corrected stresses (taking into acc
                                 sea surface elevation)
                              (3) analytical calculation (eddy viscosity neglected)

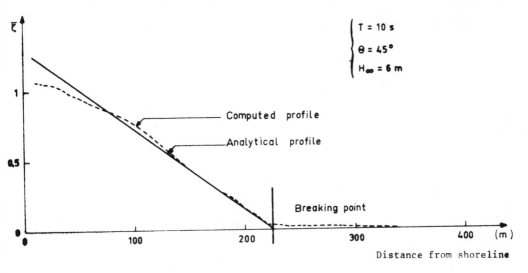

Fig. 3. Mean sea surface elevation.

The influence of some parameters has systematically been studied. We can distinguish three kinds of parameters :

- friction parameters : Chezy coefficient C_S and eddy-viscosity coefficient ν (see fig. 4, 5 and 6)
- waves parameters : period T, height H, and incidence θ (see fig. 7 and 8)
- shore parameter : bottom slope, the influence of which has not been studied (It has been chosen equal to 3 %).

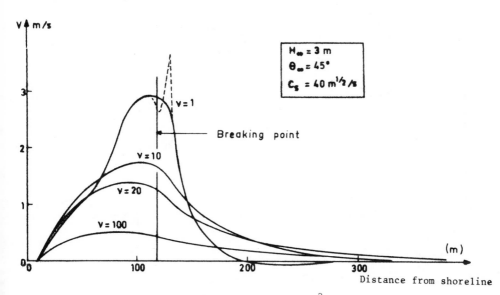

Fig. 4. Variation of velocity profile with ν (m^2/s).

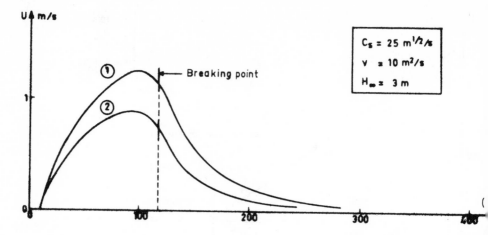

Fig. 5. Influence of bottom friction upon velocity profile

(1) Bottom friction $- \dfrac{g}{c_s^2} \vec{u} \, || \, \vec{u} \, ||$

(2) Bottom friction $- \dfrac{g}{c_s^2} \vec{u} \, || \, \vec{u} + \overrightarrow{u_{orb}} \, ||$

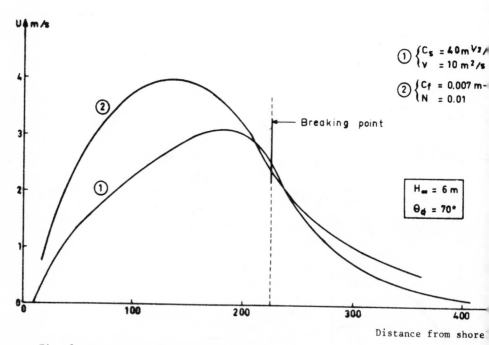

Fig. 6. Comparison with Longuet-Higgins results

(1) Computation with a bottom friction $- \dfrac{g}{c_s^2} \vec{u} \, || \, \vec{u} \, ||$

(2) Longuet-Higgins velocity profile.

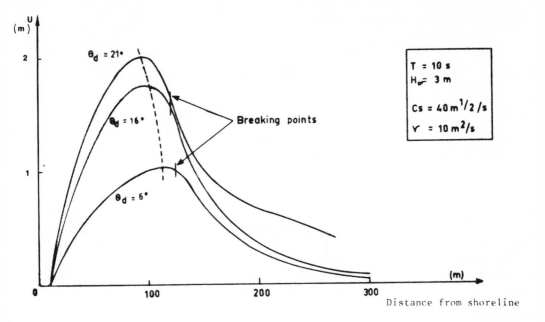

Fig. 7. Influence of wave incidence upon velocity profile.

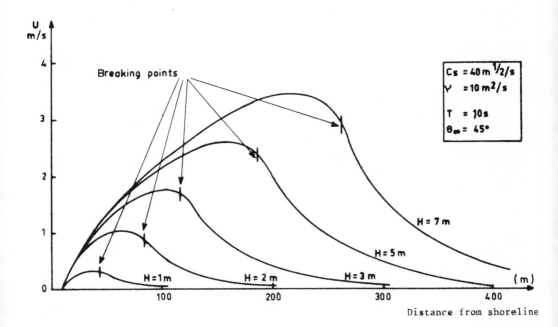

Fig. 8. Influence of wave height upon velocity profile.

Comparison with measurements

Full scale measurements are very rare. Furthermore, the few experiments are ve dissimilar and have sometimes been influenced by an other phenomena as wind for exar

The measured currents can be compared with the computed ones, but we had to chc quite different friction coefficients in order to obtain quite similar velocities.

The comparison between the full scale measurements and the computed results is hereunder discribed :

Full scale measurements		Computed results		
Conditions	measured velocity (m/s)	Parameters		Computed velocity (m/s)
		ν (m^2/s)	Cs $(m^{1/2}/s)$	
Torrey Pines Beach (1950) T = 12 s, θ_d = 6°, H_d = 1,55 m, m = 0,027	surface : 0,11 bottom : 0,06	10	10	0,08
			40	0,43
Putnam measurement T = 10 s, θ_d = 12°, H_d = 1,52 m, m = 0,31	1,10		40	0,58
			60	0,73
Putnam measurement T = 8 s, θ_d = 12°, H_d = 2,59 m, m = 0,02	0,76		25	0,69
			40	1,05

To measure the longshore currents, some experiments have been done in laborator They are mainly due to Putman (1949), Savile (1950), Brelner and Kamplanis (1963), Galoin and Eagleson (1965).

You will find hereunder some examples of the measurements made in laboratory compared with the computed results (The laboratory results are transposed to a rea case thanks to Froude similitude : geometric scale 100).

Laboratory measurements		Computed results		
Conditions	measured velocity (m/s)	Parameters ν (m^2/s)	C_s $(m^{1/2}/s)$	Computed velocity (m/s)
H = 7,3 m T = 9,5 s m = 0,1 θ = 18°	1,55	10	$\dfrac{10}{40}$	$\dfrac{1,36}{5,95}$
H = 7,0 m T = 12,0 s m = 0,1 θ = 11,5	3,87	10	40	4,39

With a constant eddy viscosity coefficient $\nu = 10$ m^2/s the value of C_s to calibrate the model with the measurements varies from 10 to 60.

Conclusion

Several cases are to be considered depending on the information we have. If any measurements have been made, it is only necessary to take them to calibrate the model. In the contrary, if the model is a predictive one, it is necessary to make a further study to point out the influence of each parameter so that the friction coefficient value could be obtained from the bottom characteristics.

SEMI CIRCULAR BAY

The advantage of a rectilinear shoreline was the possibility of calibrating the model with the measurements a ready made. On the contrary, all the possibilities of this two dimensional model have not been used. It is in a case without symetry that it will be the most usefull, because it will give the current direction.

Description of the bay

The studied case is idealized. The shoreline is circular (radius = 260 m) and the bottom slope is constant (m = 3%). The bay is opened on a rectilinear shore where the bottom slope is also 3 %.

Results

The results obtained in the case of a bay are quiete different from those obtained along a infinite shoreline :
- the current pattern shows big gires
- the strong current area is larger than the surf area.

These remarqs makes appear the importance of longshore currents in pollution problems.

The results obtained for two directions of wave (in the direction of the bay axi and edged) are presented on figures 9 and 10. In the second case, the rectilinear beach velocity profile has been imposed on the lateral boundaries.

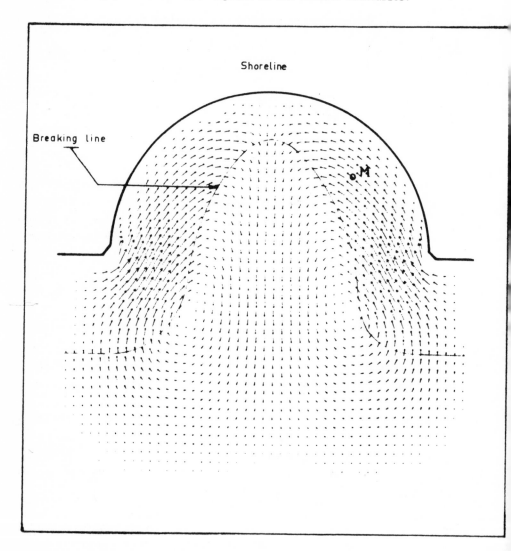

Fig. 9. Longshore currents in a circular bay (wave propagation in the direction of bay axis).

$H_\infty = 3m$

$\theta_\infty = 90°$

$T = 10$ s

Scale 1/45000

$C_S = 40$ m$^{1/2}$/s

$\nu = 50$ m^2/s

Velocity : ⊔ 1 m/s

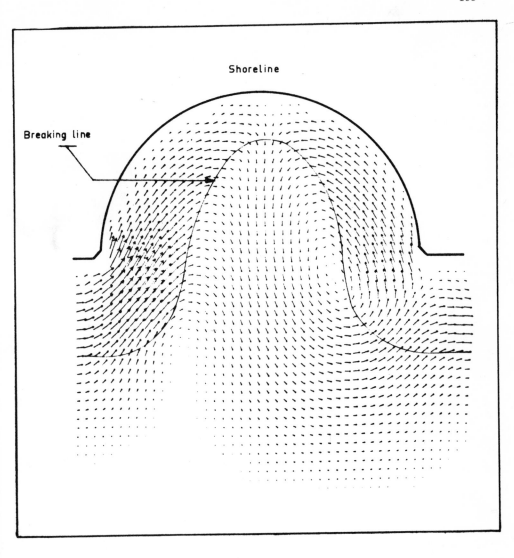

Fig. 10. Longshore currents a in circular bay (Edged wave propagation)
H∞ = 3 m c_s = 40 m$^{1/2}$/s
θ∞ = 75°   ν = 50 m^2/s
T   = 10 s
Scale 1/45000   Velocity : ⊢⊣ 1 m/s

CONCLUSION

    The numerical twodimensional model gives an estimation of the longshore currents
induced by waves fore any shape of shoreline.

TIME SERIES MODELLING OF STORM SURGES ON A MEDIUM-SIZED LAKE

W.P. BUDGELL[1] and A. EL-SHAARAWI[2]

[1]Ocean and Aquatic Sciences
Canada Centre for Inland Waters, Burlington, Ontario, Canada

[2]Applied Research Division, National Water Research Institute
Canada Centre for Inland Waters, Burlington, Ontario, Canada

INTRODUCTION

The time series modelling approach of Box and Jenkins (1970) has been extensively applied to model time series data in a variety of fields. This approach has been used by econometricians for modelling and forecasting economic data. Examples of such studies are given by Feige et al. (1974), Granger (1969) and Pierce (1977). In hydrology and water management this procedure has been used by McKerchar and Delleur (1974) and O'Connell (1971). Recently, El-Shaarawi and Whitney (1978) have adopted this approach to develop a sampling plan for the phosphorus loadings from the Niagara River into Lake Ontario.

In the present paper we describe the steps that may be followed in developing an empirical time series model within the framework of Box and Jenkins. These steps will be illustrated by modelling the dynamical behaviour of water level fluctuations of a medium-sized lake during storm events.

THE DATA SET

The data used in this study are water level measurements and wind stress estimates at Lake St. Clair, situated in the Great Lakes of North America (Fig. 1). Hourly water level measurements were obtained from a float-type gauge at Belle River. Hourly wind stress values were estimated from meteorological observations recorded at nearly Windsor Airport (Fig. 2).

The wind stress estimates were obtained from the standard formula

$$\underset{\sim}{\tau} = c_d \rho_a \underset{\sim}{v} |\underset{\sim}{v}| \tag{1}$$

198

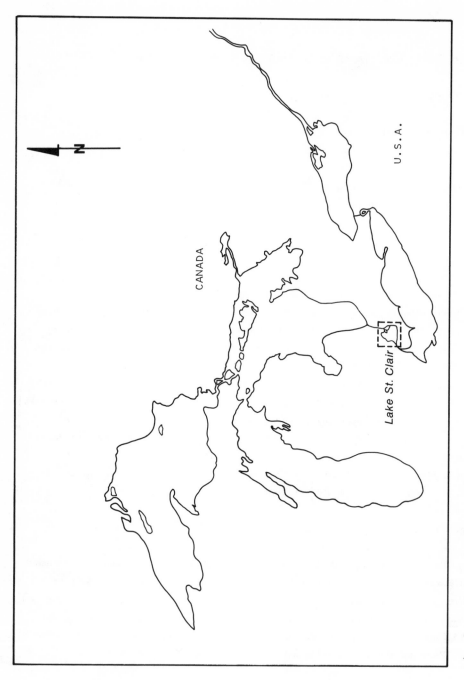

Fig. 1. Location of Lake St. Clair

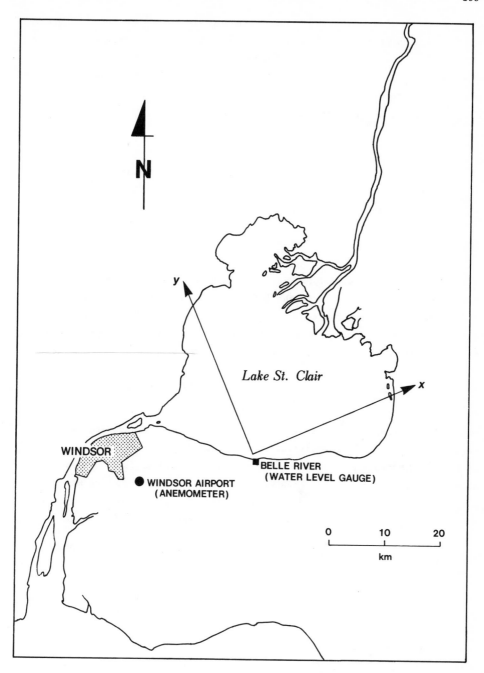

Fig. 2. Location of sampling stations

where $\underset{\sim}{\tau}$ is the wind stress, C_d, is the drag coefficient, ρ_a is the air density and $\underset{\sim}{V}$ is the observed wind vector. The drag coefficient was made dependent upon the atmospheric stability, as suggested by McClure (1970), and upon the wind speed, as suggested by Smith and Banke (1975). The exact form of the equation used for the computation of Lake St. Clair drag coefficients was that proposed by Hamblin (1978):

$$C_d = 0.001(1.0 - 0.09\ (T_a - T_w))(1.0 + 0.001048\ |\underset{\sim}{V}|) \tag{2}$$

where $T_a - T_w$ is the air-water temperature difference and CGS units are employed.

The wind stress, $\underset{\sim}{\tau}$, is resolved into τ_x and τ_y components along the x and y axes shown in Fig. 2.

THE MODEL

The water level fluctuations can be viewed as the response or output of a dynamic system which is under the influence of a set of physical factors such as the x and y components of wind stress. Suppose that the observed data on the system take the form of time series, i.e. the data were collected at N equidistant points in time. These data can be written as

$$(\eta_1, \tau_{x1}, \tau_{y1}),\ (\eta_2, \tau_{x2}, \tau_{y2}),\ \dots,\ (\eta_t,\ \tau_{xt},\ \tau_{yt}),\ \dots,\ (\eta_N,\ \tau_{xN},\ \tau_{yN})$$

where η_t, τ_{xt} and τ_{yt} are the water level measurement and the x and y components of wind stress at time t, (t = 1,2,..., N). The observed data can be regarded as a tri-variate realization or a sample from an infinite population of such realizations generated by some stochastic process. This process is assumed to be included in the general time series transfer function model proposed by Box and Jenkins (1970). This model contains two major components. The first is given by the class of the discrete transfer function model

$$\eta_t - \delta_1 \eta_{t-1} - \dots - \delta_r \eta_{t-r} = \omega_0 \tau_{x,t-b_1} + \omega_1 \tau_{x,t-b_1-1} + \dots$$

$$+ \omega_s \tau_{x,t-b_1-s} + \lambda_0 \tau_{y,t-b_2} + \lambda_1 \tau_{y,t-b_2-1} + \dots + \lambda_v \tau_{y,t-b_2-v} \tag{3}$$

for t = 1,2,...,N. The parameters b_1, b_2, δ_i (i = 1,2,...,r), ω_j (j = 1,2, ...,s) and λ_k (k = 1,2,...,v) are unknown constants which must be estimated. Writing B for the backward shift operator, where, $B^m \eta_t = \eta_{t-m}$, then equation (3) can be rewritten as

$$\delta(B)\eta_t = \omega(B)B^{b_1}\tau_{xt} + \lambda(B)B^{b_2}\tau_{yt} \tag{4}$$

where $\delta(B) = 1 - \delta_1 B - \delta_2 B^2 - \cdots - \delta_r B^r$

$\omega(B) = \omega_0 + \omega_1 B + \cdots + \omega_s B^s$

and $\lambda(b) = \lambda_0 + \lambda_1 B + \ldots + \lambda_v B^v$

Finally, equation (4) maybe rearranged as follows

$$\eta_t = \delta^{-1}(B)\omega(B)B^{b_1}\tau_{xt} + \delta^{-1}(B)\lambda(B)B^{b_2}\tau_{yt} \qquad (5)$$

The second part of the model arises by realizing that the dynamic system will be affected by disturbances, or noise, whose net effect is to corrupt the output predicted by the transfer function by an amount N_t. The combined transfer function–noise model may then be written as

$$\eta_t = \delta^{-1}(B)\omega(B)B^{b_1}\tau_{xt} + \delta^{-1}(B)\lambda(B)B^{b_2}\tau_{yt} + N_t \qquad (6)$$

When the data are available, the problem is to fit the model described by equation (6) to the data. The approach we shall employ for identifying and fitting the transfer function model differs from that given in Box and Jenkins (1970), but it can be regarded as generalization to the approach given by Haugh and Box (1977) for fitting a transfer function when only one independent variable is available. This approach can be described as follows:

(i) fit a univariate Box and Jenkins model to each of the three series η_t, τ_{xt} and τ_{yt};

(ii) estimate the residual for each univariate model;

(iii) fit a Box and Jenkins transfer function to the residuals, taking the residual of the η_t series as the dependent variable and the residuals of the τ_{xt} and τ_{yt} series as the independent variables;

(iv) substitute the expressions for the η_t, τ_{xt} and τ_{yt} residuals obtained in step (i) into the transfer function obtained in step (iii) to obtain a transfer function in terms of the variables η_t, τ_{xt} and τ_{yt};

(v) check the adequacy of fit; respecify and fit the models, if necessary.

In the next two sections the steps listed above will be described in more detail.

THE UNIVARIATE MODEL

The steps that may be followed in identifying, fitting and checking the adequacy of a univariate time series model will be described here. Suppose we have a time series of the N successive observations X_1, X_2,..., X_N taken at equidistant points in time. This series is assumed to be a realization from an infinite population of such realizations generated by some stochastic process. This process is assumed to be stationary.

Stationarity requires the absence of a trend in the mean value as well as the absence of any trend in all higher moments of the process. If the process is nonstationary, a prior transformation is performed on the data to achieve stationarity. Differencing operations are commonly used to obtain the desired transformation. A difference of order d is defined as

$$H_t = (1-B)^d X_t$$

where H_t is assumed to be a stationary series. An alternative transformation to differencing is known as the detrending procedure. This latter technique is often used to eliminate deterministic trends and seasonal effects.

Assuming that the sequence X_1, X_2, ..., X_N is stationary, the Box and Jenkins model is given as

$$X_t - \beta_1 X_{t-1} - \beta_2 X_{t-2} - \cdots - \beta_p X_{t-p} = a_t - \theta_1 a_{t-1} - \theta_2 a_{t-2} - \cdots - \theta_q a_{t-q} \quad (7)$$

where β_i (i = 1,2,...,p), θ_j (j = 1,2,...,q), p and q are unknown parameters to be estimated and a_t (t= 0,1,2,...,N) is a sequence of independent random variables with mean 0 and variance σ_a^2. The model described by equation (7) is often called the Autoregressive Moving Average Model of order p and q and is usually denoted by ARMA(p,q). The form of the model indicates that past values of the series can be used in forecasting its future values.

In order to fit the above model to an observed time series, the following steps are required:

(i) transform the series to achieve stationarity and determine the values of p and q;

(ii) given the values of p and q estimate β_1, β_2,..., β_p; θ_1, θ_2,..., θ_q and σ_a^2; and

(iii) check the adequacy of the model.

The estimation of the model parameters can be accomplished by using computer routines developed by McLeod (1977) at the University of Waterloo. Steps (i) and (iii) may be carried out by the examination of the sample autocorrelation function (ACF) and the sample partial autocorrelation function (PACF). The estimated autocorrelation coefficient at lag k is

$$r_k = \sum_{t=1}^{N-k} (X_t - \bar{X})(X_{t+k} - \bar{X}) / \sum_{t=1}^{N} (X_t - \bar{X})^2 \quad (8)$$

where
$$\bar{X} = \sum_{t=1}^{N} X_t / N$$

The value of r_k specifies how much information is contained in an observation made at time t about an observation made at time t+k. The other type of correlation which may be used in model identification is the estimated lag k partial autocorrelation, $\hat{\phi}_{kk}$. The values of $\hat{\phi}_{kk}$ can be estimated by solving the system of equations

$$r_j = \hat{\phi}_{k1} r_{j-1} + \hat{\phi}_{k2} r_{j-2} + \cdots + \hat{\phi}_{kk} r_{j-k} \tag{9}$$

$$j = 1, 2, \ldots, k$$

The value $\hat{\phi}_{kk}$ measures the dependence of X_{t+k} on X_t after eliminating the influence of $X_{t+1}, X_{t+2}, \ldots, X_{t+k-1}$ upon X_{t+k}.

The first step in identifying the model is to determine if the stochastic process is stationary. This is done by examining the ACF and PACF. If the ACF and PACF neither damp out with increasing k nor truncate, but instead remain large, then the process is nonstationary. In this case the data have to be transformed either by taking differences or by detrending. The ACF and PACF of the transformed data are then calculated and examined for stationarity. If the process is still nonstationary another transformation is attempted until stationarity is achieved.

Once the data, or transformed data, satisfy the stationarity requirement, the behaviour of the ACF and PACF determines an approximate estimate for the p and q values defined in equation (7). In the case where q=0, the ACF dies out slowly but the PACF is zero after lag p. On the other hand, when p=0, the ACF is zero after lag q but the PACF damps out slowly. When neither p nor q is equal to zero, both the ACF and PACF decay slowly. To determine when the population autocorrelations and partial autocorrelations are effectively zero, the sample estimates must be tested for significance. A test of significance of the sample autocorrelations was given by Bartlett (1935). The estimated variance of the autocorrelation, r_k, is

$$\hat{\sigma}^2_{r_k} \simeq \frac{1}{N} \{1 + 2 \sum_{i=1}^{k-1} r_i^2\} \tag{10}$$

which is calculated under the assumption that the population autocorrelation of lag greater than or equal to k is zero. The estimate of the variance for the partial autocorrelation of lag k, $\hat{\phi}_{kk}$, was given by Quenouille (1949), as

$$\hat{\sigma}^2_{\phi_{kk}} \simeq 1/N \tag{11}$$

After this preliminary identification of the model, the parameters of the
model may be estimated by using the method of maximum likelihood. The adequacy
of the fitted model is then checked by examining the autocorrelation function
of the estimated residuals $\{\hat{a}_t\}$, where $\hat{a}_t$ is estimated from equation (7) as

$$\hat{a}_t = X_t - \hat{\beta}_1 X_{t-1} - \cdots - \hat{\beta}_\rho X_{t-\rho} + \hat{\theta}_1 \hat{a}_{t-1} + \cdots + \hat{\theta}_q \hat{a}_{t-q} \tag{12}$$

where $\hat{\beta}_i$ (i=1,2,...,p) and $\hat{\theta}_j$ (j=1,2,...,q) are estimates of β_i (i=1,2,...,p)
and θ_j (j=1,2,...,q), respectively. If the model is adequate the autocorrela-
tion function of $\{\hat{a}_t\}$ will show the pattern associated with a set of independent observations (i.e., all autocorrelations at lags greater than one are
zeros).

THE TRANSFER FUNCTION MODEL

Suppose that univariate models have been fitted to the three series η_t,
τ_{xt} and τ_{yt}, and let $\hat{u}_\eta$, $\hat{u}_x$ and $\hat{u}_y$ represent the residual series obtained from
these fitted models. If it is assumed that there is no feedback, then the
residual of η_t may be expressed in terms of that of τ_{xt} and τ_{yt}. An appropriate model may be written as

$$\hat{u}_{\eta t} = \mu(B)\hat{u}_{xt} + \nu(B)\hat{u}_{yt} + \psi(B)a_t \tag{13}$$

where $\mu(B) = \mu_0 + \mu_1 B + \cdots + \mu_\ell B^\ell$,

$\nu(B) = \nu_0 + \nu_1 B + \cdots + \nu_m B^m$,

$\psi(B) = 1 + \psi_1 B + \cdots + \psi_n B^n$, and

$\{a_t\}$ represents a set of white noise with zero mean and variance σ_a^2. The
problem is then to determine the values of ℓ, m and n. The cross correlation
function (CCF) is the appropriate tool for the identification of these constants.

The cross correlation between two given time series, X_t and Y_t, can be
estimated from the equation

$$r_{xy}(k) = c_{xy}(k) \Big/ \sqrt{c_x(0)c_y(0)} \tag{14}$$

where $r_{xy}(k)$ is the estimated cross correlation at lag k,

$$c_{xy}(k) = \frac{1}{N} \sum_{t=1}^{N-k} (X_t - \bar{X})(Y_{t+k} - \bar{Y}), \quad k \geq 0$$

$$= \frac{1}{N} \sum_{t=1+k}^{N} (X_t - \bar{X})(Y_{t+k} - \bar{Y}), \quad k < 0$$

$$c_x(0) = \frac{1}{N} \sum_{t=1}^{N} (X_t - \bar{X})^2$$

$$c_y(0) = \frac{1}{N} \sum_{t=1}^{N} (Y_t - \bar{Y})^2$$

Under the assumption that X_t and Y_t are independent white-noise series, we have

$$\text{var}(r_{xy}(k)) \simeq \frac{1}{N-k} \tag{15}$$

The values of ℓ, m and n can be determined from an examination of the estimated cross correlations $r_{u_x u_\eta}(k)$, $r_{u_y u_\eta}(k)$ and $r_{u_x u_y}(k)$. Clearly, the comparison of $r_{u_x u_\eta}(k)$ and $r_{u_y u_\eta}(k)$ with their standard errors will determine the appropriate values of ℓ and m. Once ℓ and m have been determined, then the fact that $u_{\eta t}$ is a white-noise series imposes a set of restrictions on the model, the number of these restrictions will be equal to n.

After the order (ℓ, m, n) of the model has been established, it is possible to estimate the unknown parameters $\mu_0, \mu_1, \ldots, \mu_\ell, \nu_0, \nu_1, \ldots, \nu_m, \psi_1, \psi_2, \ldots, \psi_n$. One approach for obtaining approximate estimates for these parameters is the method of moments. This method might be appropriate if the number of observations is large. The other approach is to use an approximate maximum likelihood estimation technique. An algorithm for computing the estimates in this manner was made available to us through the University of Waterloo Mathematics Faculty Computer Facility.

The complete model is obtained by combining equation (7), which is used to define the residual series $u_{\eta t}$, u_{xt} and u_{yt}, with the noise transfer function model described by equation (13). The final form of the model becomes

$$\delta(B)\eta_t = \omega(B)\tau_{xt} + \lambda(B)\tau_{yt} + \theta(B)a_t \tag{16}$$

After the model parameters have been estimated, then the estimated residual

series $\{a_t\}$ can be calculated. The autocorrelation function of the series
is computed and compared with its standard error to ensure that the model
has accounted for the relationship between successive observations. In add-
ition, the cross-correlation functions between $\hat{a}_t$ and $\hat{u}_{xt}$, and between $\hat{a}_t$
and $\hat{u}_{yt}$ are calculated and compared to their standard errors to make sure
that the model has adequately described the relationship between the dependent
series, η_t, and the two independent series, τ_{xt} and τ_{yt}.

APPLICATIONS

The previous analysis has been applied to develop a transfer function
model for a storm event which took place July 10-14, 1964. The storm, which
has 91 hourly observations in η_t, τ_{xt} and τ_{yt}, will be used to illustrate the
procedures described in the previous sections.

Fig. 3 shows the time histories of the water level fluctuations, η_t, and
the normal wind stress components, τ_{xt} and τ_{yt}. The plots of the ACF (Figs.
4,6 and 8) and PACF (Figs. 5, 7 and 9) of η_t, τ_{xt} and τ_{yt} suggest that these
variables behave in a similar manner. It can be seen from these figures that
the ACF does not die out quickly, suggesting nonstationarity. The original
series are then transformed by taking the first difference. The resulting
autocorrelation functions, (Figs. 10, 12 and 14), and partial autocorrelation
functions (Figs. 11, 13 and 15) become very small, indicating that the differencing operation has induced stationarity.

Since the autocorrelation at lag one is high in all three cases, it is
assumed that a moving average model of order one might be appropriate in each
case. The structure of the models is as follows:

$$(1-B)X_t = (1-\theta_1 B)a_t \tag{17}$$

or

$$X_t - X_{t-1} = a_t - \theta_1 a_{t-1}$$

The estimates of θ_1, the standard error of the estimate and σ_a^2 for each
of the three univariate models is given in Table 1.

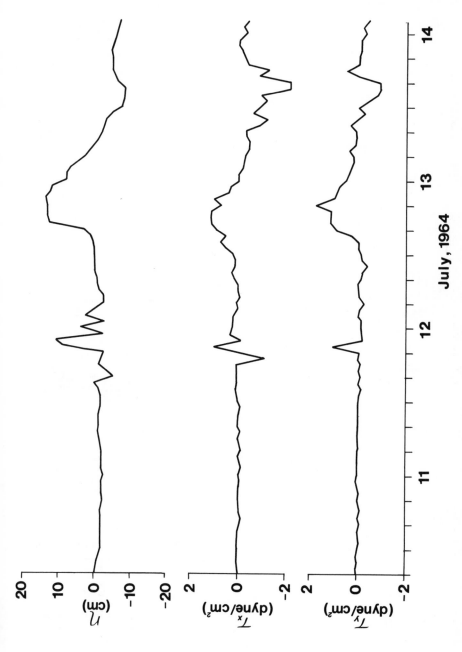

Fig. 3. Time series plots of water levels and x and y components of wind stress

208

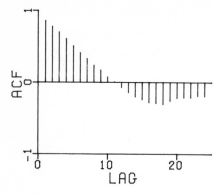

Fig. 4.   ACF of η_t

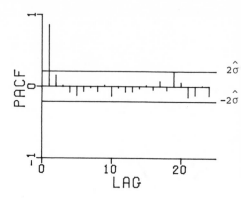

Fig. 5.   PACF of η_t

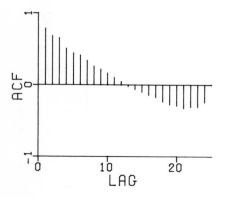

Fig. 6.   ACF of τ_{xt}

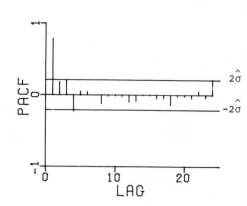

Fig. 7.   PACF of τ_{xt}

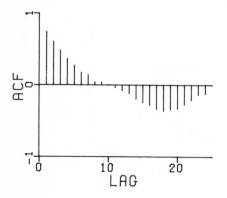

Fig. 8.   ACF of τ_{yt}

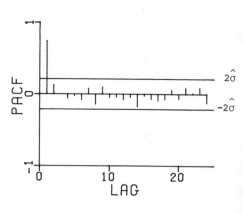

Fig. 9.   PACF of τ_{yt}

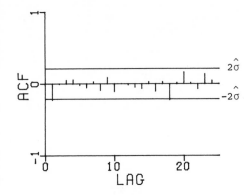

Fig. 10. ACF of $(1-B)\eta_t$

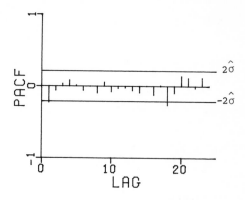

Fig. 11. PACF of $(1-B)\eta_t$

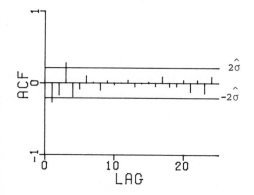

Fig. 12. ACF of $(1-B)\tau_{xt}$

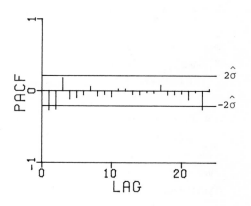

Fig. 13. PACF of $(1-B)\tau_{xt}$

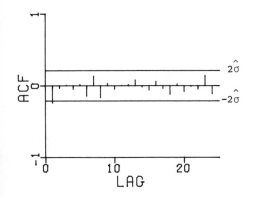

Fig. 14. ACF of $(1-B)\tau_{yt}$

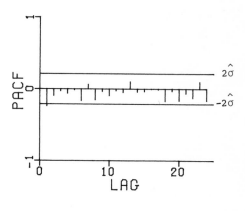

Fig. 15. PACF of $(1-B)\tau_{yt}$

TABLE I

Estimated parameters in the univariate models

Variable	$\hat{\theta}_1$	S.E.$(\hat{\theta}_1)$	$\hat{\sigma}_a^2$
η_t	0.227	0.103	6.65
τ_{xt}	0.354	0.099	0.12
τ_{yt}	0.275	0.101	0.09

Residual series were computed for each of the three variables through the use of the following equation:

$$\hat{a}_t = X_t - X_{t-1} + \hat{\theta}_1 \hat{a}_{t-1} \tag{18}$$

The residual series $\hat{u}_{\eta t}$, $\hat{u}_{xt}$ and $\hat{u}_{yt}$ were estimated for the variables η_t, τ_{xt} and τ_{yt}, respectively, and are shown in Figs. 16, 17 and 18.

To identify a residual transfer function model, the cross correlations of $\hat{u}_{xt}$ and $\hat{u}_{\eta t}$, $\hat{u}_{yt}$ and $\hat{u}_{\eta t}$, and $\hat{u}_{xt}$ and $\hat{u}_{yt}$ were computed. These cross correlation functions are shown in Figs. 19, 20 and 21, respectively. From these plots it can be seen that the cross correlations between $\hat{u}_{xt}$ and $\hat{u}_{\eta t}$ are significant at lags 0, 1 and 2, the cross correlations between $\hat{u}_{yt}$ and $\hat{u}_{\eta t}$ are significant at lags 0 and 1, and the cross correlations between $\hat{u}_{xt}$ and $\hat{u}_{yt}$ are significant at lags -2, 0 and 2. Thus the transfer function model described by equation (13) will be of the form:

$$\hat{u}_{\eta t} = (\hat{\mu}_0 + \hat{\mu}_1 B + \hat{\mu}_2 B^2)\hat{u}_{xt} + (\hat{\nu}_0 + \hat{\nu}_1 B)\hat{u}_{yt}$$

$$+ (1 - \hat{\theta}_1 B - \hat{\theta}_2 B^2)\hat{a}_t \tag{19}$$

The estimated coefficients are listed in the following table:

TABLE II

Estimated coefficients for noise transfer function model

Coefficient	Estimate
$\hat{\mu}_0$	1.087
$\hat{\mu}_1$	1.447
$\hat{\mu}_2$	-1.736
$\hat{\nu}_0$	1.892
$\hat{\nu}_1$	3.585
$\hat{\theta}_1$	0.072
$\hat{\theta}_2$	-0.047

The residual variance, $\hat{\sigma}_a^2$, was 4.47 cm^2.

The equations describing the univariate models for η_t, τ_{xt} and τ_{yt} are substituted into the noise transfer function model given by equation (19) to provide the final transfer function model described below

$$\eta_t = 1.628\eta_{t-1} - 0.725\eta_{t-2} + 0.097\eta_{t-3}$$

$$+ 1.087\tau_{xt} - 0.185\tau_{x,t-1} - 3.296\tau_{x,t-2} + 3.355\tau_{x,t-3}$$

$$- 1.069\tau_{x,t-4} + 0.108\tau_{x,t-5}$$

$$+ 1.892\tau_{yt} + 0.595\tau_{y,t-1} - 4.416\tau_{y,t-2} + 2.217\tau_{y,t-3}$$

$$- 0.288\tau_{y,t-4}$$

$$+ a_t - 0.927a_{t-1} + 0.349a_{t-2} - 0.080a_{t-3} - 0.013a_{t-4}$$

$$- 0.001a_{t-5} \tag{20}$$

To ensure that this model accounts for the autocorrelation of η_t as well as the dependence of η_t upon τ_{xt} and τ_{yt}, the autocorrelation function of $\hat{a}_t$ and the cross correlation functions of $\hat{u}_{\eta t}$ with respect to $\hat{u}_{xt}$ and of $\hat{u}_{\eta t}$ with respect to $\hat{u}_{yt}$ were calculated and plotted (Figs. 22, 23 and 24). None of the residual auto or cross correlations are more than marginally significant. This indicates that the model is adequate. A plot of the predicted and observed water level fluctuations is shown in Fig. 25.

212

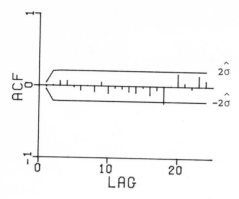

Fig. 16.   ACF of $\hat{u}_\eta$

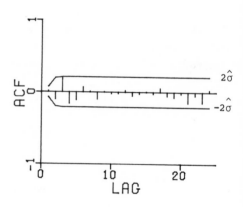

Fig. 17.   ACF of $\hat{u}_x$

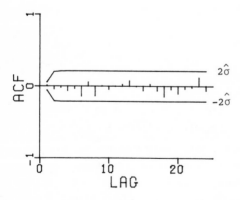

Fig. 18.   ACF of $\hat{u}_y$

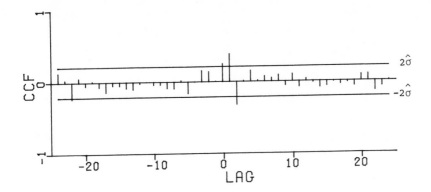

Fig. 19. $r_{u_x u_\eta}$ (CCF of $\hat{u}_\eta$ on $\hat{u}_x$)

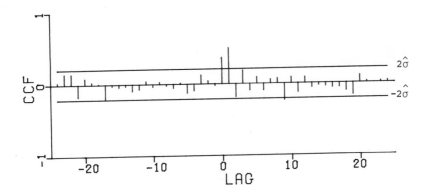

Fig. 20. $r_{u_y u_\eta}$ (CCF of $\hat{u}_\eta$ on $\hat{u}_y$)

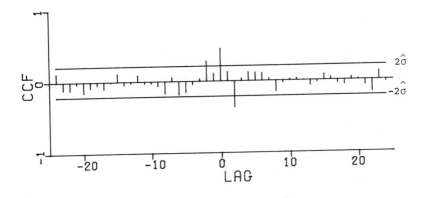

Fig. 21. $r_{u_x u_y}$ (CCF of $\hat{u}_y$ on $\hat{u}_x$)

214

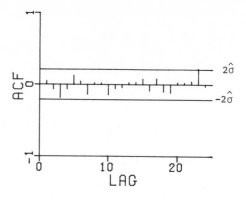

Fig. 22. ACF of $\hat{a}_t$

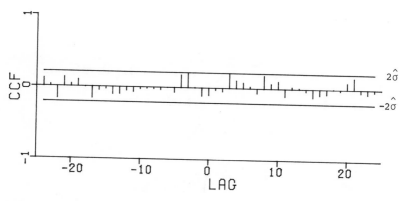

Fig. 23. $r_{u_x a}$ (CCF of $\hat{a}_t$ on $\hat{u}_{xt}$)

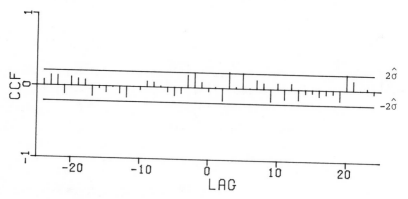

Fig. 24. $r_{u_y a}$ (CCF of $\hat{a}_t$ on $\hat{u}_{yt}$)

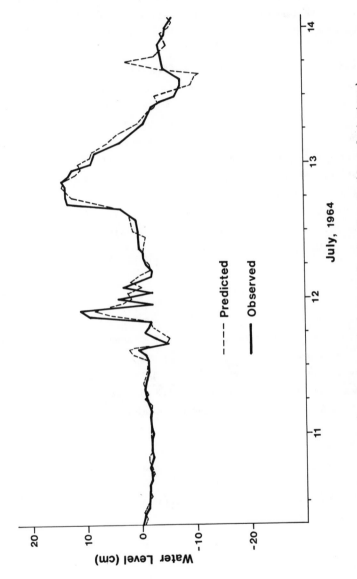

Fig. 25. One step ahead predictions and observed water levels (dependent storm)

In order to test the model under an independent set of conditions it was applied to an additional 18 storm events. On these 18 storms, the model accounted for 72 per cent of the total variation in the water level (R^2 = .72, where R is the multiple correlation coefficient). The residual variance was 16.6 cm^2. The largest storm in the data set occurred November 1-4, 1966. The predicted and observed hourly water levels for this storm are presented in Fig. 26.

Improved estimates for the model coefficients could be obtained if the estimates were based on data collected from several storm, events. This could be accomplished by "lumping" statistically similar storm data. In addition, accuracy could be enhanced if over-lake as opposed to over-land winds were used in the modelling study. Unfortunately, very few over-lake wind measurements are available for Lake St. Clair. To date, attempts to use lake-land wind ratios derived from other lakes in Lake St. Clair applications have been disappointing. Hamblin (1978) found that correcting the land wind for the over-lake effect, after Phillips and Irbe (1978), failed to improve the accuracy of Lake St. Clair water level predictions obtained from a spectral dynamical model.

CONCLUSIONS

In this paper an empirical time series model has been developed to relate water level fluctuations to normal wind stress components for storm surge events on a medium-sized lake. The model produced acceptable results when used to predict water level fluctuations on Lake St. Clair. Improved accuracy could be achieved by combining statistically similar storm data and by using over-lake as opposed to over-land wind measurements.

ACKNOWLEDGMENTS

The authors would like to thank Dr. P.F. Hamblin of the National Water Research Institute in Burlington, Ontario for many stimulating discussions and helpful suggestions.

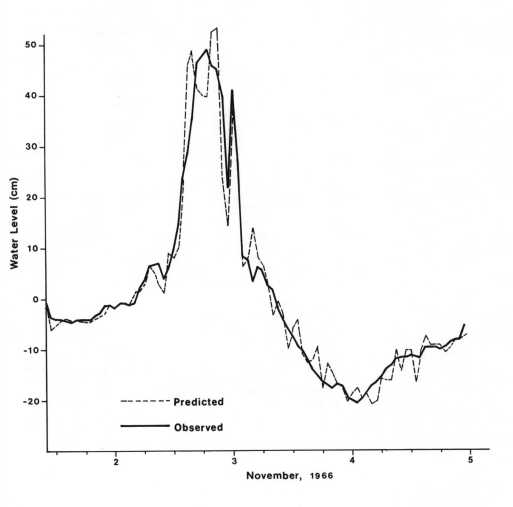

Fig. 26. One step ahead predictions and observed water levels (independent storm)

REFERENCES

Bartlett, M.S., 1935. Some aspects of the time-correlation problem in regard to tests of significance. Journal of the Royal Statistical Society, 98:536-543.

Box, G.E.P. and Jenkins, G.M., 1970. Time Series Analysis: Forecasting and Control. Holden-Day, San Francisco.

El-Shaarawi, A. and Whitney, J., 1978. On determining the number of samples required to estimate the phosphorous input contributed by Niagara River to Lake Ontario. Canada Centre for Inland Waters Report Series, in press.

Feige, E.L. and Pearce, D.K., 1974. The causality relationship between money and income: a time series approach. Presented at the Annual Meeting of Midwest Economic Association, Chicago.

Granger, C.W.J., 1969. Investigating causal relationships by econometric models and cross spectral methods. Econometrica, 37:424-438.

Hamblin, P.F., 1978. Storm surge forecasting in enclosed seas. Proc. of the 16th Conference on Coastal Engineering, ASCE, Hamburg.

Haugh, L.D. and Box, G.E.P., 1977. Identification of dynamic regression (distributed lag) models connecting two time series. Journal of the American Statistical Society, 72:121-130.

McClure, D.J., 1970. Dynamic forecasting of Lake Erie water levels. Report No. 70-250 H. Hydro-Electric Power Commission of Ontario, Research Div. Report, Toronto.

McKerchar, A.I. and Delleur, J.W., 1974. Application of seasonal parametric linear stochastic models to monthly flow data. Water Resources Research, 10:246-254.

McLeod, A.I., 1977. Improved Box-Jenkins estimators. Biometrika, 64:531-534.

O'Connell, P.E., 1971. A simple stochastic modelling of Hurst's Law. Proc. of I.A.S.H. International Symposium on Mathematical Models in Hydrology, Warsaw.

Phillips, D.W. and Irbe, J.G., 1978. Lake to land comparison of wind, temperature and humidity on Lake Ontario during the International Field Year for the Great Lakes. Report No. CL1-2-77. Atmospheric Environment, Fisheries and Environment Canada, Toronto.

Pierce, D.A., 1977. Relationships and the lack thereof - between economic time series, with special reference to money and interest rates. Journal of the American Statistical Association, 72:11-22.

Quenouille, M.H., 1949. Approximate tests of correlation in time-series. Journal of the Royal Statistical Society (B), 11:68-74.

Smith, S.D. and Banke, E.G., 1975. Variation of the sea surface drag coefficient with wind speed. Quarterly Journal of the Royal Meteorological Society, 101:665-673.

WIND INDUCED WATER CIRCULATION OF LAKE GENEVA

S.W. BAUER and W.H. GRAF

Laboratoire d'Hydraulique (LHYDREP), Ecole Polytechnique Fédérale, Lausanne
Switzerland

(Received 17 August 1978; accepted 24 August 1978)

ABSTRACT

A numerical modeling technique is used to simulate flow patterns at various
depths in the Lake of Geneva (Le Léman) for a homogeneous situation encountered
usually during winter months. Subsequently, a vertically integrated flow pattern
is obtained.

A measuring campaign is under way which provides data on the velocity, direction
and temperature of the atmosphere and the water.

Using these data, two quasi-steady state situations are compared with model
simulations for winter months of 1977 and 1978.

DESCRIPTION OF MATHEMATICAL MODEL

A system of currents in a lake may be considered as water movement on a large

scale. It can be described by the three components of the momentum equation and

the continuity equation for a homogeneous (non stratified) and incompressible fluid

(Liggett, 1970):

$$\rho[\frac{\partial u}{\partial t} + \frac{\partial}{\partial x}(u^2) + \frac{\partial}{\partial y}(uv) + \frac{\partial}{\partial z}(uw) - fv] = -\frac{\partial p}{\partial x} + \frac{\partial}{\partial z}(\eta\frac{\partial u}{\partial z}) + \frac{\partial}{\partial x}(\varepsilon\frac{\partial u}{\partial x}) + \frac{\partial}{\partial y}(\varepsilon\frac{\partial u}{\partial y}) \quad (1)$$

$$\rho[\frac{\partial v}{\partial t} + \frac{\partial}{\partial x}(uv) + \frac{\partial}{\partial y}(v^2) + \frac{\partial}{\partial z}(vw) + fu] = -\frac{\partial p}{\partial y} + \frac{\partial}{\partial z}(\eta\frac{\partial v}{\partial z}) + \frac{\partial}{\partial x}(\varepsilon\frac{\partial v}{\partial x}) + \frac{\partial}{\partial y}(\varepsilon\frac{\partial v}{\partial y}) \quad (2)$$

$$\rho g = -\frac{\partial p}{\partial z} \quad (3)$$

$$\frac{\partial u}{\partial x} + \frac{\partial v}{\partial y} + \frac{\partial w}{\partial z} = 0 \quad (4)$$

The boundary conditions applicable on solid boundaries are

$$u = v = w = 0 \quad (5)$$

and at the free surface with z = 0

$$\eta \frac{\partial u}{\partial z} = \tau_x \; ; \; \eta \frac{\partial v}{\partial z} = \tau_y \qquad\qquad\qquad (6)$$

The symbols in equation 1-6 are as follows:

u, v and w are the velocity components in the x, y and z directions respectively
where x is positive towards east, y is positive towards north and z is positive
upwards with zero at the water surface,

t is the time,

f is the Coriolis parameter,

ρ is the fluid density,

p is the local pressure,

η and ε are the vertical and horizontal components of the eddy viscosity,

g is the acceleration of gravity   and

τ_x and τ_y are the wind shear stresses on the water surface in the x and y direction
respectively.

Equation 3 expresses the hydrostatic equilibrium, which is a valid assumption
for shallow lakes, i.e.: D/L $\ll$ 1 where L and D are characteristic vertical and
horizontal dimensions (for the Léman: D/L $\sim$ 0,03).

Equations 1-3 may be further simplified according to the following assumptions:

a) stationary flow: $\frac{\partial u}{\partial t} = 0$, $\frac{\partial v}{\partial t} = 0$;
    such a situation might be imagined for a wind which blows long enough to
    establish a stationary circulation;

b) the inertia forces are small when compared with the Coriolis forces, i.e., the
    Rossby number is small. In this case equations 1-3 may be linearized:

    $\frac{\partial}{\partial x} (uv) = 0$, $\frac{\partial}{\partial y} (uv) = 0$, etc.

    (for the Léman the Rossby number $\sim$ 0,1);

c) horizontal diffusion is small compared to vertical diffusion, which can be ac-
    cepted for shallow lakes:

    $\frac{\partial}{\partial x}(\varepsilon \frac{\partial v}{\partial x}) = 0$, $\frac{\partial}{\partial y}(\varepsilon \frac{\partial u}{\partial y}) = 0$, etc.

    (for the Léman: D/L $\sim$ 0,03);

d) the vertical component of the eddy viscosity η which is not known is assumed
    to be constant over the entire lake:

    $\frac{\partial}{\partial z}(\eta \frac{\partial v}{\partial z}) = \eta \frac{\partial^2 v}{\partial z^2}$, etc.

e) the influence of external in- and outflows in the Leman on wind induced currents
    on the lake is negligible. Thus, in the immediate vicinity of river mouths,
    the model is not applicable.

With the above simplifications and conditions the system of equations describing wind induced currents in a homogeneous shallow lake is thus:

$$- \rho f v = - \frac{\partial p}{\partial x} + \eta \frac{\partial^2 u}{\partial z^2} \qquad (7)$$

$$+ \rho f u = - \frac{\partial p}{\partial y} + \eta \frac{\partial^2 v}{\partial z^2} \qquad (8)$$

$$\rho g = - \frac{\partial p}{\partial z} \qquad (9)$$

$$\frac{\partial u}{\partial x} + \frac{\partial v}{\partial y} + \frac{\partial w}{\partial z} = 0 \qquad (10)$$

The numerical solution of the model described thus far is obtained by Gallagher et al. (1973) and based on a mathematical formulation by Liggett et al. (1969).

ADAPTATION OF MATHEMATICAL MODEL FOR THE LEMAN

The mathematical model as described above has been coded and provided to LHYDREP by J.A. Liggett of Cornell University. Subsequently a suite of routines allowing graphical representation in 2 and 3 dimensions of the numerical results as well as of the geometrical representation in finite elements - see Figure 1 - of the lake geometry have been developed at LHYDREP. After several trial geometries a finite element grid consisting of 579 nodes (about one node per km^2), 1025 triangular elements and 131 boundary points was found to give reasonable results (see Bauer et al., 1977).

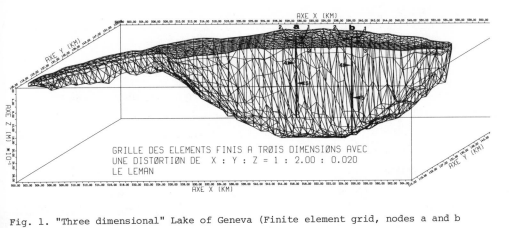

GRILLE DES ELEMENTS FINIS A TRØIS DIMENSIØNS AVEC
UNE DISTØRTIØN DE X : Y : Z = 1 : 2.00 : 0.020
LE LEMAN

Fig. 1. "Three dimensional" Lake of Geneva (Finite element grid, nodes a and b showing position of instruments).

It should be noted that compared to its width, the Leman is very shallow (0,01 < D/L < 0,03). Due to numerical reasons of the finite element approximation it is necessary that the minimum depth allowable is about 3 % of the maximum depth.

Thus there are no zero depths at the boundaries (see Figure 1) which however is not a severe limitation for the geometrical representation.

To execute the model a series of parameters (see equations 7-9) must be entere the density of water, ρ, the Coriolis parameter, f, the acceleration of gravity, g, and the eddy viscosity, η. Also, replacing in equation 6 the shear stress, τ, by equation 11, viz.

$$\tau = C_f \cdot \rho_a \cdot U_{wind}^2 \tag{11}$$

where C_f is a wind shear stress coefficient,

ρ_a is the density of air   and

U_{wind} is the free stream wind velocity.

Two further parameters, C_f and ρ_a are to be supplied. The following parameters have been taken as constants in the model simulations:

density of air   : $\rho_a = 1,2 \text{ kg/m}^3$

density of water   : $\rho = 999,9 \text{ kg/m}^3$

acceleration of gravity : $g = 9,80 \text{ m/s}^2$

Coriolis parameter   : $f = 0,000105 \text{ s}^{-1}$ (for a mean latitude of the Leman of $46°25'$)

The parameters η and C_f, which are unknown are thus subject to determination by a model calibration.

A trial simulation for a set of fixed and chosen parameters has been done and is reported by Bauer et al. (1977).

MEASURING CAMPAIGNS

To allow for comparison between the currents occurring in nature and their simulat by a model it is necessary to study the distribution of the water velocities and the winds which generate them. Therefore a measuring campaign has been started to measure simultaneously and in situ velocities and directions of the wind and currents. Furthermore, temperature profiles of atmosphere and water were measured.

Ideally, measurements should be taken by a large number of recording instrument placed according to some grid system over the entire domain. Such measurements, horizontally and vertically distributed, would allow to obtain a synoptic view of the phenomenon. Such a solution however is at present, for financial and operation reasons, impossible and it was thus necessary to arrive at a much more modest solution.

Keeping the concept of a vertical scheme, a single station is placed in the lake. Measured and recorded are:

- velocity of the atmosphere at three (3) different altitudes as well as its
  directions and the temperature;
- velocity, direction, temperature and pressure of the currents at five (5) different depths.

A description of this installation is given by Prost et al. (1977) and a schematic
view of the station is shown in Figure 2.

This station has been kept for a duration of weeks at one place. Since for the
servicing of the current meters it is necessary to lift the entire installation
out of the water, this operation is also used to change its position in the lake,
thus allowing for a coverage of different areas. During the winter months, the
installation was located close to the two nodes a and b as indicated in Figure 1.

Measurements:

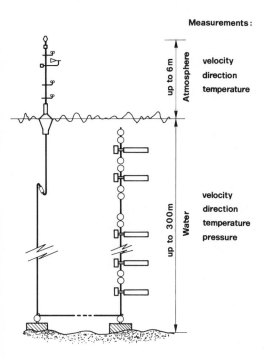

velocity
direction
temperature

velocity
direction
temperature
pressure

Fig. 2. Schematic view of measuring station

For measurements in the atmosphere, conventional cup anemometers of the type
Aanderaa (WSS 219) are being used to measure the velocities. The number of rotor
revolutions is counted electronically and transmitted in digital form to a data
logger. Wind directions are determined with the aid of a wind vane type Aanderaa
(WDS 2053) and a magnetic compass measuring the orientation of the buoy with respect to north. The air temperature sensors used were Aanderaa (1289 A).

DATA - OBTAINED SINCE FEBRUARY 1977

The periods of the data collected thus far by LHYDREP in the Lake of Geneva are summarized in Table 1.

TABLE 1

Summary of data periods collected by LHYDREP (Dates given by day, month and year)

No	Period	Position (km)		Depth (m)
		x	y	
1	1/2/77 - 1/3/77	528,28	149,18	73,0
2	1/3/77 - 25/4/77	531,24	147,84	198,9
3	16/6/77 - 24/8/77	538,39	148,68	293,4
4	31/8/77 - 8/11/77	542,02	147,48	289,9
5	9/12/77 - 15/2/78	542,02	147,48	289,9

To allow inspection of these data several computer programs permitting analog representation of the data were developed at LHYDREP. After all data were plotted, visual inspection showed that the lake has been reasonably homogeneous for the entire periods 1 and 2 and for period 5 after 14/1/78. Using average temperatures from 5,6; 10,0; 20,0; 35,0 and 55,0 m depth the overall mean lake temperature was 5,725 ^{o}C with a standard deviation of the means of 0,020 ^{o}C for period 1. Similarly for period 2, using average temperatures from 5,8; 10,2; 20,2; 84,4 and 148,9 m depth, the overall mean lake temperature was 6,143 ^{o}C with a standard deviation of 0,412 ^{o}C. For period 5 after 14/1/78 the overall mean temperature obtained from 3,9; 8,6; 18,5; 92,7 and 195,6 m depth was 5,968 ^{o}C with a standard deviation of 0,222 ^{o}C.

Typical records of the data are shown in Figures 3 and 4 for the periods 2 and 5 respectively. In Figures 3 and 4, the time axis is horizontal, whereby every 24 hours a vertical line, showing the date in years, months, days, hours and minutes indicates the start of a new day. The data represented in the topmost band are the temperatures of the water in 5 depths corresponding to the positions of each current meter. Proceeding downwards, the next band shows the wind velocities and then follow the velocity observations of the 5 currentmeters. The subsequent 6 bands show directions, the first being the direction of the wind followed by the direction of the 5 currentmeters. The scales - shown every five days - have been selected such that the temperature band extends over a range of 5-10 ^{o}C, the wind speed band over 0-10 m/s, the water velocity bands over 0-10 cm/s and the direction bands over 0-360°. (In case these ranges are exceeded by the data to be plotted, the

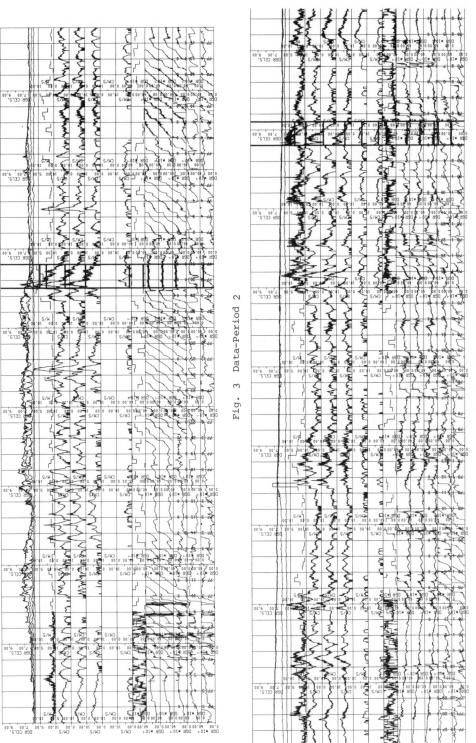

Fig. 3 Data-Period 2

Fig. 4 Data-Period 5

trace of one band continues over neighbouring bands but keeps its original scale.)
In Figures 3 and 4 it can be seen that records of wind were not always available
(Figure 3: after 9/3/77, Figure 4: between 3/1/78 and 23/1/78). Thus, wind observations of the Swiss Meteorological Service recorded at 7, 13 and 19 hours at
Lausanne are inserted in the wind bands assuming a six hours duration (straight
line segments) for each of these observations. Comparison of the LHYDREP and
"Lausanne" wind data (see Figure 3, before 9/3/77 and Figure 4 before 3/1/78 and
after 23/1/78) show reasonable agreement. Also, currentmeter 4 of Figure 3 recorded only current directions and thus no data are shown on the velocity band of
currentmeter 4 of this Figure.

SELECTION OF PERIODS USED FOR SIMULATION

It was stated above that the mathematical model is conceived for the following
limiting conditions:
(1) homogeneity of the water body, i.e., the lake must have almost no temperature
    stratification;
(2) stationarity of the velocity and direction of the wind and the water.
The manner of satisfying the above conditions is discussed in the following.
Since the "stationarity" is more restrictive we shall start with it.
    (ad 2) A climatological particularity of the Leman basin is that there exist
two important and strong winds: one north-easterly - the bise - and one south-
westerly - the vent (Primault, 1972). These winds often blow for periods of days
and, with the possible exception of the eastern end of the lake, produce probably
a constant intensity and direction of shear on the lake surface. Searching for
such situations in periods 2 and 5 of Table 1 (see Figures 3 and 4), we have
selected two time periods of data, subsequently to be used for a model simulation.
These periods are:
        bise :   29/3/77, 12h00 - 30/3/77, 24h00
        vent :   2/2/78, 00h00 - 3/2/78, 12h00
If one regards the direction of currents at these layers, one finds them to be
reasonably constant; however, the velocity of currents at the different layers
is not at all constant (!). Vectorially averaged values for the velocities of
wind and currents were calculated as indicated - Figures 3 and 4 with bold lines -
and are summarized in Tables 2 (BISE) and 3 (VENT).
    (ad 1) For the periods selected above, i.e. bise and vent, density variations
were calculated and found to be weak; results are given in Tables 2 and 3.

TABLE 2

BISE: Mean (temporal) velocities and temperatures of currents at x = 531,24 km
and y = 147,84 km between 29/3/77, 12 hours and 31/3/77, zero hours
Average wind velocity at Lausanne[1]: $(U_{wind})_x$ = - 3,5 m/s
$(U_{wind})_y$ = - 8,4 m/s
U_{wind} =   9,1 m/s = 32,8 km/h

Depth (m)	u (cm/s)	v (cm/s)	$\sqrt{u^2 + v^2}$ (cm/s)	t (°C) [2]	ρ (g/ml)	$\Delta\rho/\bar{\rho}_o$
5,6	- 14,77	- 1,60	14,86	6,287	0,999931014	$2,1101221 \cdot 10^{-7}$
10,2	- 12,84	- 1,16	12,89	6,293	0,999930803	$9,5665536 \cdot 10^{-7}$
20,2	- 9,56	0,96	9,61	6,320	0,999929846	$-1,1799582 \cdot 10^{-5}$
84,4	-	-	-	5,962	0,999941645	$-8,3563836 \cdot 10^{-6}$
148,9	- 1,1	0,02	1,10	5,664	0,999950001	
Weighted mean with 200 m max. depth	- 3,64	0,09	3,64	-	$\bar{\rho}_o =$ 0,999942124	

[1] Velocity components of wind and currents: u is positive for flow towards east
v is positive for flow towards north
[2] Subject to accuracy of ± 0,01 °C according to instrument specification

TABLE 3

VENT: Mean (temporal) velocities and temperatures of currents at x = 542,02 km
and y = 147,48 km between 2/2/78, zero hours and 3/2/78, 12 hours
Average wind velocity at station[1]: $(U_{wind})_x$ = 4,7 m/s
$(U_{wind})_y$ = 4,3 m/s
U_{wind} = 6,4 m/s = 22,9 km/h

Depth (m)	u (cm/s)	v (cm/s)	$\sqrt{u^2 + v^2}$ (cm/s)	t (°C) [2]	ρ (g/ml)	$\Delta\rho/\bar{\rho}_o$
0,0	-	-	-	6,055	0,999938764	$-3,8424000 \cdot 10^{-6}$
3,9	8,40	4,00	9,30	5,930	0,999942607	$-5,6352933 \cdot 10^{-7}$
8,6	7,92	1,16	8,00	5,911	0,999943170	0
18,5	3,92	1,08	4,07	5,911	0,999943170	$2,0113047 \cdot 10^{-6}$
92,7	- 1,29	2,66	2,96	5,978	0,999941159	$-1,2234736 \cdot 10^{-5}$
195,6	- 0,60	2,22	2,30	5,527	0,999953397	
Weighted mean with 305 m max. depth	0,21	2,20	2,21	-	$\bar{\rho}_o =$ 0,999947942	

[1] See Table 2
[2] See Table 2

CALIBRATION OF MODEL

As has been stated above, the two model parameters subject to model calibration (in this study) are the wind shear coefficient, C_f, and the eddy viscosity, η.

Let us look first at the influence of the eddy viscosity upon the simulated velocities, keeping the shear stress coefficient constant. This is done in Figure where the velocity vectors for η = 1000, 500 and 100 cm^2/s for three (3) points are shown every 2,5 m. It can be seen that in each case there is a deviation on

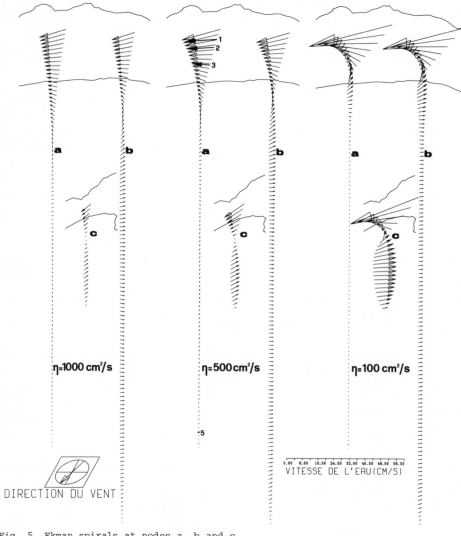

Fig. 5. Ekman spirals at nodes a, b and c

the water surface of about 45° to the right between the wind direction and the current direction. Then, proceeding downwards, the vectors diminish and turn to the right forming the Ekman spiral. At a certain depth, D_e, the vectors have turned 180° relative to the vectors on the water surface. For an idealized ocean of infinite dimensions this depth, D_e, is given by (Dietrich et al., 1975)

$$D_e = \pi \sqrt{\frac{2 \eta}{f}} \tag{12}$$

Thus, the depth, D_e, is proportional to the square root of η as seen in Figure 5. Furthermore, since the mass transport over the depth, D_e, is independent of the eddy viscosity (Dietrich et al., 1975) - i.e. $\int_o^{D_e} v dz = \text{constant}$ - a change of D_e produces a change in the vertical distribution of the velocities. This is also evident in Figure 5.

As for the influence of the C_f values on the simulation, it is to be noted that this factor acts merely like a scale factor on the velocity scales.

Concluding from the above remarks the following procedure is proposed for calibration of the model:

(1) trying different values of η; select η such that reasonable directional agreement between observations and model simulation is obtained (greater importance must be attached to close agreement in the top layers);

(2) check the thus obtained η value against values found in the literature;

(3) for the chosen value of η adjust the wind shear stress coefficient, C_f, such that the magnitudes of simulated and observed velocities agree reasonably well (again, greater importance must be attached to close agreement in the top layers);

(4) check if the selected value of C_f agrees with the ones calculated from direct observations or found in the literature.

Following the above procedure, the bise (for description see Table 2) was investigated first and is discussed herewith (the same was also done for the vent):

(ad 1) Trying various values of $50 \text{ cm}^2/s \le \eta \le 2000 \text{ cm}^2/s$, it was found that $\eta = 500 \text{ cm}^2/s$ gave directions of vectors that agreed well with the observations, as it is to be seen in Figure 5.

(ad 2) If the η value is derived from a well-accepted equation (Neumann and Pierson, 1966):

$$\eta = 0{,}1825 \cdot 10^{-4} \, u_{wind}^{5/2} / \rho \tag{13}$$

where U_{wind} is in cm/s,

ρ is in g/cm^3 and

η is in cm^2/s

with a wind velocity of 32,7 km/h, an eddy viscosity of $460 \text{ cm}^2/s$ is obtained.

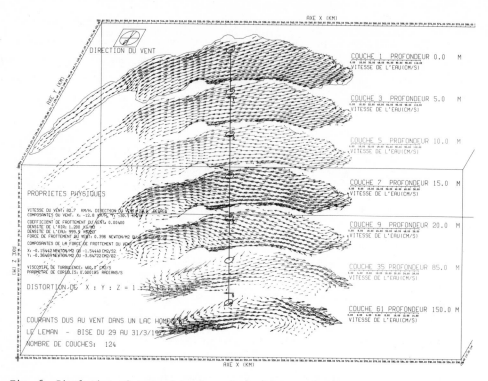

Fig. 6. Simulation of current vectors for a Bise, and observations.

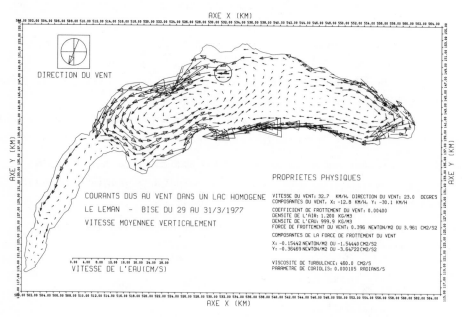

Fig. 7. Simulation of vertically integrated current vectors for a Bise, and observations.

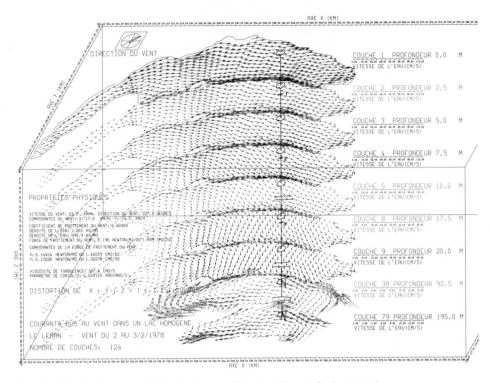

Fig. 8. Simulation of current vectors for a Vent, and observations

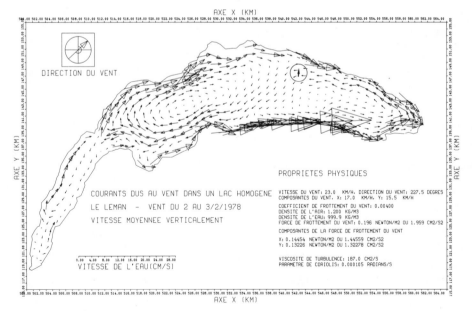

Fig. 9. Simulation of vertically integrated current vectors for a Vent, and observations.

232

This is thought to be reasonably close to 500 cm^2/s, and a value of 460 cm^2/s was adopted for further calculations.

(ad 3) In order to obtain agreement of the magnitudes of the vectors, the wind shear stress coefficient, C_f, was taken as $C_f = 0,004$.

(ad 4) C_f values have been calculated at LHYDREP by Prost (personal communication, 1978) for our measuring campaign and do indeed corroborate with our chosen values.

The results of this model simulation (calculation) are shown for the bise in Figure 6 and for the vent in Figure 8. Drawn are the computer calculated circulation patterns at different layers and the in situ measured velocity vectors at respective depths. Considering the assumptions made in the mathematical simulation and the difficulties encountered in an in situ measuring campaign, the agreement is considered to be reasonably good.

Furthermore, in Figure 7 (bise) and Figure 9 (vent) we show a comparison of depth-average velocities calculated and measured. Agreement is extremely encouraging for the bise, but certainly less good for the vent. (We de not exclude that further calibration could be applied to reach better agreement for the vent as well). Interestingly enough the model circulation pattern leads us to agree with a conclusion recently drawn by Hamblin (1976):"Currents demonstrate the general tendency to follow the wind in the nearshore region, whereas the return current opposed to the wind direction is situated in the central portion of the lake".

ACKNOWLEDGEMENT

This work was partially sponsored by the Swiss National Science Foundation (FNSRS) under its special program "Fundamental problems of the water cycle in Switzerland".

REFERENCES

Bauer, S.W., Graf, W.H. and Tischer E., 1977. Les courants dans le Léman en saison froide. Une simulation mathématique. Bull. Techn. Suisse Romande, 103:239-243.

Dietrich, G., Kalle, K., Krauss, W. and Siedler, G., 1975. Allgemeine Meereskunde, Gebrüder Bornträger, Berlin-Stuttgart.

Gallagher, R.H., Liggett, J.A. and Chan, S.K.T., 1973. Finite Element Shallow Lake Circulation. Proc. Am. Soc. Civ. Engs., Vol. 99, No HY7

Hamblin, P.F., 1976. Seiches, circulation, and storm surges of an ice-free Lake Winnipeg. J. Fish. Res. Board Can., 33:2377-2391.

Liggett, J.A., 1970. Cell Method for computing lake circulation. Proc. Am. Soc. Civ. Engs., Vol. 96, No HY3.

Liggett, J.A. and Hadjitheodorou C., 1969. Circulation in shallow homogeneous lakes. Proc. Am. Soc. Civ. Engs., Vol. 95, No HY2.

Neumann, G. and Pierson, W.H. Jr., 1966. Principles of physical oceanography.
   Prentice Hall, Englewood Cliffs, N.J.
Primault, B., 1972. Etude mésoclimatique du Canton de Vaud, Cahier de l'aménage-
   ment régional 14, Office cantonal vaudois de l'urbanisme, Lausanne.
Prost, J.-P., Bauer, S.W., Graf, W.H. and Girod, H., 1977. Campagne de mesure
   des courants dans le Léman. Bull. Techn. Suisse Romande, 103:243-249.

NON-LINEAR THREE-DIMENSIONAL MODELLING OF MESOSCALE CIRCULATION IN SEAS AND LAKES

Jacques C.J. NIHOUL[1], Y. RUNFOLA and B. ROISIN

Mécanique des Fluides Géophysiques, Université de Liège, Sart Tilman B6, B-4000 Liège (Belgium)

[1]Also at the Institut d'Astronomie et de Géophysique, Université de Louvain (Belgium)

ABSTRACT

Two-dimensional and one-dimensional models of mesoscale hydrodynamics are discussed with particular emphasis on the possibility of combining them to obtain a three-dimensional description of the currents in seas and lakes.
A particular attention is paid to the variable eddy viscosity multimode model developed by Nihoul (1977) and a numerical generalization of this model is presented in which, by successive iterations at each time step, the non-linear advection terms are taken into account in the one-dimensional Ekman model while the parameterization of the bottom stress is, if necessary, revised in the depth-averaged two-dimensional model according to the vertical structure.
An example of application to the North Sea is given in illustration.

INTRODUCTION

Mesoscale phenomena in seas and great lakes are characterized by time scales ranging from hours to days. They encompass inertial oscillations, tides, wind induced currents, storm surges and diurnal thermally induced fluctuations.

Their governing equations are obtained from the general Navier-Stokes equations by application of the Boussinesq approximation and the quasi-hydrostatic approximation (e.g. Nihoul, 1976). In this context, taking into account the different orders of magnitude of the horizontal and vertical velocity and length scales, one can generally neglect also the components of the Coriolis force where the horizontal component of the Earth's rotation vector appears, and the terms of horizontal turbulent diffusion as compared with the vertical turbulent diffusion.

The Boussinesq approximation is tantamount to assuming that the specific mass of sea water is a constant while its specific weight may be variable; small deviations of specific mass being there multiplied by the acceleration of gravity g , much larger than typical accelerations of the fluid.

The variations of the specific weight appear, in the hydrodynamic equations, as a vertical force, the "buoyancy", the magnitude of which must be regarded as an additional variable for which a supplementary equation is required.

In the scope of the Boussinesq approximation, one can relate buoyancy to temperature, salinity and turbidity variations and it is generally assumed that the three governing equations for temperature, salinity and turbidity can be combined in a single equation for buoyancy - Although this is obviously feasible when only one of the three variables (usually temperature) plays a significant role in the density variations, in the general case, it constitutes an additional approximation requiring further assumptions on the turbulent diffusion coefficients and the possibility of expressing volume sources of buoyancy (like the effect of radiation) in terms of buoyancy alone.

The system of equations governing mesoscale circulations, even in the simplest case of a single equation for buoyancy, is still a formidable problem. It is a system of five non-linear partial differential equations. In real situations, the boundaries may be very irregular (coasts, sea floor,...) and the boundary conditions are often partly inadequate (in particular, along open sea boundaries). The equations contain eddy diffusion coefficients which are unknown functions of space and time (and, possibly, of the velocity and buoyancy fields) and one of the first problem is to establish an adequate parameterization for them.

The solution of the three-dimensional time dependent equations of the mesoscale circulation does not seem to be possible, at this stage, without rather severe simplifications. If one takes, for instance, the paper by Freeman et al (1972) - one of the very few attempting to solve the complete three-dimensional model -, one finds dangerously restrictive hypotheses such as constant eddy viscosity, zero bottom stress, uniform depth (when buoyancy is taken into account), zero buoyancy (when depth's variations are included) and a constant wind stress whose relation with the wind velocity, incidently, is not correct.

Confronted with the complexity of the three-dimensional model, one naturally tries to reduce its size and one turns to situations which can be described by two-dimensional or one-dimensional models.

With their main interest in the vertical structure of currents and density, several authors, advocating the small value of the Rosby number in mesoscale flows $\left(0(10^{-1})\right)$, have neglected the non-linear advection terms. Since the horizontal diffusion terms are also negligible, the resulting equations known as the Ekman equations contain no derivative with respect to the horizontal coordinates x_1 and x_2 , except for the pressure gradient which appears as an unknown forcing term related essentially to the atmospheric pressure gradient and the sea surface slope.

Some authors (e.g. Welander, 1957) have attempted to find an analytical solution of the Ekman equations where forcing functions like pressure gradient and wind stress appear as kernels of convolution integrals.

Others have tried to eliminate the pressure gradient by considering not the horizontal current, but its deviation from either a geostrophic current (defined as to be driven by the pressure gradient) or a depth-averaged current.

Most of the models of the diurnal thermocline fall in this category (e.g. Niiler, 1977 ; Phillips, 1977 ; Kitaigorodskii, 1979).

Along a rather similar line, one can also differentiate with respect to the vertical coordinate x_3 and derive, from the Ekman equations, a complete set of three equations for the vertical shear

$$\underset{\sim}{\omega} = \frac{d\underset{\sim}{u}}{dx_3}$$ (where $\underset{\sim}{u}$ is the horizontal current vector) and buoyancy.

More interested in the general circulation pattern of a continental sea or a lake, many authors have restricted their attention to the horizontal distribution of surface slope and depth-averaged currents. When the water column is well mixed and buoyancy can be ignored, integration is carried from the bottom to the surface. In more complicated cases, several layers are treated separately and characterized by their depth-averaged properties. Depth-averaged models have been extensively applied in the recent years and detailed references can be found in numerous reviews and books (e.g. Nihoul, 1975 ; Cheng et al, 1976 ; Nihoul and Ronday, 1976).

The two kinds of models, local one-dimensional models and depth-averaged two-dimensional models, have their obvious limitations. The one-dimensional Ekman models are not applicable in certain regions (like the vicinity of tidal amphidromic points or in coastal zones) where the non-linear advection terms are not negligible (e.g. Ronday, 1976).

One can also show that these terms must be retained everywhere if the mesoscale circulation model is to be exploited to compute the residual macroscale circulation in tidal seas like the North Sea (Nihoul and Ronday, 1975).

The depth-averaged models allow only for a crude representation of stratification and give no information on the vertical profile of the horizontal current which may be rather essential in such fields as sediments transport, off-shore engineering, current meters data interpretation...

Moreover, neither the Ekman equations nor the depth-averaged equations constitute a closed system. At one stage or another, one-dimensional Ekman models cannot be pursued without a knowledge of surface elevation, geostrophic or mean current, bottom stress,... to materialize the results of an analytical solution or to formulate the boundary conditions, at the bottom for instance. Two-dimensional depth-averaged models, on the other hand, require a parameterization of the bottom stress (introduced in the equations by the vertical integration) and classical empirical formulas in terms of the depth-averaged velocity may not be entirely satisfactory, especially in particular situations like the reversal of tides in weak wind conditions (Nihoul, 1977).

In fact, it is obvious that the two types of models are complementary and should be run in parallel, following some appropriate iteration procedure.

MESOSCALE HYDRODYNAMIC EQUATIONS

In the scope of the approximations described in section 1, the equations of mesoscale marine hydrodynamics can be written

$$\frac{\partial \underset{\sim}{u}}{\partial t} + \nabla \cdot \underset{\sim}{u}\,\underset{\sim}{u} + \frac{\partial}{\partial x_3}\,(u_3 \underset{\sim}{u}) + f\,e_3 \wedge \underset{\sim}{u} = -\,\nabla q + \underset{\sim}{\psi} + \frac{\partial}{\partial x_3}\left(\tilde{\nu}\,\frac{\partial \underset{\sim}{u}}{\partial x_3}\right) \tag{1}$$

$$\nabla \cdot \underset{\sim}{u} + \frac{\partial u_3}{\partial x_3} = 0 \qquad ; \qquad \frac{\partial q}{\partial x_3} = -\,a \tag{2)(3}$$

$$\frac{\partial a}{\partial t} + \nabla \cdot (ua) + \frac{\partial}{\partial x_3}\,(u_3 a) = P_a + \frac{\partial}{\partial x_3}\left(\tilde{\lambda}\,\frac{\partial a}{\partial x_3}\right) \tag{4}$$

$$\frac{\partial \zeta}{\partial t} + \underset{\sim}{u} \cdot \nabla \zeta = u_3 \qquad \text{at} \qquad x_3 = \zeta \tag{5}$$

$$\underset{\sim}{u} = 0 \qquad , \qquad \left(\frac{\partial h}{\partial t} + \underset{\sim}{u} \cdot \nabla h = -\,u_3\right) \qquad \text{at} \qquad x_3 = -\,h \tag{6}$$

where the e_3-axis is vertical, pointing upwards with its origin at the reference sea level and where

$u = u_1 \, e_1 + u_2 \, e_2$ is the horizontal current velocity vector

u_3 is the vertical component of the three-dimensional velocity vector

$\nabla = e_1 \frac{\partial}{\partial x_1} + e_2 \frac{\partial}{\partial x_2}$ is the horizontal Nabla operator

f is the Coriolis parameter, twice the vertical component of the
  Earth's rotation vector

$q = \frac{p}{\rho_o} + g \, x_3$ (where p is the pressure, ρ_o the constant reference density and g the acceleration of gravity)

Ψ is the astronomical tidal force[*]

$\tilde{\nu}$ is the vertical eddy viscosity

a is the buoyancy $\left(a = - g \, \dfrac{\rho - \rho_o}{\rho_o}\right)$

P_a is the rate of buoyancy production

$\tilde{\lambda}$ is the vertical eddy diffusivity for buoyancy

ζ is the surface elevation

h is the depth

$h + \zeta = H$ is the water height

DEPTH-INTEGRATED AND MULTI-LAYER MODELS

The difficulty of solving the three-dimensional system of equations (1) to (4) has already been pointed out.

In the case of shallow well-mixed seas and lakes, assuming negligible buoyancy and renouncing the determination of the vertical variations, one usually integrates the equations over depth and restrict attention to the computation of the surface elevation and of the depth-averaged velocity field $\bar{u}$. Integration over depth, however, introduces the bottom stress into the equations.

The bottom stress (per unit mass of sea water) is defined as

---

[*]In seas and lakes, the astronomical tides are usually negligible with respect to wind induced currents and inertial oscillations, or incoming long waves produced by perturbations of the sea surface due to oceanic tides and atmospheric depressions (e.g. Ronday, 1976) Ψ is then neglected. (In the North Sea, for instance, surface elevations due to incoming ocean tides may range to several meters while astronomical tidal elevations never exceed a few centimeters).

$$\tau_{\underset{\sim}{b}} = \left(\underset{\sim}{\tilde{\nu}} \frac{\partial \underset{\sim}{u}}{\partial x_3} \right)_{x_3=-h}$$

and must be parameterized in terms of the mean velocity $\overline{\underset{\sim}{u}}$ although, from a physical point of view, it should really be expressed in terms of bottom currents.

Two-dimensional depth-integrated models can be improved to give some indications of vertical variations by considering different layers. Multi-layer models determine the depth-mean velocity of each layer and thus provide a staircase approximation of the velocity profile. They allow a parameterization of the bottom stress in terms of the mean current in the bottom layer but they introduce additional approximations such as interfacial friction coefficients in the boundary conditions at the interfaces between layers (e.g. Leendertse et al, 1973).

Moreover the volume of computation involved, when the number of layers increases, severely limits this number (or equivalently the number of vertical grid points in a three-dimensional attempt) and the variations in the vertical are very crudely represented.

Being limited in the number of layers, it seems reasonable to define them in relation with the vertical buoyancy structure (for instance a well-mixed layer above the diurnal thermocline and a stratified layer below). Unfortunately, with this definition, the interfaces between layers are not fixed and how they vary in time is very often poorly known and, in any case, very difficult to take into account (e.g. Cheng et al, 1976).

In most mesoscale phenomena, one can regard the non-linear advection terms as small except, probably, in localized regions where exceptionnally high velocities or rapid spatial variations significantly increase their order of magnitude.

If these terms are neglected, the mesoscale hydrodynamic equations can be transformed in many different ways into a one-dimensional system.

Some typical one-dimensional models are briefly[*] discussed in the following with particular emphasis on the possibility of

---

[*]The review has no ambition of being exhaustive and, far from drawing up an inventory, one intends to focus one's attention on a limited number of papers which serve to illustrate the discussion. In particular, studies of diurnal thermal fluctuations and the dynamics of the thermocline, which are not the main concern of this paper, will not be considered. An extensive review of them is presented in this volume by Kitaigorodskii.

combining such models with two-dimensional ones to obtain the full
three-dimensional picture.

VERTICAL SHEAR MODELS

Differentiating eq.(1) with respect to x_3 and neglecting the astronomical force Ψ^*, one gets

$$\frac{\partial \underset{\sim}{\omega}}{\partial t} + f \underset{\sim}{e}_3 \wedge \underset{\sim}{\omega} = \underset{\sim}{\nabla}a + \frac{\partial^2}{\partial x_3^2} (\tilde{\nu}\underset{\sim}{\omega}) \tag{8}$$

where

$$\underset{\sim}{\omega} = \frac{\partial \underset{\sim}{u}}{\partial x_3} \tag{9}$$

is the vertical shear vector.

Eq.(8) and eq.(4) (linearized) constitute a closed system for $\underset{\sim}{\omega}$ and a.

If one excepts estuarine and similar regions where horizontal gradients of buoyancy (related to horizontal salinity gradients, for instance) may play an important part, it is customary to neglect the horizontal gradient of a , regarding locally the marine system as "horizontally homogeneous".

In that case, eq.(8) can usually be solved for $\underset{\sim}{\omega}$ and eq.(4) for a , the eddy diffusivity $\tilde{\lambda}$ being eventually a function of $\|\underset{\sim}{\omega}\|$.

The velocity field u can then be derived from $\underset{\sim}{\omega}$ within a "constant of integration" (actually a function of x_1 , x_2 and t) which depends on the general circulation in the area.

The same result can be obtained by considering a "geostrophic current" $\underset{\sim}{u}_g$ independent of depth and solution of the equation

$$\frac{\partial \underset{\sim}{u}_g}{\partial t} + f \underset{\sim}{e}_3 \wedge \underset{\sim}{u}_g = - \underset{\sim}{\nabla}q \tag{10}$$

where $\underset{\sim}{\nabla}q$, in the hypothesis of horizontal homogeneity and after, integration of eq.(3), is given by

$$\underset{\sim}{\nabla}q = \underset{\sim}{\nabla} (\frac{p_a}{\rho_o} + g \, \zeta) \tag{11}$$

where p_a is the atmospheric pressure.

The unknown forcing term $\underset{\sim}{\nabla}q$ can then be eliminated by studying the velocity difference $\underset{\sim}{u} - \underset{\sim}{u}_g$.

Obviously the geostrophic current plays the same role as the "constant of integration" mentioned above.

_____
[*] Vertical variations of $\underset{\sim}{\Psi}$ are, in any case, entirely negligible (e.g. Ronday, 1976).

This type of approach has been used extensively in thermocline models (e.g. Niiler, 1977 ; Phillips, 1977 ; Kitaigorodskii, 1979).

The difficulty here resides in the expression of the boundary conditions. For instance, the value of ω at the bottom is related to the bottom stress which is either unknown or parameterized in terms of the depth-averaged velocity $\bar{u}$. The no-slip condition at the bottom will require that $\underset{\sim}{u}$ be zero, i.e. $\underset{\sim}{u} - \underset{\sim}{u}_g = - \underset{\sim}{u}_g$. Apart from writing a formal analytical solution, in both cases, there is an unknown function of x_1, x_2 and t ($\bar{u}$ or $\underset{\sim}{u}_g$) to determine separately.

ANALYTICAL MODELS

Assuming pseudo-horizontal homogeneity and neglecting the non-linear advection terms, one can, with more or less reasonable assumptions on the expression of the eddy viscosity $\tilde{\nu}$, derive an analytical solution of eq.(1) in terms of the unknown forcing term q (and Ψ^* if needed).

If Ψ^* is neglected, eq.(1) takes, in these conditions, the simple form

$$\frac{\partial \underset{\sim}{u}}{\partial t} + f \underset{\sim}{e}_3 \wedge \underset{\sim}{u} = - \underset{\sim}{\nabla}(\frac{p_a}{\rho_o} + g\zeta) + \frac{\partial}{\partial x_3} (\tilde{\nu} \frac{\partial \underset{\sim}{u}}{\partial x_3}) \qquad (12)$$

where eq.(11) has been used.

Eq.(12) is known as the Ekman equation.

A well-known solution of this type is the model of Welander (1957) Neglecting the spatial variations of a and the forcing function Ψ and assuming constant vertical eddy viscosity $\tilde{\nu}$, Welander seeks an analytical solution of eq.(12), with the initial conditions $\underset{\sim}{u} = 0$ for t = 0 and the boundary conditions

$$\begin{cases} \tilde{\nu} \frac{d \underset{\sim}{u}}{d x_3} = \underset{\sim}{\tau}_s & \text{at the surface} \qquad (13) \\ \\ \underset{\sim}{u} = 0 & \text{at the bottom} \qquad (14) \end{cases}$$

where $\underset{\sim}{\tau}_s$ is the wind stress (per unit mass of sea water).

Obtained by superposition of elementary solutions corresponding to Heaviside step functions forcing terms (e.g. Hidaka, 1933), the final solution of Welander appears as the sum of two convolution integrals with respective kernels q(t) and $\underset{\sim}{\tau}_s(t)$.

The velocity profile determined by Welander depends thus on the time history of the atmospheric pressure, the wind stress and the surface elevation. Obviously, the latter must be determined, some way or another, before one can exploit the result in practical applications.

In reality, the analytical solution of the Eckman equation must be regarded as a first step, paving the way to accurate two-dimensional modelling. Jelesnianski, for instance, (Jelesnianski, 1970) obtains the same solution as Welander by application of the Laplace transform. From the velocity profile, he derives expressions for the mean velocity $\bar{u}$ and the bottom stress τ_b . Eliminating the convolution integral with unknown kernel q , one can derive an expression for τ_b in terms of $\bar{u}$ and τ_s which can be used in a subsequent depth-averaged model.

As shown by Forristall, using a simplified version of Jelesnianski's integrals (Foristall, 1974), the two-dimensional depth-averaged model gives $\bar{u}$ and q and the latter can be substituted in Jelesnianski's formula to determine the velocity profile $u(x_3)$.

The main shortcoming of the Welander-Jelesnianski-Foristall approach is the severe hypothesis made on the vertical eddy viscosity ($\tilde{\nu}$ = constant) to obtain an analytical solution. This hypothesis is in contradiction with the observations and indeed the models fail to reproduce correctly the bottom boundary layer characteristics and the bottom stress turns out to be a linear function of the mean velocity and not a quadratic one, as it should.

The same objection can be made to all models constructed along the same line : too severe hypotheses, made at the beginning, handicap the whole model and the exploitation of the results in realistic situations. One can refer here to the model of Liggett and Hadjitheodorou (1969) (rectangular lake, constant eddy viscosity, steady state, no atmospheric pressure gradient), Gedney and Lick (1972) an improved version of the preceding one (constant eddy viscosity, steady state, no atmospheric pressure gradient), Witten and Thomas (1976) (steady state, no atmospheric pressure gradient and an unrealistic and rather unfortunate exponential form for the eddy viscosity).

MULTI-MODE MODELS

Generalizing the concept of vertical integration, Heaps (1972) suggested that the vertical variations could be elegantly taken into account by expanding the velocity u in series of eigenfunctions of the turbulent operator

$$\frac{d}{dx_3} \left(\tilde{\nu} \frac{d}{dx_3} \right)$$

The philosophy of this approach can be summarized as follows : substituting the series expansion in the Ekman equation, one gets a system of equations for the coefficients, each of which is a function of x_1, x_2 and t and satisfies a two-dimensional equation, in many ways similar to the two-dimensional equation of a depth-averaged model.

The final result is obtained by superposing the solutions of two-dimensional models, each of which corresponds to a different vertical mode (as compared with the multi-layer model where several two-dimensional models are solved simultaneously, each of them corresponding to a different vertical layer).

This method was applied to the Irish Sea by Heaps and Jones (1975). A slightly more sophisticated version was used by Davies (1977) in a numerical experiment (rectangular sea basin of constant depth) with the object of testing the sensitivity of the results to various parameters.

The difficulty here resides in the parameterization of the eddy viscosity $\tilde{\nu}$ and the determination of the corresponding eigenfunctions. Heaps and his co-workers were forced to introduce several simplifications regarding especially the eddy viscosity $\tilde{\nu}$ (assumed independent of x_3 and a function of x_1, x_2 and t such that the ratio $\tilde{\nu}/h$ is a constant) and the bottom stress (taken as a linear function of the bottom velocity assumed different from zero, in contradiction with the obvious no-slip condition)[*].

VARIABLE EDDY VISCOSITY MULTI-MODE MODEL

Most models described sofar can be labelled "1D + 2D" in the sense that they seek, by a preliminary one-dimensional model, some appropriate formulation of the vertical dependence of the velocity field to use subsequently in a complementary two-dimensional model. The almost inevitable hypothesis of constant eddy viscosity, in the preliminary analytical calculations, turns out to be a terrible embarrassment when exploiting the results of the one-dimensional model in two-dimensional modelling as, with an initially incorrect representation of the bottom boundary layer and of the bottom stress, one may doubt whether there will be,ultimately,real improvements in the two-dimensional forecast.

---

[*] In his numerical experiment, Davies (1977) was able to reduce the severity of some of the assumptions but then, even with a simple piecewise linear eddy viscosity, his solution is entirely numerical and the physical insight of the eigenfunction expansions is somewhat obscured

In a recent paper on tides and storm surges in well-mixed seas, Nihoul (1977) approached the problem in a contrary direction and recognizing that two-dimensional depth-integrated models, carefully calibrated, have been successfully applied to many lakes and seas and that the results are available, he suggested a locally one-dimensional multi-mode model using the predictions of the preexisting two-dimensional depth-integrated model to provide the local values of the forcing terms and of the boundary conditions.

Indeed, a depth-integrated model provides, at any point where one might desire the vertical current profile, the local surface elevation, mean velocity and associated bottom stress.

The only difficulty here is that, in two-dimensional models, the bottom stress is parameterized by a quadratic formula assuming that the stress is (except for a small wind stress correction) in the direction of the mean velocity. This type of parameterization is one of the things one would like to verify by a three-dimensional model. Although the quadratic law may be generally applicable, one suspects that it could be faulty in certain situations, such as tide reversals, when the mean current becomes very small.

However, it is readily seen that, the local wind stress being known, the two-dimensional model provides two additional boundary conditions instead of one for the second order Ekman equation (the velocity and the stress at the bottom). Nihoul (1977) thus derives an analytical solution of the local Ekman equation as a series of eigenfunctions of the turbulent operator (15) using the surface stress $\underset{\sim}{\tau}_s$ and the bottom stress $\underset{\sim}{\tau}_b$ as boundary conditions. The velocity deviation $\hat{\underset{\sim}{u}} = \underset{\sim}{u} - \bar{\underset{\sim}{u}}$ is obtained as a functional of $\underset{\sim}{\tau}_s$ and $\underset{\sim}{\tau}_b$ and the additional boundary condition ($\hat{\underset{\sim}{u}} = - \bar{\underset{\sim}{u}}$ at the bottom) is used to determine the relationship between $\underset{\sim}{\tau}_b$ and $\bar{\underset{\sim}{u}}$ and to verify the two-dimensional parameterization. In this model, which could be labelled "2D + 1D", the necessary matching of the one-dimensional and the two-dimensional models requires a more realistic parameterization of the eddy viscosity which is taken as a function of t, x_1, x_2 and x_3 and, in particular, respects the observed asymptotic form of the eddy coefficient in the bottom boundary layer.

The application of the model to the North Sea shows that, for typical values of tides and storm surges, the classical bottom friction law is valid over most of the tidal cycle but fails, in magnitude and in direction, during a comparatively short period of time, at tide reversal. This may be regarded as a validation of the depth-integrated model, the result of which can be used to determine the velocity profile at any grid point where the information might be requested (off-shore structure, current-meters mooring, sediment transport problem,...).

There is however, in principle, no difficulty in setting, for improved numerical forecasting, an iteration process by which the corrected relationship between $\bar{\underline{u}}$ and $\underline{\tau}_b$ is introduced in the two-dimensional model ; the two-dimensional model being run a second tim to feed back better values of ζ , $\bar{\underline{u}}$ and $\underline{\tau}_b$ into the one-dimensiona model, etc...

One can go a step further and include the non-linear advection terms in the iteration process. In this way, the combined 2D + 1D model is applicable everywhere and results in a truly non-linear three-dimensional model of mesoscale circulation.

This model is described in the next section.

NON-LINEAR THREE-DIMENSIONAL MODEL OF MESOSCALE CIRCULATION IN WELL MIXED SHALLOW SEAS

For simplicity, one shall restrict attention to well-mixed shallo seas like the North Sea and assume that buoyancy is negligible[*]. Th astronomical tidal force $\underline{\Psi}$ will also be neglected. (The justification is given in section 2).

In the integration of eq.(1) over depth, the non-linear advection term gives two contributions related respectively to the product of the mean values and the mean product of the deviations around the means. The second contribution is responsible for an enhanced horizontal diffusion (comparable to turbulence but generally more efficient) called "shear effect" diffusion (e.g. Nihoul, 1975).

Shear effect diffusion is essential in the depth-integrated dispersion equation of a passive scalar but, although it is larger than its turbulent equivalent, it remains generally negligible in the momentum equation as compared with the effects of the Coriolis force and the surface gradient (e.g. Ronday, 1976). There is no difficulty in including the shear effect in a 2D + 1D model. It is excluded in the following, for the sake of simplicity.

---

[*] The extension of the model to stratified seas is now under investigation. It requires a more subtle parameterization of the eddy viscosity which must be allowed to vary with the Richardson number and, at least, one more iteration loop introducing the effect of buoyancy in the pressure gradient and in the two-dimensional model.

Substracting the depth-averaged equation from eq.(1), one can derive an equation for the velocity deviation $\hat{\underset{\sim}{u}}$. It is convenient, at this stage, to change variable from x_3 to

$$\xi = \frac{x_3 + h}{H} \tag{16}$$

The variable ξ is taken to vary from 0 to 1, provided this does not create a singularity at the bottom in which case one must take the rugosity length z_O explicitly into account and set the lower limit at $\xi_O = \frac{z_O}{H} \ll 1$ ($\ln \xi_O \sim - 10$ in the North Sea).

The equation for $\hat{\underset{\sim}{u}}$ can then be written (Nihoul, 1977)

$$\frac{\partial \hat{\underset{\sim}{u}}}{\partial t} + f \underset{\sim}{e}_3 \wedge \hat{\underset{\sim}{u}} + \underset{\sim}{n} = \frac{\partial}{\partial \xi} \left(\frac{\tilde{\nu}}{H^2} \frac{\partial \hat{\underset{\sim}{u}}}{\partial \xi} \right) - \frac{\underset{\sim}{\tau}_s - \underset{\sim}{\tau}_b}{H} \tag{17}$$

where

$$\underset{\sim}{n} = \underset{\sim}{u} . \nabla \underset{\sim}{u} - \overline{\underset{\sim}{u}} . \nabla \overline{\underset{\sim}{u}} + H^{-1} \frac{\partial \underset{\sim}{u}}{\partial \xi} (1 - \xi) (\underset{\sim}{u} . \nabla h + u_3)$$

$$+ H^{-1} \frac{\partial \underset{\sim}{u}}{\partial \xi} \xi \left((\underset{\sim}{u}_s - \underset{\sim}{u}) . \nabla \zeta - (u_{3s} - u_3) \right) \tag{18}$$

the subscript s denoting surface values.

In his study of tides and storm surges in the North Sea, Nihoul (1977) has shown that the non-linear terms $\underset{\sim}{n}$ were almost everywhere negligible and that, according to observations, the eddy viscosity was well represented by an equation of the form

$$\tilde{\nu} = \kappa \| \tau_b \|^{1/2} H \lambda (\xi) \tag{19}$$

where κ is the Von Karman constant and λ a non-dimensional function of ξ behaving asymptotically as ξ for small ξ. (The eddy viscosity in the bottom layer reduces then to the well-known boundary layer expression $\tilde{\nu} = \kappa u_\ast z$ where $u_\ast$ is the friction velocity and $z = x_3 + h$).

Assuming a parabolic law for $\lambda^\ast$, Nihoul (1977) derived an analytical solution in the form of a multi-mode expansion giving $\hat{\underset{\sim}{u}}(\xi)$ as a functional of $\underset{\sim}{\tau}_s$ and $\underset{\sim}{\tau}_b$.

Using the results of the two-dimensional depth-integrated model of the North Sea (Ronday 1976), he computed the evolution with time of the velocity profile $\underset{\sim}{u}(t, x_1, x_2, x_3)$ at a series of representative horizontal grid points in the Southern Bight. The extra condition

$$\overline{\underset{\sim}{u}} = - \hat{\underset{\sim}{u}}(\xi_O) \tag{20}$$

[*] It was shown by Roisin (1977) that the final results were rather insensitive to the exact form of λ ; the cogent factor being its asymptotic behaviour $\lambda \sim \xi$ for $\xi \ll 1$.

was used as a test of consistency ; a significant discrepancy indic
ting a local influence of the non-linear terms or a temporary failu
of the parameterization of τ_b in the depth-integrated model (memo
effects for small $\bar{\underset{\sim}{u}}$).

At all the grid points where the study was conducted, Nihoul (19
found very little difference between the mean velocity calculated b
the depth-integrated model and calculated by the local multi-mode s
lution.

However, no point was taken in danger zones near coasts or tidal
amphidromic points and the model was not tested for other regions
than the Southern Bight and the North Sea or for different types of
mesoscale circulations or climatic conditions. The necessity was
thus felt of a generalization of the model based on an iteration pr
cedure by which the non-linear terms could be included in the one-
dimensional local Ekman model while the parameterization of the bot
stress could be revised in the two-dimensional model according to t
calculated vertical structure.

The two-dimensional model gives the components $\bar{u}_1$ and $\bar{u}_2$ of
the mean velocity and the surface elevation ζ at the points of a
staggered horizontal grid

$$\bar{u}_{2+}$$
$$.$$

$$\bar{u}_{1-} \qquad \zeta \qquad \bar{u}_{1+}$$
$$\text{o} \qquad \text{x} \qquad \text{o}$$

$$\bar{u}_{2-}$$
$$.$$

The integration of the equation for the velocity deviation $\hat{\underset{\sim}{u}}$ i
made at the ζ-point . At that point, the two components of the mea
velocity $\bar{\underset{\sim}{u}}$ are calculated by linear interpolation $\left(\text{the error is i}\right.$
$(\Delta x)^2$ and consistent with the order of the two-dimensional scheme$\left.\right)$
To compute the non-linear terms which contain horizontal derivatives
of $\hat{\underset{\sim}{u}}$, the velocity deviation must be calculated simultaneously at
nine ζ-points of the two-dimensional grid. (In the application to
the North Sea, these points form a rectangle of 20 km x 26 km) .

At each time step $t + \Delta t$, a first approximation $\hat{\underset{\sim}{u}}^1$ is calcula
ted numerically setting $\underset{\sim}{n} = \underset{\sim}{n}^1(t)$.

This approximation, combined with the results of the two-dimensio
model, is then used to compute $\underset{\sim}{n}^1(t + \Delta t)$. A second approximation
$\hat{\underset{\sim}{u}}^2$ is then calculated setting $\underset{\sim}{n} = \frac{1}{2}\left(\underset{\sim}{n}^1(t + \Delta t) + \underset{\sim}{n}^1(t)\right)$. With t

econd approximation, $n^2(t + \Delta t)$ is computed and the iteration con-
inues.

At the initial time, the first approximation is obtained taking
$= 0$ and the result of the analytical model can be used.

The numerical method is an extension of the compact hermitian me-
hod developed by Adam (1976).

As boundary conditions for the determination of $\hat{u}$, one imposes
he surface stress and the bottom stress (the latter obtained from
he two-dimensional model). The consistency condition

$$\bar{u} = - \hat{u} \qquad \text{at the bottom}$$

indicates if the model can proceed to the next time step or if a preliminary iteration must be performed on the parameterization of the
bottom stress.

As shown by Nihoul (1977), the classical bottom friction law constitutes the algebraic part of a differential relation where the additional terms (containing derivatives with respect to the time) are
negligible as long as the mean velocity $\bar{u}$ and the associated bottom
friction are sufficiently large. At small values of the mean velocity
however the correction may become comparatively important and, for
instance, the resulting bottom stress may have a different direction
from the mean velocity vector, even in the absence of wind.

At small velocities, the correction for the non-linear terms n
is not necessary and the result of the analytical multimode model can
be used to correct for the bottom stress parameterization.

The relationship between $\bar{u}$ and τ_b given by the multi-mode model
is rather complicated but Roisin (1977) has shown that it could be
written in much more convenient and simple forms and that, with a
good approximation, one could take

$$\tau_b = D^{1/2} \| \tau_b \|^{1/2} \bar{u} - m \tau_s + \gamma H \left(\frac{\partial \bar{u}}{\partial t} + f \, e_3 \wedge \bar{u} \right) \tag{21}$$

where D and m are the coefficients of the classical bottom friction law (D is the drag coefficient) and where γ is a numerical
factor.

If the consistency condition, $\bar{u} = - \hat{u}$ at the bottom, is not satisfied at a given time step t, eq.(21) can be introduced in a iteration scheme where the corrected value of τ_b at time t is computed in terms of the simultaneous value of $\bar{u}$ and τ_s and the imme-
diate past history of $\bar{u}$ $\left(\bar{u}(t-\Delta t) \right)$.

APPLICATION TO THREE-DIMENSIONAL MODELLING OF TIDES AND STORM SURGE
IN THE NORTH SEA

The three-dimensional model was applied to the calculation of t
and storm surges in the North Sea. The two-dimensional depth-integ
model was run for the whole North Sea with the horizontal grid use
Ronday (1976). The vertical profile of the velocity vector was ca
culated at a series of points in the Southern Bight. The test poi
were selected for their role in the existing management program of
North Sea and corresponded to places where current meters were to
moored or where the vertical variations of the current were desire
to study the deposition of silt, the distribution of eggs and larv
etc...

Although some of these points were close to the coast or in rela
vely shallow irregular depths, very few iteration stages were foun
necessary in most cases.

In many places, there is no significant improvement in taking th
non-linear advection terms into account in the calculation of $\tilde{u}$
and the variable eddy viscosity multi-mode model (Nihoul 1977) appe
to give very satisfactory results.

This is illustrated, in the following, by the results of the com
tation at the point 52°30'N 3°50'E. Fig. 1 shows the numerical gr
for the two-dimensional model of the Southern Bight of the North Se
($\Delta x_1 = 10^4$m ; $\Delta x_2 = 1.395 \ 10^4$m). The test point is indicated by a cir

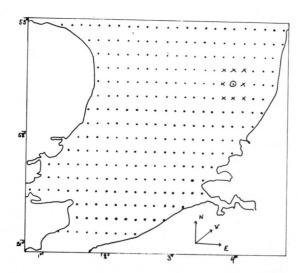

Fig. 1. Numerical grid for the two-dimensional depth integrated mode
showing the test point and the wind direction.

he depth there is 22 m   the rugosity length 1.10^{-3} m ($\xi_o \sim 5 \ 10^{-5}$).
he wind stress is oriented to the North-East as indicated in the
ower right-hand corner of the figure and it has a maximum magnitude
f 2 $10^{-4} m^2 sec^{-2}$.

In this exercise, the iteration was restricted to the introduction
f the non-linear advection terms. No correction was made on τ_b to
mphasize the comparison with the analytical multi-mode model (Nihoul,
977). Figs. 2-7 show the results of the model in which the non-
inear advection terms were included by successive iterations at each
ime step. One notices on figs. 6 and 7 that, near tide reversal,
ottom currents and surface currents may flow in opposite directions
nd that the bottom stress maintained by the current in the bottom
ayer is then difficult to relate to the mean current which may become
ery small.

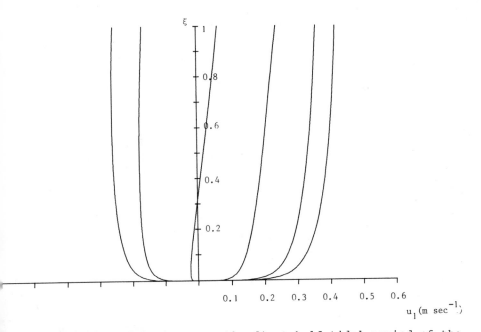

Fig.2. Evolution with time over the first half tidal period of the
       eastern component of the horizontal velocity vector (u_1) .
       The curves from right to left are vertical profiles computed
       at 54' interval.

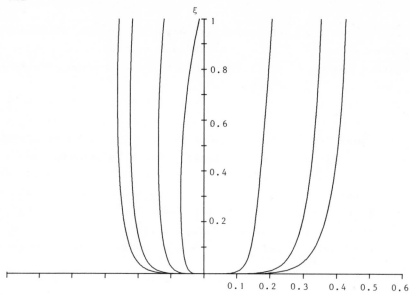

Fig. 3. Evolution with time, over the second half tidal period, of
the eastern component of the horizontal velocity vector (
The curves from right to left are vertical profiles comput
at 54' interval.

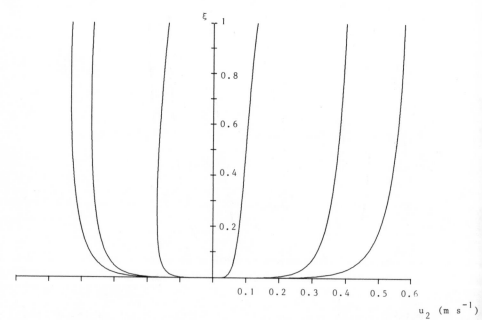

Fig. 4. Evolution with time, over the first half tidal period of th
northern component of the horizontal velocity vector (u_2) .
The curves from right to left are vertical profiles compute
at 54' interval.

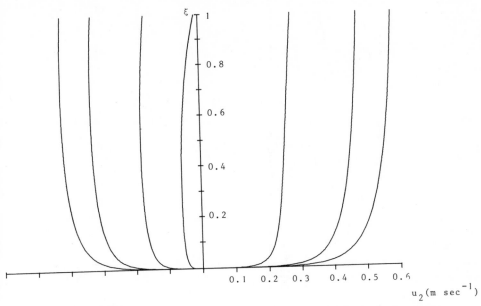

Fig. 5. Evolution with time, over the second half period of the
northern component of the horizontal velocity vector (u_2).
The curves from right to left are vertical profiles computed
at 54' interval.

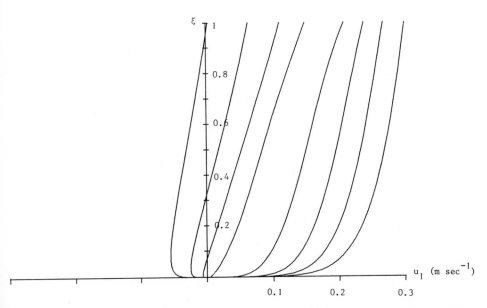

Fig. 6. Evolution with time, at tide reversal of the eastern component of the horizontal velocity vector (u_1) . The curves
from right to left are vertical profiles computed at 18'
interval.

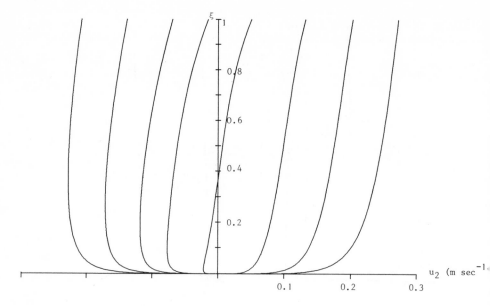

Fig. 7. Evolution with time, at tide reversal of the northern component of the horizontal velocity vector (u_2) . The curves
from right to left are vertical profiles computed at 18'
interval.

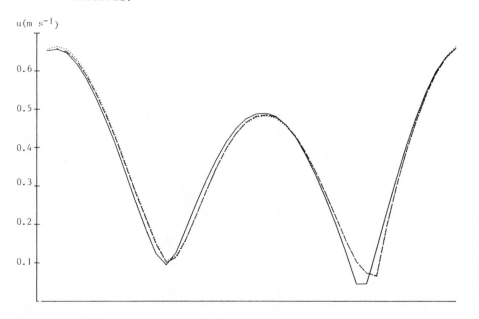

Fig. 8. Evolution with time over one tidal period of the magnitude
of the mean velocity $\underset{\sim}{u}$ computed respectively by the uncorrected two-dimensional depth integrated model (full line
——), the linear local model (dash line ---) and the non-
linear local model (dots ...) .

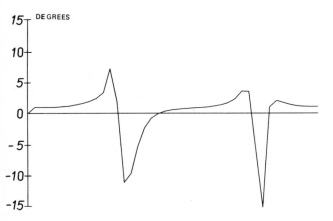

Fig. 9. Evolution with time over one tidal period of the difference
between the directions of the mean velocity computed respectively by the uncorrected two-dimensional model and by the
non linear model.

Figs. 10 and 11 show the veering of the horizontal velocity vector
over the water column. It must be noted here that most of the veering
occurs above $\xi \sim 0.1$. This is not obvious on the figures because
there is about the same number of points in the lower layer $\xi \leq 0.1$
and the upper layer $0.1 \leq \xi \leq 1$. This is actually an artifice of
the numerical method which, for increased accuracy, introduces the
change of variable $y = \ln\left(\frac{\xi}{\xi_o}\right)$ in the bottom layer $(\xi_o \leq \xi \leq 0.15)$
where the vertical gradients are large. The same remark applies to
fig. 12.

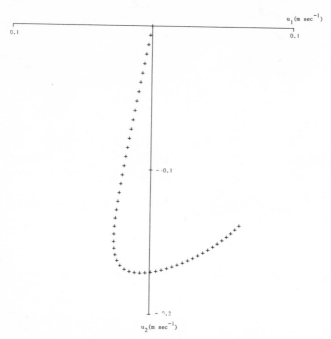

Fig. 10. Vertical veering of the horizontal velocity vector at tide reversal

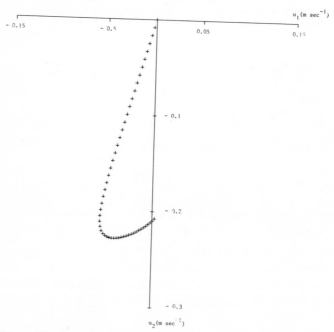

Fig. 11. Vertical veering of the horizontal velocity vector 18' after tide reversal

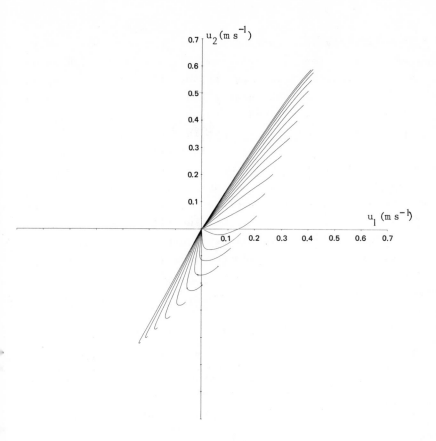

Fig. 12. Evolution with time over the first half tidal period of the
Ekman diagram showing the vertical veering of the horizontal
velocity vector. The separation between two successive curves is 18' .

Fig. 12 shows that, as long as the mean velocity is sufficiently
large, the horizontal velocity vector keeps the same direction from
the bottom to the surface. In that case, the classical bottom friction law applies and it may be assumed that the bottom stress is
roughly (except for a small wind effect) parallel to the mean velocity. On the other hand, when the mean velocity becomes small, near
tide reversal, there is a noticeable rotation of the velocity vector
between its directions in the bottom and the surface layers.

During a comparatively short period of time, one may thus expect the classical parametrization of the bottom stress to fail both in magnitude and direction. As described above, this may be corrected at each time step in the critical interval by an iteration procedure.

The iteration on the non-linear terms does not, on the other hand appear to bring significant improvement in the example considered. The effect may of course be larger at other grid points but, in general, it seems to constitute only a minor correction.

In large-scale practical applications of the three-dimensional model, one would thus be wise to define realistically limiting values of $\bar{u} + \hat{u}(\xi_o)$ and $\underset{\sim}{n}$ which one is willing to tolerate and instruct the model to proceed without any vainly expensive iteration whenever the calculated values do not exceed the limits of tolerance.

REFERENCES

Adam, Y., 1977. Highly accurate compact implicit methods and boundary conditions. J. Comput. Phys., 24:10-22.

Cheng, R.T., Powell, T.M. and Dillon, T.M., 1976. Numerical models of wind driven circulation in lakes, Appl. Math. Modelling, 1:141-159

Davies, A.M., 1977. The numerical solution of the three-dimensional hydrodynamic equations, using a B-spline representation of the vertical current profile. In: J.C.J. Nihoul (Editor), Bottom Turbulence. Elsevier Publ. Co, Amsterdam, 1-48.

Forristall, G.Z., 1974. Three-dimensional structure of storm generated currents. J. Geophys. Res., 79:2721-2729.

Freeman, N.G., Hale, A.M. and Danard, M.B., 1972. A modified sigma equations'approach to the numerical modelling of great lakes hydrodynamics. J. Geophys. Res., 7:1050-1060.

Gedney, R.T. and Lick, W., 1972. Wind-driven currents in lake Erie. J. Geophys. Res., 77:2714-2723.

Heaps, N.S., 1972. On the numerical solution of the three-dimensional hydrodynamical equations for tides and storm surges. Mem. Soc. R. Sci. Liège, 2:143-180.

Heaps, N.S., 1975. Storm surge computations for the Irish Sea using a three-dimensional numerical model. Mem. Soc. R. Sci. Liège, 7: 289-333.

Hidaka, K., 1933. Non-stationary ocean currents. Part I. Mém. Imp. Mar. Obs. Kobe, 5:141-266.

Jelesnianski, C.P., 1970. Bottom stress time-history in linearized equations of motion for storm surges. Mon. Weather Rev., 98:462-478.

Kitaigorodskii, S.A., 1979. Review of the theories of wind-mixed layer deepening. In: J.C.J. Nihoul (Editor), Marine Forecasting. Elsevier Publ. Co., Amsterdam, 1-33.

Leendertse, J.J., Alexander, R.C. and Liu, S., 1973. Rand Report R-1417 - OWRR.

Liggett, J.A. and Hadjitheodorou, C., 1969. Ann. Soc. Civil Eng. J. Hydr. Div., 95:609-617.

Niiler, P.P., 1977. One-dimensional models of the seasonal thermocline In: E.D. Goldberg, I.N. McCave, J.J. O'Brien, J.M. Steele (Editors) The Sea. Wiley Interscience Publ., New-York, 6:97-115.

Nihoul, J.C.J., 1975. Modelling of marine systems. Elsevier Oceanography Series 10, Elsevier Publ. Co., Amsterdam, 272 pp.
Nihoul, J.C.J., 1976. Modèles mathématiques et Dynamique de
  l'Environnement. Ele Publ. Co., Liège, 198 pp.
Nihoul, J.C.J., 1977. Three-dimensional model of tides and storm
  surges in a shallow well-mixed continental sea. Dyn. Atmos. Oceans,
  2:29-47.
Nihoul, J.C.J. and Ronday, F.C., 1975. The influence of the tidal
  stress on the residual circulation. Tellus, 27:484-489.
Nihoul, J.C.J. and Ronday, F.C., 1976. Hydrodynamic models of the
  North Sea, Mém. Soc. R. Sc. Liège, 10:61-46.
Phillips, O.M., 1977. Entrainment. In: E.B. Kraus (Editor), Modelling
  and Prediction of the Upper Layer of the Ocean. Pergamon Press,
  Ch.7.
Roisin, B., 1977. Modèles tri-dimensionnels des courants marins.
  Rep. ACN 3. Ministry for Science Policy Brussels, 124 pp.
Ronday, F.C., 1976. Modèles hydrodynamiques. In: J.C.J. Nihoul (Editor),
  Publ. Ministry for Science Policy Brussels, 3, 270 pp.
Welander, P., 1957. Wind action on a shallow sea : some generalizations
  of Ekman's theory, Tellus, 9:45-52.
Witten, A.J. and Thomas, J.H., 1976. Steady wind-driven currents in
  a large lake with depth-dependent eddy viscosity. J. Phys. Oceanogr.
  6:85-92.

IRREGULAR-GRID FINITE-DIFFERENCE TECHNIQUES FOR STORM SURGE CALCULATIONS FOR CURVING
   COASTLINES

W. C. THACKER

Atlantic Oceanographic and Meteorological Laboratories, National Oceanic and Atmospheric Administration, Miami, Florida, 33149, U.S.A.

ABSTRACT

   Finite-difference computations on irregular grids offer the advantages of resolving
the coastal curvature and of allowing for an explicit time step of optimal size. A
further advantage is the flexibility associated with editing the grid, which allows
for improvements in its design. The techniques are based on the approximation of partial derivatives with slopes of planar surfaces associated with triangular components
of the irregular grid. Simulations of several surges associated with various hypothetical hurricanes passing through Mobile Bay are realistic and relatively noise free.
Because these techniques are computationally efficient, they provide a practical tool
for forecasting storm surges.

INTRODUCTION

   Previous approaches to the problem of including curving coastlines into calcula-

tions of storm surges have involved either the use of transformed coordinates (e.g.,

Wanstrath et al., 1976) or the use of finite-element techniques (e.g., Pinder and

Gray, 1977). The method suggested here is to perform finite-difference calculations

directly on an irregular grid that can smoothly represent the coastline. This offers

the advantages of greater flexibility than the coordinate transformation method and

greater economy than the finite-element method.

   The plan of this presentation is, after first discussing features of the irregular

computational grid, the approximation of derivatives, and the equations used for cal-

culating storm surges, to present the results of computations for several hypotheti-

cal storms in the forms of sea surface elevation contour maps, flow vector maps, and

graphs of water elevations along the coastline, with concluding remarks concerning

the use of irregular grids in conjunction with the problem of inundation.

THE IRREGULAR GRID

   The computational grid for Mobile Bay, shown in Fig. 1, illustrates some of the

features that make irregular-grid finite-difference techniques attractive for use in

forecasting storm surges. Most obvious is the excellent representation for the curv-

ing coastline by the irregular grid; this is a significant improvement over the

MOBILE BAY AND VICINITY

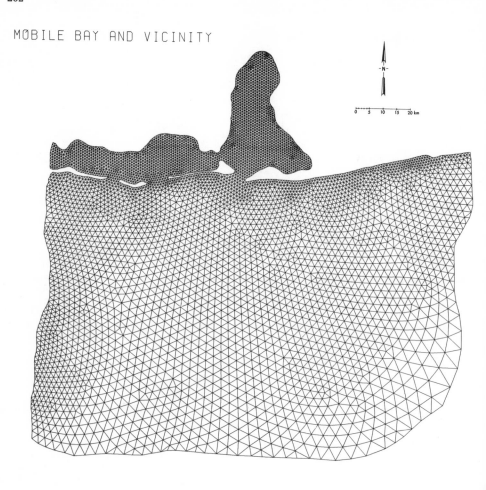

Fig. 1. Irregular grid for computing storm surges in Mobile Bay.

"stair-step" boundary of conventional uniform grids because it is at the coastline
that the forecast value of the water surface elevation is most needed. Since one
continuous grid serves for both the inland waters and for the open sea, the calculations can be made simultaneously for both regions, thereby avoiding the question of
how properly to specify boundary conditions where the inland waters meet the open
sea. The fact that the grid spacing is roughly proportional to the square root of
the basin depth (Fig. 2) reflects the fact that the storm surge is essentially a
shallow water wave with phase velocity proportional to the square root of the water
depth. It is the ratio of the grid spacing to the basin depth which limits the size
of the explicit computational time step in order to guarantee numerical stability.
Unlike the uniform grid, for which the time step is determined in the deepest water
(where this ratio is minimum and which is the least important region for the storm

DEPTH CONTOURS
(Meters)

Fig. 2. Contours of stillwater depth values used in storm surge calculations.

surge forecast), the irregular grid has a time step which is no smaller than necessary for calculations in inland waters. This distribution of grid points has the
additional advantage that it gives the highest density of points along the shoreline,
aiding in the resolution of coastal curvature.

A piecewise uniform spliced grid, such as the one for the Elbe Estuary (Ramming,
1976) shown in Fig. 3, has to some extent the same attractive features. It allows
for calculations simultaneously for inland waters and for the open sea, and the larger
grid spacing in deeper water has a beneficial effect on the time step; however, the
grid refinement scheme dictates the density of grid points in each portion of the
grid, and the "stair-step" boundary provides a rough representation of the shoreline.
Such a spliced grid (Thacker, 1976) provided the motivation for the irregular-grid
finite-difference techniques. Just as linear interpolation can be successfully used

to calculate derivatives at the "extra" points along the splices, it should also pr
vide a means for calculating derivatives at points on an irregular grid.

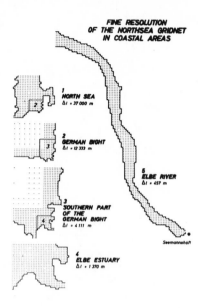

Fig. 3. Piecewise uniform spliced grid for the Elbe Estuary (Ramming, 1975).

The fact that the grid points are connected by line segments to form a mosaic of
triangular elements (Fig. 1) is reminiscent of similar grids used in finite-element
calculations (see, for example, Pinder and Gray, 1977). This similarity is due to
the fact that the techniques discussed here as well as those of the finite-element
method involve linear interpolation over triangular elements. The fundamental distinction is that the finite-element method is based directly upon approximation of
the functions, whereas the finite-difference method is based upon approximation of
the derivatives. The practical distinction is that the finite-difference technique
provide greater computational economy. The spatial averages (Thacker, 1978a and
1978b) that result from the finite-element method necessitate a matrix inversion
at each time step. In addition to this computationally expensive matrix inversion,
these averages lead to greater storage requirements, to a greater number of arithmet
operations per time step, and to a smaller value for the length of the time step tha
required by the corresponding finite-difference calculations.

Because the computational grid is irregular, only one index is used to specify th
grid points rather than two indices corresponding to distances along coordinate axes
as for the conventional uniform grids. Since the grid point index is neither simply
related to the coordinates of the grid point nor to the indices of neighboring point
that provide values necessary for evaluating derivatives, this information must be

tabulated for computation. Also, the differentiation coefficients are not simply the inverse of the grid spacing as for uniform grids. Since they vary from grid point to grid point, either they must be tabulated or they must be calculated from the tabulated values of the coordinates and the indices of neighboring points each time they are needed.

Because the manner in which the grid points are indexed is unimportant, it is a simple matter to alter the grid in order to add additional points, to remove points, or to respecify neighbors. After editing the grid, it is also a simple matter to sort and renumber the grid points for computational efficiency. The scheme used here assigns indices to the interior points first, the lowest for interior points with six neighbors, next for those with five, and then for those with seven, and assigns indices to the boundary points last, also according to the number of neighboring points. Additional editing (Thacker, 1977) guarantees that each interior grid point is situated at the geometric center of the polygon formed by the neighboring grid points. These editing procedures can also be used for finite-element grids so long as the matrix inversions are calculated by an iterative technique, but if direct inversion techniques are used, finite-element grids should be numbered so that the differences between the indices of neighboring points be as small as possible. For storm surge calculations the previous time step provides excellent values for initializing the iterative techniques, so they should be efficient as well as flexible.

APPROXIMATION OF DERIVATIVES

The slope of the spinnaker-shaped surface in Fig. 4 can be approximated by the slope of the planar surface determined by points a, b, and c. Of course, for smaller curvature the approximation is better. The planar surface is a linear interpolating function, and its derivatives provide approximations of the function specifying the curved surface,

$$\frac{\overline{\partial f}}{\partial x} = \frac{f_a(y_b - y_c) + f_b(y_c - y_a) + f_c(y_a - y_b)}{\Delta}$$

$$\frac{\overline{\partial f}}{\partial y} = -\frac{f_a(x_b - x_c) + f_b(x_c - x_a) + f_c(x_a - x_b)}{\Delta}$$

$$\Delta = x_a(y_b - y_c) + x_b(y_c - y_a) + x_c(y_a - y_b).$$

In the storm surge calculation the function f can represent the x- and y-components of the vertically integrated horizontal velocity, $\vec{U}$, and the surface elevation, H.

Since the dynamical variables are calculated at the grid points which are vertices of triangles, there is no reason for preferring the approximations corresponding to one adjacent triangle over those corresponding to any other. For this reason, the

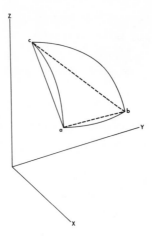

Fig. 4. The slope of the plane passing through points a, b, and c approximates
the slope of the curved surface. The plane represents the interpolating
function with derivatives that approximate the derivatives of the curved
surface.

Interior

N = 5   N = 6   N = 7

Boundary

N = 3   N = 4   N = 5   N = 6

Fig. 5. The approximations for derivatives at points on the irregular grid are
averages of the approximations obtained from the adjacent triangles.
For interior points, the approximations are centered, involving only
values associated with the N neighboring points and not the value at
the point for which the derivative is evaluated. For points on a
boundary, the approximations are "one-sided", with the values at the
grid point contributing to the evaluation of the derivative.

derivatives at a grid point are approximated by averages of the contributions from all adjacent triangles weighted according to their area, Fig. 5. The resulting N-point formulas (Thacker, 1977) are equally as simple as the three-point formulas for the slope of the surface in Fig. 4. For example, if there are five points contributing to the approximation, then the formulas are,

$$\frac{\overline{\partial f}}{\partial x} = \frac{f_a(y_b-y_e) + f_b(y_c-y_a) + f_c(y_d-y_b) + f_d(y_e-y_c) + f_e(y_a-y_d)}{\Delta}$$

$$\frac{\overline{\partial f}}{\partial y} = -\frac{f_a(x_b-x_e) + f_b(x_c-x_a) + f_c(x_d-x_b) + f_d(x_e-x_c) + f_e(x_a-x_d)}{\Delta}$$

$$\Delta = x_a(y_b-y_e) + x_b(y_c-y_a) + x_c(y_d-y_b) + x_d(y_e-y_c) + x_e(y_a-y_d).$$

In every case the numerators are given by cyclic sums of products of the values of the function at the grid points with the differences of coordinates at adjacent points, and the denominator is twice the area of the polygon formed by the N points. For regular polygons, such as the square and the hexagon shown in Fig. 6, the formulas reduce to the familiar expressions,

$$\frac{\overline{\partial f}}{\partial x} = \frac{f_a-f_c}{x_a-x_c}$$

$$\frac{\overline{\partial f}}{\partial x} = \frac{1}{6}\left[4\left(\frac{f_a-f_d}{x_a-x_d}\right) + \frac{f_b-f_c}{x_b-x_c} + \frac{f_f-f_e}{x_f-x_e}\right]$$

and

$$\frac{\overline{\partial f}}{\partial y} = \frac{f_b-f_d}{y_b-y_d}$$

$$\frac{\overline{\partial f}}{\partial y} = \frac{1}{2}\left[\frac{f_b-f_f}{y_b-y_f} + \frac{f_c-f_e}{y_c-y_e}\right].$$

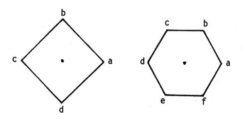

Fig. 6. For uniform grids, with points in square or hexagonal arrays such as these, the N-point formulas for approximating derivatives reduce to simple, recognizable expressions.

When the shallow water wave equations are discretized to obtain equations for the values of $\vec{U}_i^{n+\frac{1}{2}}$ and H_i^n, corresponding to the transport and surface elevation for grid point i and time level n, the partial derivatives are approximated by the appropriate N-point formulas. Only at the boundary (see Fig. 5) is the point at which the derivative is approximated also one of the N points contributing to the approximation.

GOVERNING EQUATIONS

The hydrodynamic equations governing the storm surge,

$$\frac{\partial \vec{U}}{\partial t} + \vec{\nabla} \cdot \frac{\vec{U}\,\vec{U}}{D} = - gD\vec{\nabla}H - f\hat{k}x\vec{U} + \vec{T} - \vec{B}$$

$$\frac{\partial H}{\partial t} = - \vec{\nabla} \cdot \vec{U} \; ,$$

account for the atmospheric forcing through the term, $\vec{T}$, and for the bottom friction through $\vec{B}$. The term involving the Coriolis parameter, f, and unit vector in the vertical direction, $\hat{k}$, account for the earth's rotation which has a relatively small influence on the storm surge. The term involving the gravitational acceleration, g, and the water depth, D, accounts for flow in response to slope in the sea surface. The flow accelerates in response to these forces and the sea surface rises as the flow converges.

The wind velocity and pressure gradient fields for the hurricane forcing are taken to be the same as those used by Overland (1975) for Apalachicola Bay,

$$\vec{W} = \frac{2rR}{r^2+R^2} W_{max} \hat{\phi} + \frac{rR}{r^2+R^2} \vec{S}$$

$$\vec{\nabla}P = - \Delta P \frac{R}{r^2} \exp\left(-\frac{R}{r}\right)\hat{r} \; .$$

The velocity field has two components; one is circularly symmetric with maximum value, W_{max}, at distance, r = R, from the storm's center and with inflow angle specified by the unit vector, $\hat{\phi}$, and the other approximates the assymmetry of the storm associated with its translational velocity, $\vec{S}$. The value of W_{max}, depending upon the values of the radius, R, and of the pressure drop, ΔP, used to specify the storm is determined (see Fig. 7) as in the SPLASH model (Jelesnianski, 1967) used by the National Weather Service for forecasting storm surges. The symmetric part of the wind speed, the inflow angle with maximum of 22° at 3R and 17° at large r, and the pressure gradient inward along the radial direction vary as indicated in Fig. 8. The hurricane forcing associated with these fields is given by

$$\vec{T} = C_d \vec{W}|\vec{W}|\rho_a/\rho_w - D\vec{\nabla}P/\rho_w$$

where ρ_a and ρ_w are the densities of air and water and where the drag coefficient has the value used in the SPLASH model, $C_d = 2.4 \times 10^{-3}$ at all wind speeds.

Whereas the SPLASH model uses time-history bottom stress, the more conventional quadratic stress is used here,

$$\vec{B} = g\vec{U}|\vec{U}|/C_H^2 D^2 \; ,$$

with the Chezy coefficient, $C_H = 62 \; m^{\frac{1}{2}}/sec$.

The mathematical specification is completed by the boundary conditions requiring

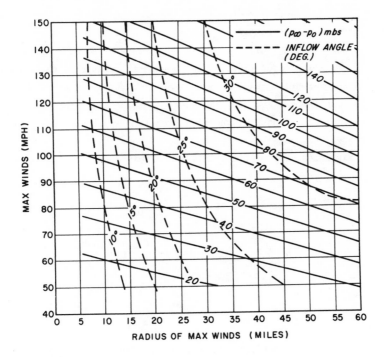

Fig. 7. This nomogram (Jelesnianski, 1967) can be used to obtain the value of
the maximum hurricane wind velocity from the values of the radius to
maximum winds and the pressure drop. Tabulated values as used by SPLASH
were used for computation.

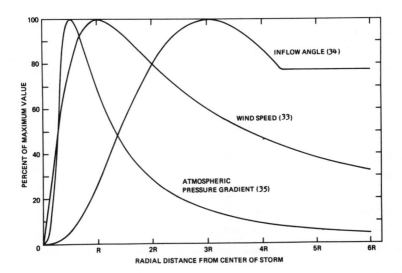

Fig. 8. Variation of hurricane wind speed, inflow angle, and pressure gradient
with radial distance from center of storm (Overland, 1975).

that there be no flow normal to the shoreline and that the surface elevation along boundaries separating the portion of the sea included in the computation from that which is excluded be that height of water supported by the atmospheric pressure drop

The finite-difference equations, which govern the values of the dynamical variabl at points on the irregular grid, have a "leap-frog" time structure with values for t transport vectors and surface elevation corresponding to different time levels separ ated by $\tau/2$, where the length of the time step is $\tau = 2.5$ minutes. Except for point on the boundary, which must satisfy the imposed boundary conditions, the values of the dynamic variables at the grid points are obtained from the equations

$$\frac{1}{\tau}\left(\vec{U}_i^{n+\frac{1}{2}}-\vec{U}_i^{n-\frac{1}{2}}\right) + \frac{\vec{U}_i^{n+\frac{1}{2}}}{D_i^n} \cdot \left(\overline{\vec{\nabla}\,\vec{U}}\right)_i^{n-\frac{1}{2}} + \vec{U}_i^{n+\frac{1}{2}}\left(\overline{\vec{\nabla}\cdot\frac{\vec{U}_i^{n-\frac{1}{2}}}{D^n}}\right)_i$$

$$= - gD_i^n\left(\overline{\vec{\nabla} H}\right)_i^n - f\hat{k}x\vec{U}_i^{n+\frac{1}{2}} + T_i^n - \frac{g\vec{U}_i^{n+\frac{1}{2}}|U_i^{n-\frac{1}{2}}|}{\left(C_H D_i^n\right)^2}$$

$$\frac{1}{\tau}\left(H_i^{n+1}-H_i^n\right) = -\left(\overline{\vec{\nabla}\cdot\vec{U}}\right)_i^{n+\frac{1}{2}} \;.$$

For those points corresponding to the coastline, the momentum equation must be al- tered to prevent flow normal to the coastline. The right-hand side, which represent the forcing, must be projected onto the line tangent to the boundary determined by the unit vector $\hat{b}_i = (\vec{x}_a-\vec{x}_c)/|\vec{x}_a-\vec{x}_c|$, where $\vec{x}_a$ and $\vec{x}_c$ are the coordinates of the point which are neighbors of point i = b lying on the boundary (see Fig. 9). This is done by taking the inner product of the right-hand side with the dyadic, $\hat{b}_i\hat{b}_i$. For those points on the computational boundary not corresponding to a coastline, the atmospheric pressure determines the value of the surface elevation at each time step

The position of the storm at the nth time step and the velocity of the storm are calculated from specified coordinates for the center of the storm at two different times, which might correspond to the forecast value for the storm to reach a desig- nated point in the vicinity of the bay and the time that the forecast is issued. From the position of the center, the values of the distances to each grid point and the values of the wind velocity and pressure gradient can be calculated in order to evaluate the storm forcing, $\vec{T}_i^n$, at the nth time step. The computations are initial- ized with a flat motionless sea, and the storm remains at the initial position for one hour (24 time steps) as it grows linearly to full strength.

For numerical stability the Coriolis term is evaluated at the time level, $n+\frac{1}{2}$. The bottom friction and advection terms involve both levels, $n+\frac{1}{2}$ and $n-\frac{1}{2}$. However, the equations can easily be rewritten with the values at step $n+\frac{1}{2}$ given explicitly in terms of values at steps n and $n-\frac{1}{2}$. Also in order to guarantee numerical stability, the advection terms are omitted for points on the open boundaries.

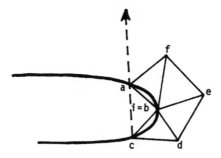

Fig. 9. Points c, d, e, f, and a are neighboring points for boundary point i = b.
The direction of flow at this point is parallel to the unit vector along
a line through the neighbors a and c which are also on the boundary.

COMPUTATIONAL RESULTS

Storm surges which might result from various hypothetical storms have been simulated in order to ascertain that the irregular-grid finite-difference techniques are
indeed capable of producing reasonable results for realistic circumstances. Four
cases are considered, corresponding to the series of Figs., 10, 11, 12, and 13.

The first case corresponds to a hurricane with R = 56 km and ΔP = 100 mb, moving
from (29°30'N, 88°30'W) to (30°40'N, 88°W) in three hours and continuing inland on
the same course for two additional hours. Although this pressure drop might be reasonable for extremely strong hurricanes in this area, the radius and forward speed
are both larger than would be expected. Calculations for this case neglected the
Coriolis and advection terms and used the stillwater rather than the total depths.

Fig. 10 shows a contour map of sea surface elevation after three hours of simulation with the storm just north of Mobile Bay. Since the line segments composing these
contours represent linear interpolations of the computed values at the grid points
with no smoothing, the smoothness of the contour lines accurately reflects the low
level of noise in the computations. The corners of the islands, which correspond to
curvature that is too great to be resolved by the boundary points of the grid, might
be responsible for some of this computational noise. This level of noise is typical
of these calculations and does not seem to increase on the time scale of these simulations.

The second case corresponds to a large hurricane which might be expected to hit
Mobile Bay. It has R = 24 km and ΔP = 100 mb, and it moves due north from (29°30'N,
88°W) to (30°30'N, 88°W) in five hours and continues north for two more hours. For
this case, the Coriolis term is included and the total depths are used. The Coriolis
term has little effect and can be dropped for computational efficiency. The total
depth does give somewhat diferent results than the stillwater depth, especially for

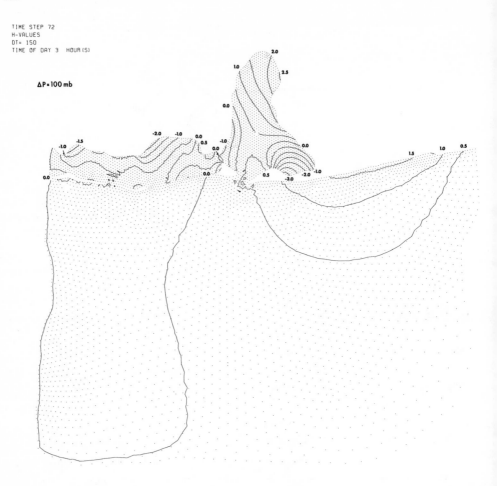

Fig. 10. The smoothness of these surface elevation contours is indicative of the low level of noise in these computations.

extreme water levels. No attempt has been made to guarantee that the total depth is always positive, but this is usually the case, even though the calculations can lead to negative elevations greater in magnitude than the stillwater depth of the basin.

Figs. 11a through 11d show vector maps of depth-averaged velocity ($\vec{U}_i/D_i$) for times corresponding to 2, 4, 5, and 7 hours of simulations for case two. The symbol § marks the position of the hurricane on the map, but it is absent from Fig. 11d because the hurricane is too far north to be represented. The smooth variation of the vectors from grid point to grid point are another indication of the low noise level in spite of the high curvature of the islands and the constraint that the flow vectors be parallel to the computational boundary, the flow through the inlets is well represented. The flow through the open computational boundary also seems to be qui-

reasonable. The fact that the corner points of the grid, where land and sea boundaries meet, are taken to be points in the sea with no flow restriction accounts for the unusual behavior of the velocity vectors at these points. This is easily corrected but should have little influence on the rest of the calculation. Figs. 11e through 11h show the variation in computed water levels for points within Mobile Bay and Mississippi Sound and along the Gulf of Mexico coast. Points on the map indicated by letters, A-X, are also indicated on the horizontal axes of the graphs.

It is interesting to note in Fig. 11a that the hurricane has moved ahead of the oceanic gyre. A corresponding gyre is not evident in subsequent maps because of the strong effect of the land boundaries on the flow. Another interesting point (Fig. 11g) is that when the storm is directly over the bay, the water levels are negative in most of the bay, because the storm had pushed the water into the Gulf and Mississippi Sound (Figs. 11a and 11b).

Case three corresponds to exactly the same storm as for case two. For this case the Coriolis term is neglected and the stillwater depth is used, just as for case one, but the advection terms are also included. The advection terms seem to introduce some noise into the computation, but they are small and have little other effect, so it seems best to neglect these terms. For this case the water elevations at the coastline are shown in Figs. 12a through 12d, which can be compared with Figs. 11e through 11h.

Case four corresponds to a storm of the same size and strength as for case two and to the same terms used in the computations. For this case the storm moves towards the east from $(30^{\circ}15'N, 88^{\circ}45'W)$ to $(30^{\circ}15'N, 87^{\circ}45'W)$ in four hours (Fig. 13a). Figs. 13a-13c show that as much as 5 m of storm surge might be expected within Mobile Bay and Mississippi Sound for such a storm.

CONCLUSION

The irregular-grid finite-difference techniques presented here provide a simple and economical means for forecasting storm surges in bays, estuaries, and lakes, so long as the elevation of the adjacent land is sufficiently high that inundation is unimportant. It might be further anticipated that an irregular grid might also be advantageous if inundation is included in the calculations. Rather than for flooding to proceed square by square as for a uniform grid, it would proceed triangle by triangle. Since the grid points can be positioned according to the elevation of the land, the triangles should more nearly approximate areas being flooded. For situations in which roads through low-lying areas are built up, forming barriers to the intruding water, the triangular grid might be especially appealing. How to proceed with such calculations, including inundation, remains to be studied.

274

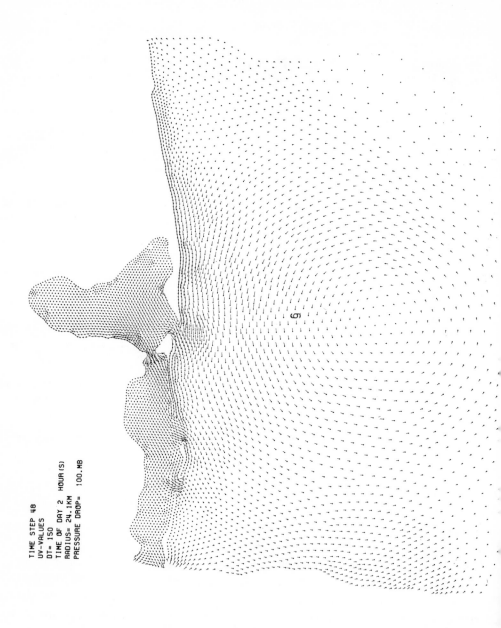

TIME STEP 48
UV-VALUES
DT= 150
TIME OF DAY 2 HOUR(S)
RADIUS= 24.1KM
PRESSURE DROP= 100.MB

Fig. 11a. Flow vectors after two hours for a large hurricane which might be
expected to move from south to north through Mobile Bay. These
results were obtained neglecting the advection terms in the momentum
equations.

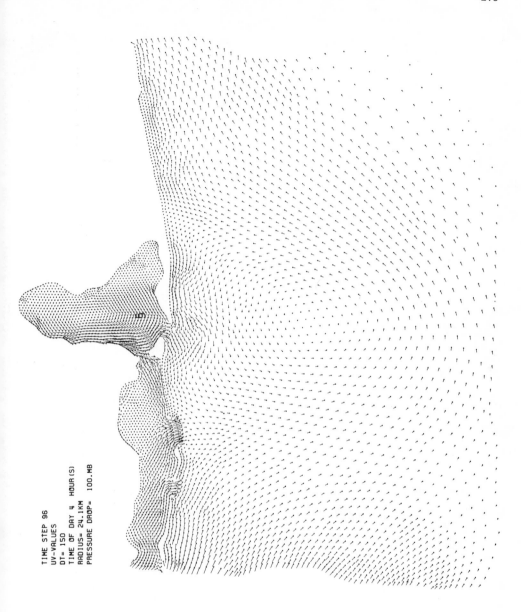

TIME STEP 96
UV-VALUES
DT= 150
TIME OF DAY 4 HOUR (S)
RADIUS= 24.1KM
PRESSURE DROP= 100.MB

Fig. 11b. After four hours.

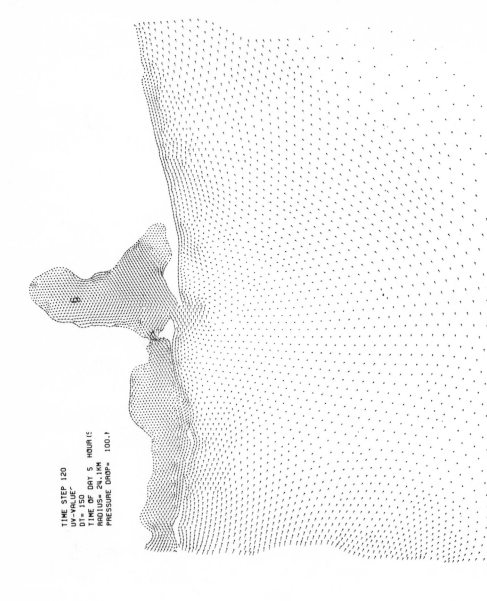

Fig. 11c. After five hours.

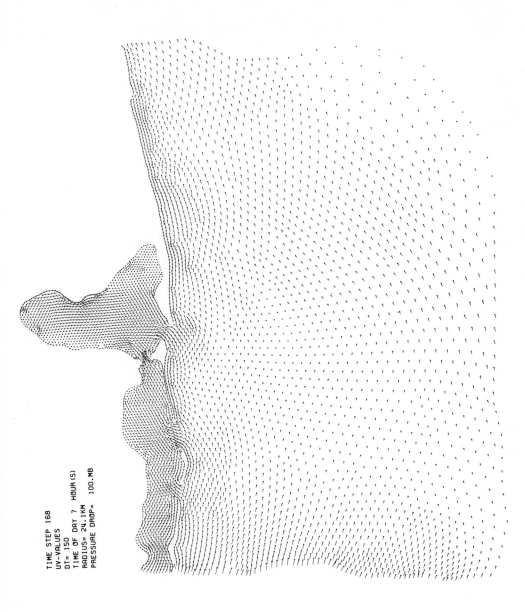

TIME STEP 168
UV-VALUES
DT= 150
TIME OF DAY 7 HOUR (S)
RADIUS= 24.1KM
PRESSURE DROP= 100.MB

Fig. 11d. After seven hours.

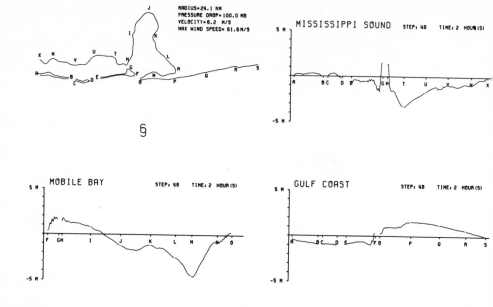

Fig. 11e. Shoreline water elevations after two hours.

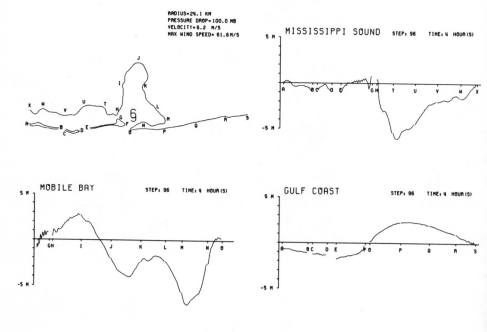

Fig. 11f. After four hours.

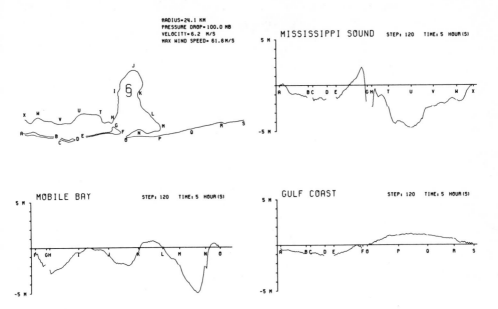

Fig. 11g. After five hours.

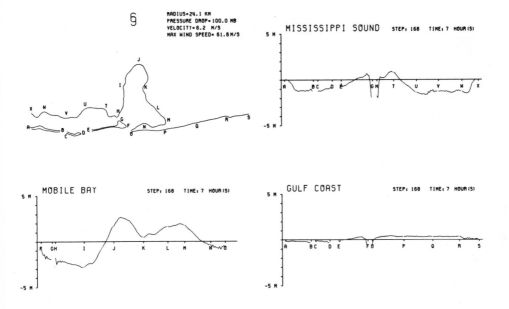

Fig. 11h. After seven hours.

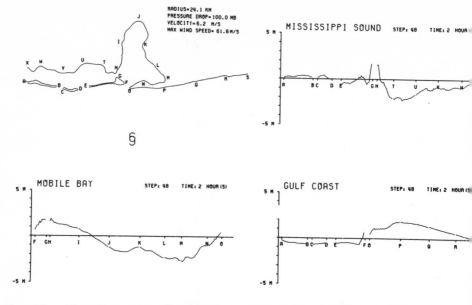

Fig. 12a. Shoreline water elevations to be compared with those in Fig. 11e. Although these results include the advection terms and neglect the Coriolis terms, the principal differences in the two sets of results are due to the fact that these involve the approximation of the total depth by the stillwater depth.

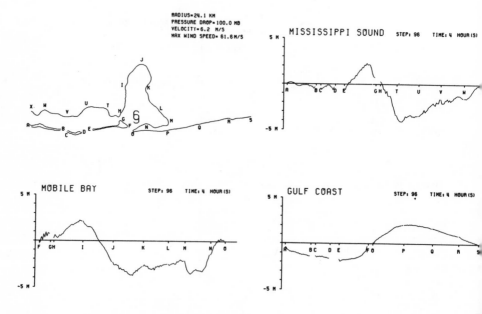

Fig. 12b. Compare with Fig. 11f.

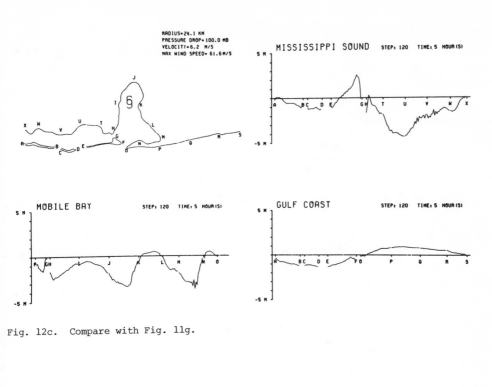

Fig. 12c. Compare with Fig. 11g.

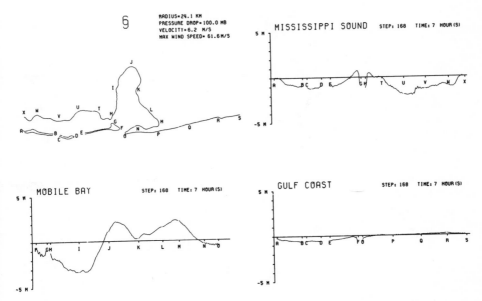

Fig. 12d. Compare with Fig. 11h.

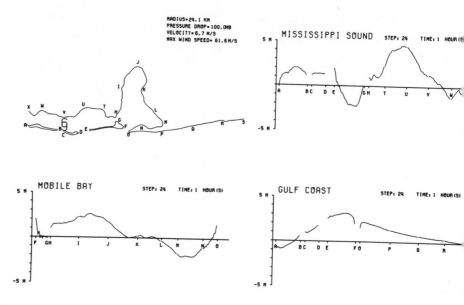

Fig. 13a.   Shoreline water elevations after one hour for a large hurricane which might be expected to move from west to east through Mississippi Sound and Mobile Bay.

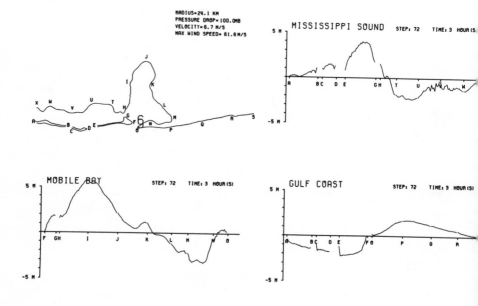

Fig. 13b.   After three hours.

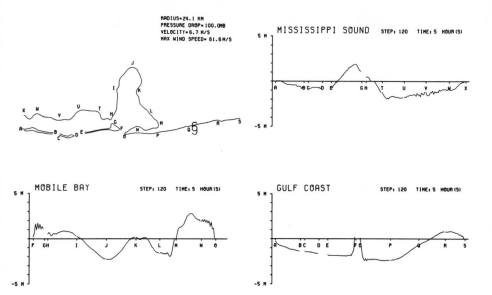

Fig. 13c.   After five hours.

ACKNOWLEDGEMENTS

Thanks are due to Gerald Putland and to Alicia Gonzales for their enthusiastic assistance, without which this presentation would not have been possible, and to Joan Wagner for her expert typing and retyping.

REFERENCES

Jelesnianski, C. P., 1967. Numerical computations of storm surges with bottom stress. Mon. Wea. Rev., 95:740-756.

Overland, J. E., 1975. Estimation of hurricane storm surge in Apalachicola Bay, Florida. NOAA Tech. Rep. NWS 17, U. S. Dept. of Commerce, 66 pp.

Pinder, G. F. and Gray, W. G., 1977. Finite Element Simulation in Surface and Sub-surface Hydrology. Academic Press, London, 295 pp.

Ramming, H-G., 1976. A nested north sea model with fine resolution in shallow coast-al areas. (Seventh Liège Colloquium on Ocean Hydrodynamics) Mémoires de la Société Royale des Sciences de Liège, X:9-26.

Thacker, W. C., 1976. A spliced numerical grid having applications to storm surge. NOAA Tech. Memo. ERL AOML-26, U. S. Dept. of Commerce, 19 pp.

Thacker, W. C., 1977. Irregular grid finite-difference techniques: simulations of oscillations in shallow circular basins. J. Phys. Oceanogr., 7:284-292.

Thacker, W. C., 1978a. Comparison of finite-element and finite-difference schemes. Part 1: One-dimensional gravity wave motion. J. Phys. Oceanogr., 8:676-679.

Thacker, W. C., 1978b. Comparison of finite-element and finite-difference schemes. Part 2: Two-dimensional gravity wave motion. J. Phys. Oceanogr., 8:680-689.

Wanstrath, J. J., Whitaker, R. E., Reid, R. O. and Vastano, A. C., 1976. Storm surge simulation in transformed coordinates. Vol. I. Theory and Application. U. S. Army, Corps of Engineers Tech. Rept. No. 76-3, 166 pp.

RECENT STORM SURGES IN THE IRISH SEA

N.S. HEAPS and J.E. JONES
Institute of Oceanographic Sciences, Bidston Observatory,
Birkenhead, Merseyside, England.

ABSTRACT

   The tidal and meteorological conditions associated with some recent very large
storm surges in the Irish Sea are described. Surges generated during a period of
ten days in November 1977 are investigated dynamically using a vertically-integrated
finite-difference model of the Irish Sea. Deductions are made concerning the possibilities of surge prediction for this area using a numerical model.

INTRODUCTION

   Major storm surges occurred in the Irish Sea in January 1976 and, more recently,

in November 1977. During each period, exceptionally high water levels were

experienced at Liverpool and at other coastal locations in the north-eastern Irish

Sea (figure 1). As a result, serious coastal flooding occurred in parts of that

region - particularly so in November 1977. The present paper describes the meteoro-

logical and tidal conditions under which these surges were generated. Emphasis is

placed on the 1977 surge events and a two-dimensional numerical model of the Irish

Sea is used to examine them in detail. An earlier study (Heaps and Jones, 1975)

investigated the large Irish Sea storm surges of January 1965 using a three-

dimensional model based on the same grid network as that used here.

   Lennon (1963) showed that major storm surges on the west coast of the British

Isles can be associated with Atlantic secondary depressions which move in towards

the coast crossing eastwards over the British Isles at a critical speed of about

40 knots. The depression tracks for a number of large surges at Avonmouth and

Liverpool were plotted and shown to lie within quite distinctive approach zones

corresponding respectively to these ports. Using an analytic continental shelf

model, Heaps (1965) found that the area of the Celtic Sea to the south of Ireland

is an important area for the generation of West Coast surges, particularly those

which influence the Bristol Channel with Avonmouth at its head (figure 1). The wind fields of the responsible secondary depressions, sweeping landwards across the Celtic Sea, force water towards the coast where a consequent rise in sea level occurs producing a surge. There is evidence for a quarter-wave tidal resonance across the Celtic Sea into the Bristol Channel which may explain the occurrence of especially large surges there when the meteorological system moves at the critical speed of 40 knots (Fong and Heaps, 1978).

The shallow north-eastern area of the Irish Sea, including Liverpool Bay, is particularly susceptible to storm surges. Heaps and Jones (1975) have shown that winds over the interior of the Irish Sea, also externally-generated surge disturbances passing into the Sea through St George's Channel and the North Channel, both significantly affect surge levels in this north-eastern region - in which the Port of Liverpool is situated. The present paper takes the study of Irish Sea surges a stage further by considering the recent large surges of November 1977, investigating their predictability in terms of a dynamical model. Comparisons are made with the surges of January 1976 and January 1965.

TIDES AND SURGES AT LIVERPOOL

A major surge peak of 1.42 m occurred at Liverpool at 01.00 h on 12 November 1977, two hours after predicted tidal high water. The superposition of surge and tide to give the elevation of the sea surface (total water level) is shown for several hours covering this event in figure 2. A maximum elevation of 6 m above ODN (Ordnance Datum Newlyn), consisting of 5 m of tide and 1 m of surge, was attained on the tidal high water at 23.09 h on 11 November.

An even larger surge peak of 1.47 m occurred at 19.00 on 14 November. However as is apparent in figure 2, this happened an hour or so before tidal low water and therefore the event posed no threat of coastal flooding. Nevertheless, a rather high water level of 5.65 m was registered on the preceding tidal high water.

For comparison, figure 3 indicates that a very large surge peak of 2.14 m occurred at 23.00 h on 2 January 1976, one hour before tidal high water. But in this case the tides were smaller and while maximum water level (at 23.20 h) was certainly high, 5.73 m, it was perhaps not as high as one might have expected with such a large surge.

Figure 4 shows that a large surge peak of 1.77 m occurred almost on tidal low water at 02.00 h on 14 January 1965; the neighbouring maximum water level reached 4 m. As indicated in the figure, there was another significant surge peak of 1.43 m at 17.15 on 17 January 1965 about one hour before tidal low water. On this occasion, the water level reached 5.11 m on the preceding tidal high water.

All the water levels quoted above are measured with respect to ODN, 0.27 m below mean sea level at Liverpool. Concern is with positive (rather than negative) surges and their effectiveness in contributing to the realisation of abnormally

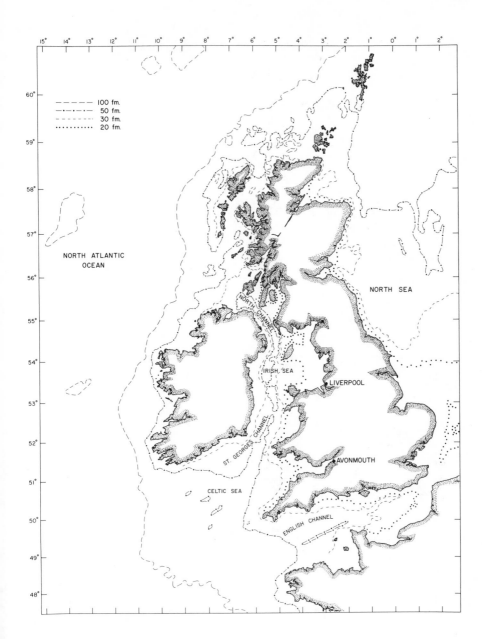

Fig. 1. Sea areas on the West Coast of the British Isles.

288

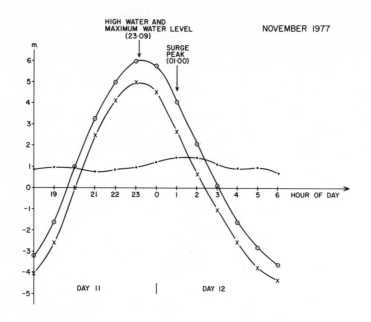

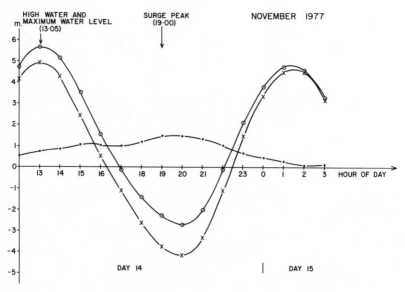

Fig. 2. Sea levels at Liverpool for 11-12 November and 14-15 November 1977 indicating times of occurrence of surge peaks and associated high waters and maximum water levels;

———X——X——X— tide, ——•——•——•— surge,

——⊙——⊙——⊙— total water level.

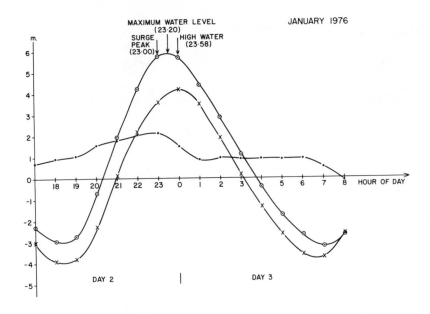

Fig. 3. Sea levels at Liverpool for 2-3 January 1976 indicating times of occurrence of the surge peak and the associated high water and maximum water level. Notation as in figure 2.

high sea levels.

It is clear from the above discussion that the tidal conditions prevailing at the time of a surge are as important as surge height itself in determining an abnormally high water level. Figures 5, 6 and 7 plot tidal high and low water levels at Liverpool throughout November 1977, January 1976 and January 1965, the times of surge peaks being indicated by arrows. The surges of November 1977 occurred at a time of high spring tides, a factor contributing to the exceptionally high water level attained on the 11th of the month. On the other hand, the surge of 2 January occurred at lower spring tides and near a tidal high water reduced by the diurnal inequality so that even though the surge peak was larger than that on 12 November the maximum water level attained was not as great. The large surge peaks on 14 November 1977 and 17 January 1965 came at spring tides, but both peaks occurred near to tidal low water and therefore could in no way contribute to the development of an especially high sea level.

METEOROLOGICAL CONDITIONS

Figure 8 shows the tracks of the centres of the depressions which gave rise to the Irish Sea surges described in the preceding section. The tracks lie within approach zones defined by Lennon (1963) - marked out by dashed lines in the figure.

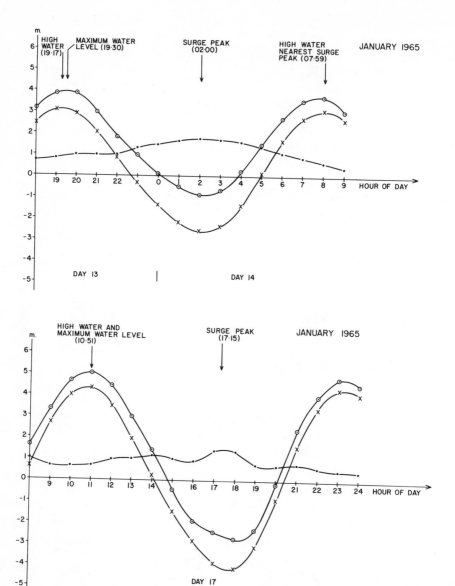

Fig. 4. Sea levels at Liverpool for 13-14 January and 17 January 1965 indicating the times of occurrence of surge peaks and associated high waters and maximum water levels. Notation as in figure 2.

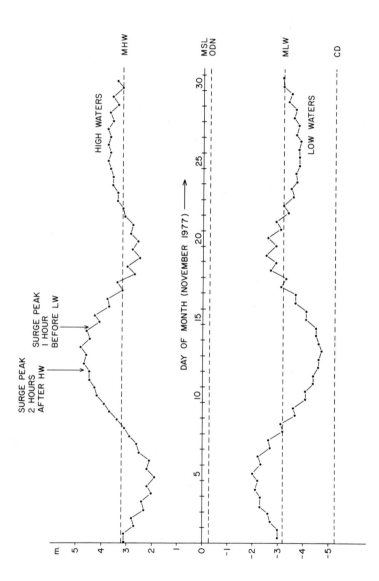

Fig. 5. Heights of high and low water at Liverpool relative to mean sea level (MSL), for November 1977, indicating times of occurrence of surge peaks. MHW = mean high water, MLW = mean low water, ODN = ordnance datum Newlyn, CD = chart datum.

292

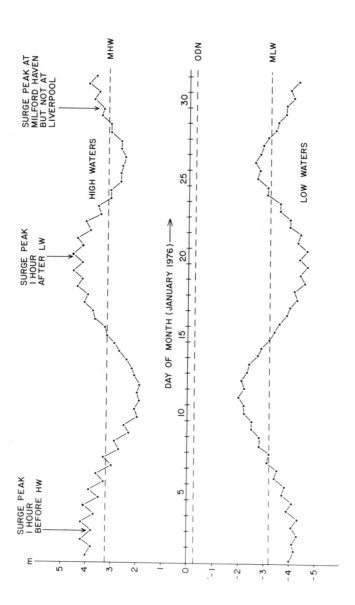

Fig. 6. Heights of high and low water at Liverpool relative to MSL, for January 1976, indicating times of occurrence of surge peaks. Notation as in figure 5.

293

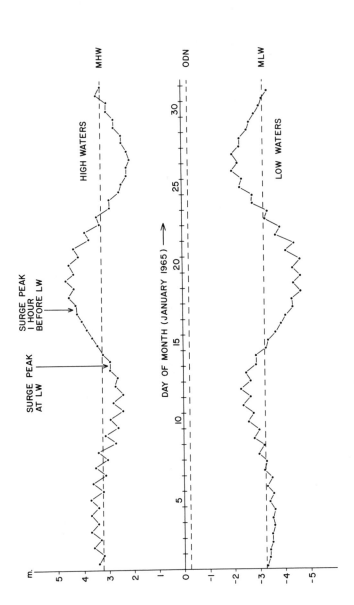

Fig. 7. Heights of high and low water at Liverpool relative to MSL, for January 1965, indicating times of occurrence of surge peaks. Notation as in figure 5.

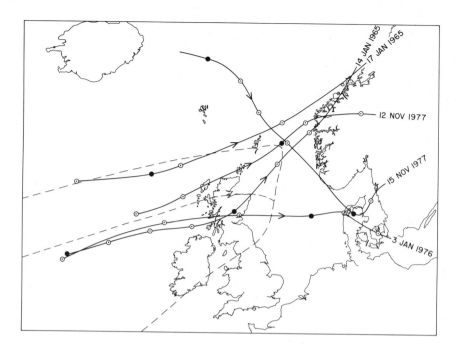

Fig. 8.   Depression tracks for five recent large surges at Liverpool;
● position at 0000 hr, ⊙ position at 0600 hr intervals.

However the track associated with the surge of 14 November 1977 is an exception
and follows a south-easterly course between Iceland and Denmark rather than an
easterly to north-easterly course over the British Isles.

The weather charts of figures 9, 10 and 11 illustrate the developing storm
patterns associated with the large surges recorded in the Irish Sea on 12 November
1977, 14 November 1977 and 2 January 1976. The secondary depression which brought
strong westerly-type winds to bear on the Irish Sea during 11 and 12 November 1977
was a poorly-defined feature (figure 9) but nevertheless a powerful surge-producing
agent. It contrasts with the larger and more clearly-defined cyclone which passed
across Scotland into the North Sea on 2 and 3 January 1976 (figure 11) again
bringing very strong westerly-type winds to the Irish Sea. The rather different
synoptic charts of 13 and 14 November 1977 (figure 10) show a frontal system and
wind fields sweeping over the British Isles from the north-west, some of the
strongest winds affecting the Irish Sea.

Figures 12, 13 and 14 plot recorded wind speed and direction, along with barometric pressure, at Ronaldsway in the Isle of Man (a central location in the
northern Irish Sea) for periods which include the large Irish Sea surges of
November 1977, January 1976 and January 1965. The times of surge peaks are

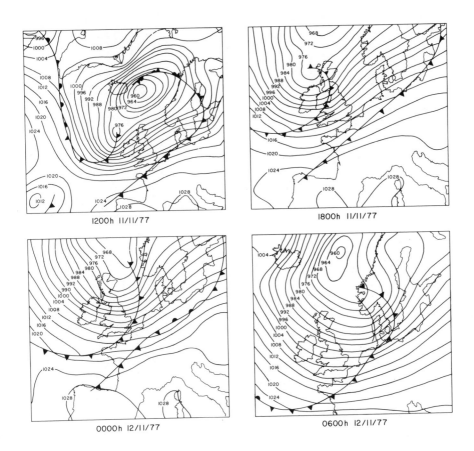

1200h 11/11/77

1800h 11/11/77

0000h 12/11/77

0600h 12/11/77

Fig. 9. Weather charts for the storm surge of 11 to 12 November 1977.

indicated. The barometric pressure variations are shown with respect to a mean
of 1012 mb. Wind angle θ in degrees is measured clockwise from the south. It is
apparent from the figures that the major surges of 12 November 1977, 2 January
1976 and 14 January 1965 were each preceded by falling barometric pressure and
rapidly strengthening winds veering from south-west to west. These characteristics
reflect the influence of an intense depression moving quickly eastwards across
the northern part of the British Isles (figures 8, 9, 11 here, also figure 1 given
by Heaps and Jones (1975)). Manifestly the surge of 14 November 1977 was associated
with strong west north-west winds maintained for over twelve hours as the result
of a northerly depression entering the North Sea (figures 8, 10). The surge of
17 January 1965 can obviously be linked to exceptionally strong west south-west
winds again maintained for half a day or so: the effect of a large depression
moving eastwards to the north of the British Isles (figure 8 here and figure 2 given
by Heaps and Jones (1975)). An overall examination of figures 12, 13 and 14 shows

296

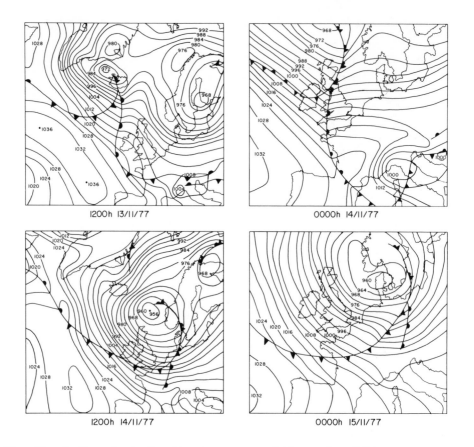

| 1200h 13/11/77 | 0000h 14/11/77 |
| 1200h 14/11/77 | 0000h 15/11/77 |

Fig. 10. Weather charts for the storm surge of 14 November 1977.

that the winds of January 1965 considerably exceeded those of January 1976 and als
those of November 1977.

IRISH SEA MODEL

To simulate the storm surges of November 1977 a two-dimensional numerical model
of the Irish Sea was formulated on the grid network shown in figure 15. The grid
has a square mesh of side 7.5 nautical miles and is constructed with reference to
a central x-directed line along the parallel of latitude $53°20'N$ and a central
y-directed line along the meridian of longitude $4°40'W$. The x coordinate increase
to the east and the y coordinate to the north. Surface elevation ζ is evaluated
at the central point of each elemental box, current u (in the x-direction) at the
mid-point of each y-directed box side, and current v (in the y-direction) at the
mid-point of each x-directed box side. Averaging u and v across an elemental box

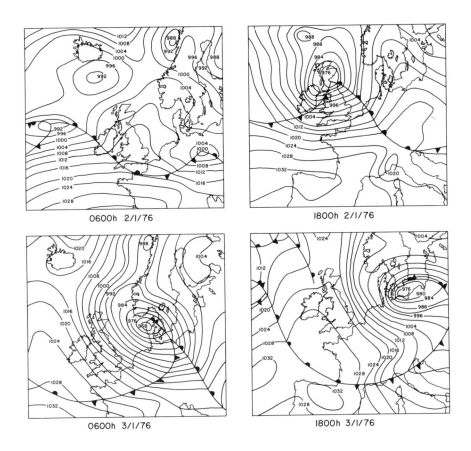

Fig. 11. Weather charts for the storm surge of 2 to 3 January 1976.

yields the current components at its centre. The model has open boundaries across
the North Channel in the north and across St George's Channel in the south.

The hydrodynamic equations of the model are:

$$\frac{\delta \zeta}{\delta t} = - \frac{\delta}{\delta x} \{ (h + \zeta) u \} - \frac{\delta}{\delta y} \{ (h + \zeta) v \}, \tag{1}$$

$$\frac{\delta u}{\delta t} = \gamma v - g \frac{\delta \zeta}{\delta x} - \frac{ku(u^2 + v^2)^{\frac{1}{2}}}{h + \zeta} + \frac{F_{sx}}{\rho (h + \zeta)} - \frac{1}{\rho} \frac{\delta p_a}{\delta x}, \tag{2}$$

$$\frac{\delta v}{\delta t} = -\gamma u - g \frac{\delta \zeta}{\delta y} - \frac{kv(u^2 + v^2)^{\frac{1}{2}}}{h + \zeta} + \frac{F_{sy}}{\rho (h + \zeta)} - \frac{1}{\rho} \frac{\delta p_a}{\delta y}, \tag{3}$$

giving the variations of ζ, u, v with respect to time t in terms of the Coriolis
effect (coefficient γ), sea surface gradients (factored by g the acceleration of
the Earth's gravity), quadratic bottom friction (coefficient k), components of
wind stress on the sea surface (F_{sx}, F_{sy}), and gradients of atmospheric pressure p_a
over the sea surface. Here: $\gamma = 1.1667 \times 10^{-4}$ s^{-1}, g = 9.81 ms^{-2} and k = 0.0026.
Also ρ = 1025 kg m^{-3}, the water density.

298

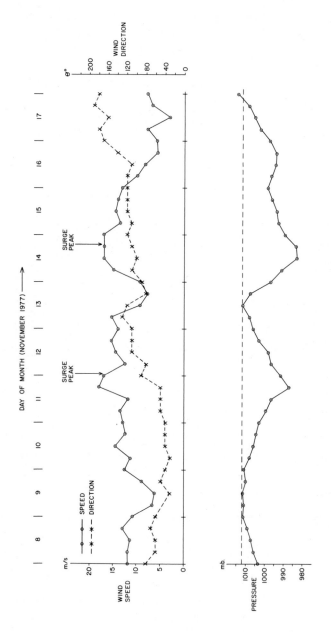

Fig. 12. Recorded wind speed and direction, and barometric pressure, at Ronaldsway (Isle of Man):
8-17 November 1977.

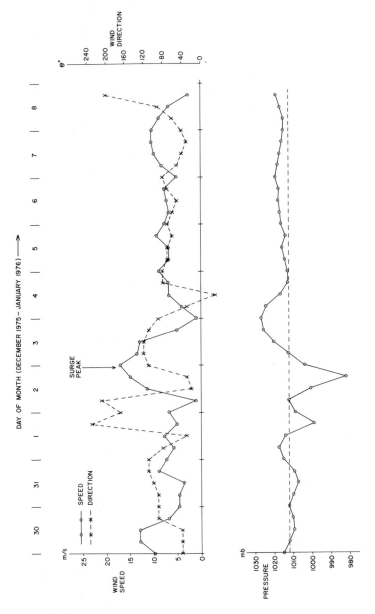

Fig. 13. Recorded wind speed and direction, and barometric pressure, at Ronaldsway: 30 December 1975 – 8 January 1976.

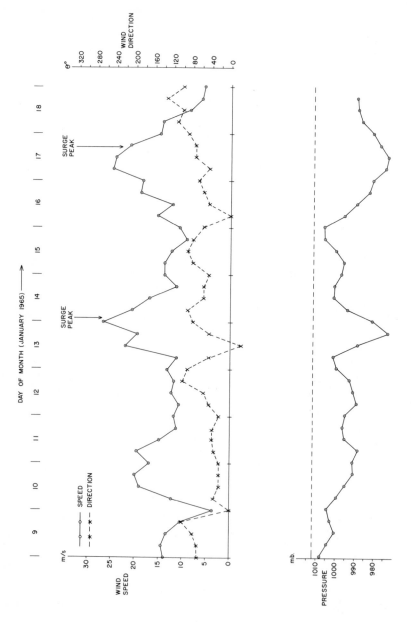

Fig. 14. Recorded wind speed and direction, and barometric pressure, at Ronaldsway: 9-18 January 1965.

In the equations, h denotes the undisturbed depth of water, prescribed realistically over the grid at the mid-points of the box sides. The total water depth
at any time is h + ζ, determined at the mid-point of a box side with ζ an average
of the values taken from the centres of the adjacent boxes.

An explicit finite difference scheme was used to develop solutions of the dynamical equations, yielding elevation ζ and depth-mean currents u, v through time
over the Irish Sea. The scheme is basically similar to that used by Heaps and
Jones (1975) with the frictional term and the total depth h + ζ treated as in the
paper by Flather and Heaps (1975). In generating solutions through time, starting
from a state of rest with ζ = u = v = 0 everywhere, the ζ, u, v are incremented
from values at t to values at t + Δt over successive time intervals Δt. In this
procedure, elevation ζ is prescribed at the open boundaries as time advances, also
wind stress components F_{sx}, F_{sy} and atmospheric pressure gradients $\delta p_a/\delta x$, $\delta p_a/\delta y$
at the u and v points of the model. Zero normal flow is postulated at the land
boundaries. Having regard to numerical stability, it was found convenient to take
Δt = 120 s.

TIDAL COMPUTATIONS

Tides were generated in the model for the whole of November 1977 in response to
specified open boundary tides consisting of the M_2 and S_2 constituents - the
principal harmonic components. Amplitudes and phases of the tidal input, applied
at the elevation points adjacent to the northern and southern open boundaries, are
given in table 1. Basically this input comes from cotidal charts and from a
numerical tidal model of the sea areas on the west coast of the British Isles.

The tides generated in the model were analysed to yield the M_2 and S_2 components
at Port Patrick, Belfast, Douglas, Workington, Heysham, Liverpool, Hilbre Island,
Holyhead, Dublin and Fishguard (see figure 15 for these locations). In table 2
the results of this analysis are compared with corresponding results derived from
the analysis of observations. There is satisfactory agreement, with discrepancies
in tidal amplitude for the most part being less than 0.13 m and discrepancies in
tidal phase not exceeding 6° for M_2 and 16° for S_2. A similar comparison is also
made in table 2 for tidal flows through the North Channel across section C_6C_7 in
figure 15. The agreement between model and observation is again quite good. Here,
the observational results come from measurements of voltage across the North Channel
by Prandle and Harrison (1975) with conversion from voltage to flow using a calibration factor due to Hughes (1969).

The model tide, limited to M_2 and S_2 by our restricted knowledge of the open
boundary tides, obviously differs from the predicted tide based on a comprehensive
set of harmonic constants. Some measure of this difference can be gained from the
tidal curves of figures 19 and 20 for Workington and Liverpool: deviations of

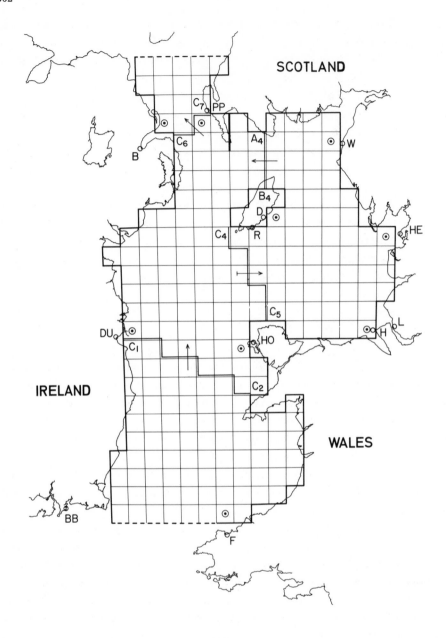

Fig. 15. Irish Sea model: ——————— land boundary; — — — — — — open sea boundary;
═══════ flow section; ○ tide gauge and ⊙ corresponding elevation point
of the model. Key: PP = Port Patrick, B = Belfast, D = Douglas, R = Ronaldsway,
W = Workington, HE = Heysham, L = Liverpool, H = Hilbre Island, HO = Holyhead,
DU = Dublin, F = Fishguard, BB = Baginbun.

TABLE 1

Amplitude H (metres) and phase g (degrees) of the input tides at the elevation points adjacent to the northern and southern open boundaries

Northern boundary

		West				East	
M_2	H	0.71	0.76	0.84	0.90	0.94	
	g	321	329	336	343	347	
S_2	H	0.16	0.18	0.21	0.25	0.27	
	g	7	13	19	23	27	

Southern boundary

		West					East
M_2	H	0.92	0.94	0.98	1.06	1.14	1.22
	g	168	178	186	194	201	208
S_2	H	0.39	0.40	0.41	0.43	0.44	0.46
	g	216	221	228	233	237	241

roughly 10 to 17 per cent in range are evident between the model tide and the tide from a full prediction. As it turns out, the accurate reproduction of tide in the model is not essential for our purposes with the emphasis on storm surge computation. Thus, with meteorological as well as tidal forcing included in the model, the total motion of tide and surge is computed from which the regime of tide alone is subtracted. The result gives the computed surge - conditioned by interaction with the tide. It is assumed that an approximate tide is sufficient for the satisfactory determination of this interaction.

STORM SURGE COMPUTATIONS

The model was run to simulate the regime of tide and surge in the Irish Sea for the period 00.00 h 8 November - 23.00 h 17 November 1977. In this, surge elevation was added to tidal elevation along the open boundaries. Simultaneously, fields of wind stress and horizontal atmospheric pressure gradient were applied to the sea surface.

At each elevation point adjacent to the northern open boundary a surge elevation was prescribed (at hourly intervals) equal to that observed at Port Patrick. At each elevation point adjacent to the southern open boundary a surge elevation was prescribed (also at hourly intervals) from a linear interpolation, with respect to distance, between the observed surge at Fishguard and that at Baginbun. The surges observed at Port Patrick, Fishguard and Baginbun (locations shown in figure 15) are plotted through time in figure 16.

The changing fields of wind stress and atmospheric pressure gradient were evaluated at three-hourly intervals over six rectangular sub-areas of the Irish

TABLE 2

Amplitude (in metres) and phase (in degrees) of the M_2 and S_2 surface tides at various Irish Sea ports, comparing results from the numerical model with those derived from observation. A similar comparison is made for the M_2 and S_2 tidal flows (in 10^5 m^3/s units) through the North Channel, section C_6C_7.

		Amplitude		Phase	
		Model	Observed	Model	Observed
M_2	Port Patrick	1.36	1.34	329	333
	Belfast	1.14	1.20	321	315
	Douglas	2.43	2.31	326	327
	Workington	2.68	2.72	331	334
	Heysham	3.08	3.15	325	326
	Liverpool	2.95	3.08	316	322
	Hilbre Is.	2.95	2.92	316	318
	Holyhead	1.55	1.79	287	292
	Dublin	1.42	1.34	321	326
	Fishguard	1.32	1.36	214	208
	C_6C_7	28.9	24.0	42	43
S_2	Port Patrick	0.35	0.38	5	16
	Belfast	0.27	0.29	357	352
	Douglas	0.70	0.72	359	7
	Workington	0.77	0.90	6	14
	Heysham	0.91	1.01	0	8
	Liverpool	0.87	1.00	350	5
	Hilbre Is.	0.87	0.95	350	0
	Holyhead	0.48	0.59	312	328
	Dublin	0.38	0.40	346	357
	Fishguard	0.48	0.53	246	247
	C_6C_7	8.9	8.2	68	79

Sea region following a method used by Heaps and Jones (1975). Pressure gradients $\delta p_a/\delta x$, $\delta p_a/\delta y$ and geostrophic wind (R_G, θ_G) were evaluated uniformly over each rectangle in terms of differences of observed barometric pressures taken over distances of approximately 60 nautical miles. Surface wind (R, θ) was then deduced from the empirical relations:

$$R = 0.56R_G + 0.24, \quad \theta = \theta_G - 22. \tag{4}$$

Here: R_G, R denote wind speeds in m/s and θ_G, θ wind angles in degrees measured clockwise from the south. Resultant wind stress, F dynes/cm^2 in the direction θ, was subsequently evaluated using the square law:

$$F = 12.5cR^2 \tag{5}$$

with the drag coefficient c given by

$$10^3 c = 0.554, \quad R < 4.917$$
$$= -0.12 + 0.137R, \quad 4.917 < R < 19.221 \tag{6}$$
$$= 2.513, \quad R > 19.221$$

Then, components of wind stress were determined from:

$$F_{sx} = F \sin\theta, \ F_{sy} = F \cos\theta. \tag{7}$$

Subjected to open-boundary elevations of tide and surge, wind stresses and atmospheric pressure gradients, the model yielded the combined motion of tide and surge in the Irish Sea through the period 8-17 November. The tidal motion alone, determined separately by the model as prescribed in the preceding section, was subtracted from the combined motion to yield the storm surge. Surge levels (model) are compared with surge levels (observation) for a number of Irish Sea ports in figures 17 and 18. The locations of all these ports are indicated in figure 15. On the observational side, the surge level at a place is obtained by taking the difference between the observed and the tidally-predicted water levels there, hour by hour. There is thus a correct correspondence between this procedure and the modelling one for the computation of surges.

An examination of the residuals in figures 17 and 18 shows that the large semi-diurnal-type fluctuations observed during 8-11 November at Workington, Heysham and Hilbre Island are quite nicely reproduced by the model. The fluctuations are somewhat overestimated at Douglas, and at Liverpool their phasing is in error due, no doubt, to the inability of the model to reproduce the influence of the Mersey Estuary. At Holyhead and Dublin the fluctuations are smaller and reasonably well reproduced. They are present at Belfast but, again as at Liverpool, their phasing comes out incorrect due presumably in this case to the omitted influence of Belfast Lough.

The main surge peak which occurred near midnight on 11 November is predicted quite well by the model at Workington, Heysham and Hilbre Island. A magnified diagram of the Workington residuals near the maximum is shown in figure 19. Note from this diagram that the peak occurred on the rising tide, a feature common to all the other ports apart from Belfast and Liverpool. The Liverpool residuals near the surge maximum are shown in figure 20. It can be seen from this figure that, while the model surge maximum occurs on the rising tide, the observed maximum was higher and occurred five to six hours later. In effect, there is a significant contribution missing from our Liverpool surge prediction on 12 November. The source of this error is suggested by figure 21 showing wind speeds recorded at Liverpool on 11 and 12 November. A rapid fall followed by a rapid rise evident in the recorded speed between 18.00 h and 23.00 h on the 11th is clearly not represented in the wind field used for the model computations. Other anemometer observations around the coastline of Liverpool Bay indicate that this fall and rise in speed was fairly local to Liverpool - at least in its intensity. We propose, therefore, that the surge contribution missing from our Liverpool prediction was generated by local wind variations which could not be accounted for by the barometric pressure differences on which the model winds were based. Figure 22 indicates that observed surge peaks at Liverpool on the 14th not reproduced by the model might also be

306

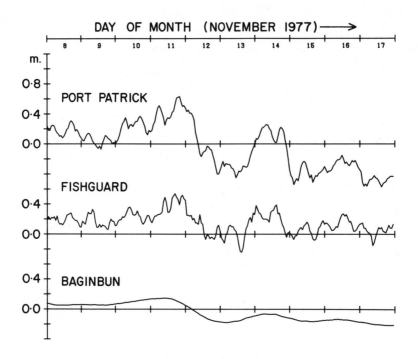

Fig. 16.   Observed residual elevations at Port Patrick, Fishguard and Baginbun
(derived for the first two of these ports on the basis of tidal predictions and
for the third on the basis of the X_o-filter).

attributed to local variations in wind speed not accounted for by the larger-scale
model winds. A finer resolution of the wind structure over the Irish Sea is
clearly required for input to the model to improve its performance at Liverpool -
and quite possibly at other places.

Returning to consideration of figures 17 and 18, it should be pointed out that
the observed surge profiles at Heysham and Hilbre Island terminated prematurely at
the end of 11 November due to the failure of the tide gauges at those locations
under storm conditions. Moreover, shifts in datum in the Workington and Holyhead
tide gauges occurred on the 11th due to slippage of their recording mechanisms when
high water levels were attained. In the surge profiles shown for Workington and
Holyhead, adjustments in datum have been made in an attempt to minimise these obser
vational errors.

In a repeat run with the model for the period 8-17 November, wind stresses and
atmospheric pressure gradients were set to zero. Motion in the Irish Sea was there
obtained due solely to tide and surge on the open boundaries. Subtracting the mode
tide then gave the externally-generated surge in the Irish Sea. Associated residua

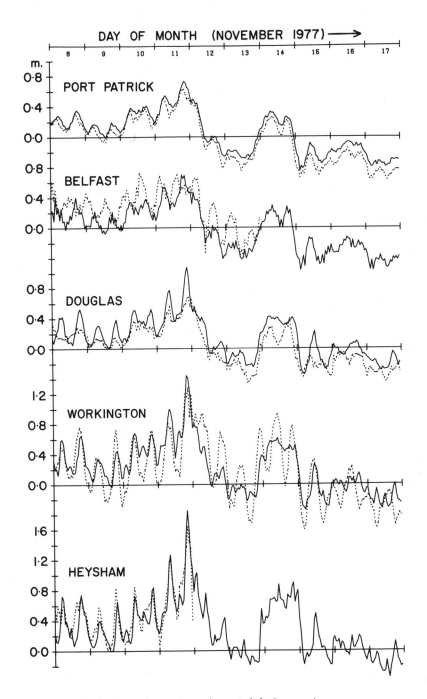

Fig. 17. Residual elevations at various Irish Sea ports: ——————— from
the numerical model, ------ from observation.

308

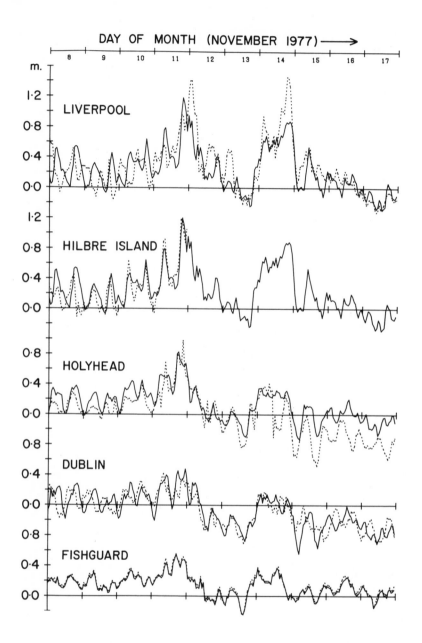

Fig. 18. Residual elevations at various Irish Sea ports:
——— from the numerical model, ‑ ‑ ‑ ‑ ‑ ‑ ‑ ‑ ‑ ‑ from observation.

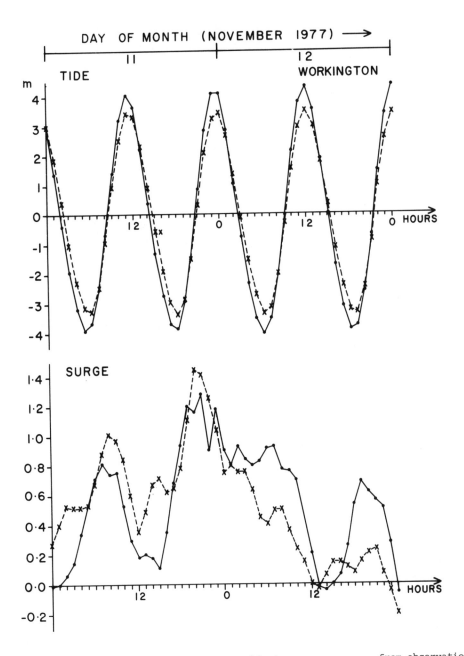

Fig. 19. Tide and surge elevations at Workington, ———————— from observations and a full harmonic tidal prediction based on observations, — — — — — — — from the model. Hourly values are plotted. Tidal heights are given to mean sea level datum.

310

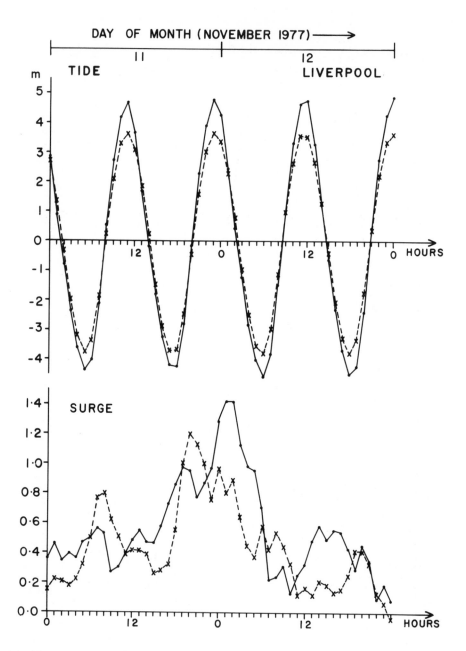

Fig. 20. Tide and surge elevations at Liverpool, ————— from observations and a full harmonic tidal prediction based on observations, — — — — — — from the model. Hourly values are plotted. Tidal heights are given to mean sea level datum.

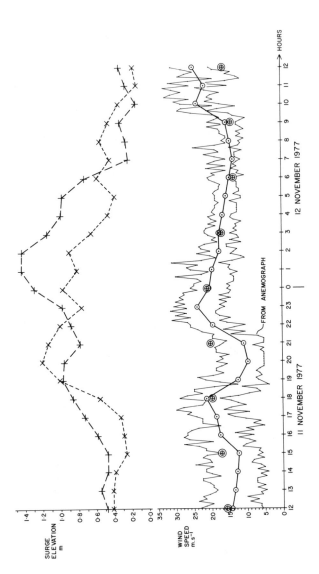

Fig. 21. Wind speeds recorded by the anemometer at Seaforth, Liverpool, 11-12 November 1977. Limits of the anemograph record are shown together with hourly means (—○——○——○——). Wind speeds used in the model computations, for Liverpool Bay, are denoted by ⊕ . Surge elevations at Liverpool are shown, as observed (—+—+—+——) and as determined from the model (—·—✕—·—✕—·—).

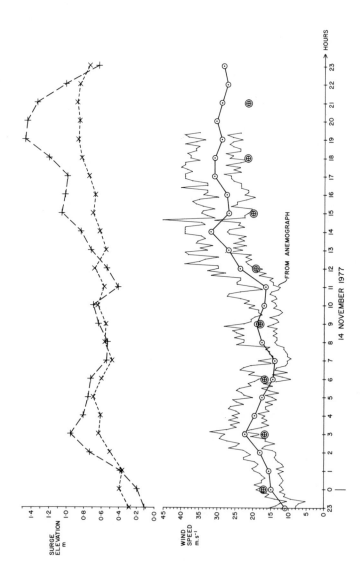

Fig. 22. Wind speeds (and their hourly means) recorded by the anemometer at Seaforth, Liverpool, 14 November 1977. Wind speeds used in the model computations, for Liverpool Bay, are also shown. The observed and computed surges at Liverpool are plotted. Notation as in figure 21.

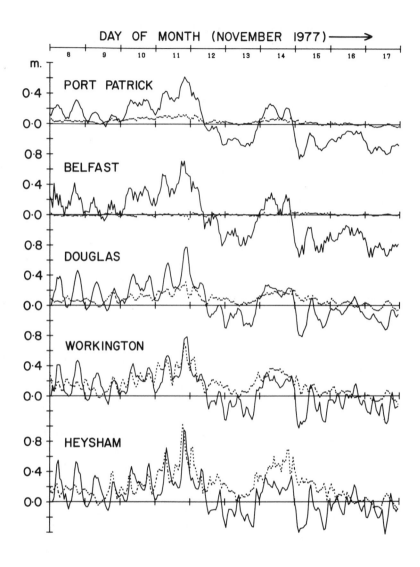

Fig. 23. Residual elevations from the model resolved into a part due to
disturbances entering across the open boundaries (────────) and a part
due to wind and atmospheric pressure gradients over the model area (- - - - - - -).

314

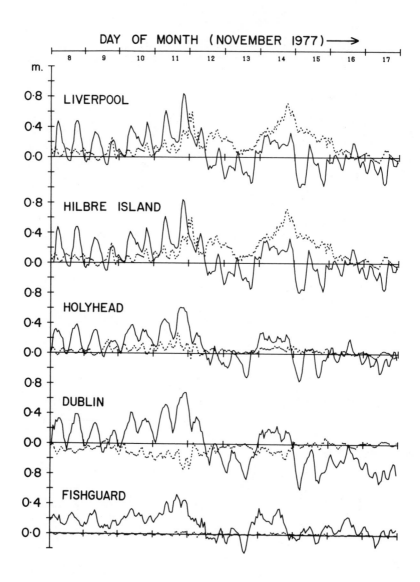

Fig. 24. Residual elevations from the model resolved into a part due to
disturbances entering across the open boundaries (———————) and a part due
to wind and atmospheric pressure gradients over the model area (– – – – – – –).

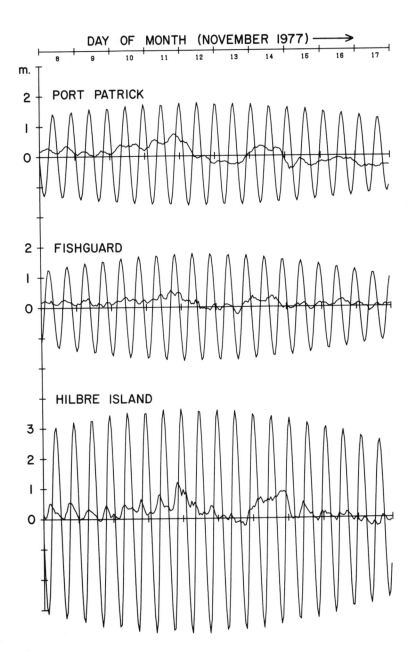

Fig. 25. Computed tides ($M_2 + S_2$) and residuals (the smaller variations shown) for (a) Port Patrick, (b) Fishguard and (c) Hilbre Island.

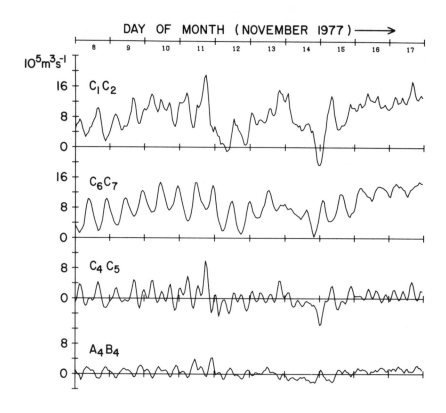

Fig. 26. Residual flows (in $10^5 m^3/s$ units) across sections C_1C_2, C_6C_7, C_4C_5 and A_4B_4: from the numerical model. Positive flow directions are shown in figure 15.

elevations are plotted through time in figures 23 and 24. Also plotted are the residual elevations due to the direct action of wind and atmospheric pressure over the Irish Sea (elevations obtained by subtracting the external surge from the total surge determined originally). Thus, figures 23 and 24 show the total surge of figures 17 and 18 resolved into an external part coming from the open boundaries and an internal part coming from the effects of wind and atmospheric pressure (essentially wind) over the Irish Sea. It is evident from these figures that the open-boundary influence generally predominates. Clearly, however, the wind effect can be equally important at Workington, Heysham, Liverpool and Hilbre Island along the north-eastern coast. Of special interest is the fact that at Workington and Heysham on 11 November the two surge components were of similar magnitude and were directly superimposed to produce the high surge peaks observed. There was a some-what less effective superposition at Liverpool and Hilbre Island on the same day.

Evidently the large surges at Heysham, Liverpool and Hilbre Island on 14 November were mainly generated by winds over the interior of the Irish Sea.

Figures 23 and 24 indicate that the large semidiurnal-type surge fluctuations during 8-11 November originated mainly from the open boundaries. Such fluctuations are also evident at Heysham and Workington as the result of meteorological forcing over the Irish Sea. This suggests that the Irish Sea basin has a natural mode of oscillation of near-semidiurnal period which may be excited by external surges on the open boundaries and, to a lesser extent, by wind stress and atmospheric pressure acting on the surface of the basin. The magnification of the tides in the Irish Sea may well depend on the existence of this mode which would seem to have a maximum amplitude in the neighbourhood of Heysham.

Figure 25 compares the tidal and total surge profiles from the model at Port Patrick, Fishguard and Hilbre Island. The semidiurnal fluctuations discussed above are shown to occur with their peaks consistently on the rising tide, which suggests that they are primarily the product of surge-tide interaction on the open boundaries which propagates (with the tide) into the interior of the Irish Sea region. There may be further interaction within the region itself but, more likely, the main internal modifications come from a magnification due to the existence of a natural basin-mode of approximately semidiurnal period. The dynamics of surge-tide interaction in the Irish Sea requires further detailed study.

Surge flows across sections C_1C_2, C_6C_7, C_4C_5 and A_4B_4 of the Irish Sea (figure 15), as derived from the model, are plotted through time in figure 26. These plots complement the results for surface elevation given in figures 17 and 18. The C_1C_2 and C_6C_7 flows show an average transport from south to north through the Irish Sea, over the period 8-17 November, of around $8 \times 10^5 \mathrm{m}^3 \mathrm{s}^{-1}$. This must be largely due to a southerly wind component between the 8th and the 11th (figure 12) but subsequently, with west to north-west winds, due to a generally downward gradient of residual sea-surface elevation from south to north between the opposite open ends of the Irish Sea (compare the surge elevation at Port Patrick with that at Fishguard in figures 17 and 18). When this gradient is small on the 12th and on the 14th, the flow is also small. Comparatively little of the sustained south to north transport appears to pass through the eastern part of the Irish Sea across C_4C_5 and A_4B_4. Main features of the transports shown in figure 26 are the semidiurnal-type fluctuations representing, particularly during 8-11 November, a succession of flow pulses directed alternately in and out of the northern Irish Sea. These pulses may be associated with the similar fluctuations of surface level already discussed. In an inward pulse, water passes northwards across C_1C_2 and (simultaneously) southwards across C_6C_7, turning eastwards across C_4C_5 and A_4B_4 into the eastern region of the Irish Sea. In the following outward pulse the flow directions are reversed. Fluctuations in flow of approximately quarter-diurnal frequency are strongly evident across C_4C_5. The flows across A_4B_4 are smaller and also exhibit these higher-frequency oscillations.

318

CONCLUDING REMARKS

1. Recent large storm surges in the Irish Sea (during November 1977, January 1976 and January 1965) may be associated with the type of weather conditions identified by Lennon (1963) as being relevant to the generation of major surges on the west coast of the British Isles. An exception was the surge of 14 November 1977, caused by a depression which followed a track between Iceland and Denmark rather than one which passed from west to east across the British Isles.

2. An examination of tide, surge and total water level at Liverpool during the recent surge events has emphasised the point that tidal conditions prevailing at the time of a surge may be just as important as surge height itself in determining an abnormally high water level. Thus, a moderately large surge on a very high tide might raise sea level to a greater extent than a major surge on a somewhat lower tide.

3. A two-dimensional numerical model of the Irish Sea was able to reproduce the main features of the surges observed at a number of Irish Sea ports during the period 8-17 November 1977. External surges entering the Irish Sea through the North Channel and St George's Channel had a substantial effect on the interior surge levels. Meteorological forces acting over the Irish Sea itself were responsible for important surge contributions at ports such as Workington, Heysham and Liverpool in the north-eastern region.

4. Local variations in wind strength appear to be able to generate significant surges at Liverpool not accounted for by the model with surface winds assessed on the basis of barometric pressure differences taken over distances of about 60 nautical miles. Presumably, therefore, the model's performance could be usefully improved by running it with a more detailed wind structure over the sea surface.

5. Large semidiurnal-type fluctuations were a feature of the surges in the Irish Sea during the period 8-11 November 1977. The model reproduced them quite well and indicated that they originated mainly from variations of surge level on the open boundaries, possibly exciting a natural mode of oscillation of the Irish Sea basin of near-semidiurnal period. Semidiurnal-type fluctuations of surge level on the open boundaries, most likely arising from surge-tide interaction, were influential in producing the internal fluctuations.

6. A model for forecasting storm surges in the Irish Sea needs to be larger in area than the research model of the present paper. For forecasting purposes a model is required which does not depend quite so critically as the present one on open-boundary surge conditions. A new model satisfying this requirement, covering all the sea areas on the west coast of the British Isles, is under development (Owen and Heaps, 1978).

ACKNOWLEDGEMENTS

The authors are grateful to a number of colleagues at I.O.S. Bidston for advice and assistance in this study. Members of the Tidal Computation Section determined most of the observed residual elevations shown and those for Baginbun came from work by Dr D.T. Pugh.

Thanks are due to Mr R.A. Smith for preparing the diagrams and to Miss Barker and Mrs Young for typing the manuscript.

The work described in this paper was funded by a Consortium consisting of the Natural Environment Research Council, the Ministry of Agriculture, Fisheries and Food, and the Departments of Industry and Energy.

REFERENCES

Flather, R.A. and Heaps, N.S., 1975. Tidal computations for Morecambe Bay. Geophys. J. R. astr. Soc., 42: 489-517.

Fong, S.W. and Heaps, N.S., 1978. Note on quarter-wave tidal resonance in the Bristol Channel. Institute of Oceanographic Sciences Report No. 63.

Heaps, N.S., 1965. Storm surges on a continental shelf. Phil. Trans. R. Soc., A,257: 351-383.

Heaps, N.S. and Jones, J.E., 1975. Storm surge computations for the Irish Sea using a three-dimensional numerical model. Mém. Soc. r. sci. Liège, ser. 6, 7: 289-333.

Hughes, P., 1969. Submarine cable measurements of tidal currents in the Irish Sea. Limnol. Oceanogr., 14: 269-278.

Lennon, G.W., 1963. The identification of weather conditions associated with the generation of major storm surges on the west coast of the British Isles. Q. Jl. R. met. Soc., 89: 381-394.

Owen, A. and Heaps, N.S., 1978. Some recent model results for tidal barrages in the Bristol Channel. Proceedings of the Colston Research Symposium 1978, University of Bristol (in press).

Prandle, D. and Harrison, A.J., 1975. Recordings of potential difference across the Port Patrick-Donaghadee submarine cable. Institute of Oceanographic Sciences Report No. 21.

RESULTS OF A 36-HOUR STORM SURGE PREDICTION OF THE NORTH SEA FOR
3 JANUARY 1976 ON THE BASIS OF NUMERICAL MODELS 1)

G.FISCHER

Meteorologisches Institut der Universität Hamburg

ABSTRACT

Within the "Sonderforschungsbereich 94" of the University of Hamburg and in
collaboration with the "Deutsches Hydrographisches Institut" and "Deutscher Wet-
terdienst", a group has been established a few years ago with the aim to explore
the feasibility of forecasting North-Sea storm surges by integrating numerically
a combined atmospheric-oceanographic physical model.
A first step into this direction is the simulation of the severe storm and the
resulting water levels occuring on 3 January 1976. For this purpose, the atmospheric
model was run with a resolution of 8 levels in the vertical and a horizontal grid
spacing of 1.4° in latitude and 2.8° in longitude on the northern hemisphere. The
initial conditions are based upon observations of 2 January 1976, 12^{h} GMT, i.e.
about 24 hours before the storm reached its greatest intensity in the southern
parts of the North-Sea.
The surface geostrophic wind predicted by the atmospheric model was converted
into stress values through a bulk formula which then entered the North-Sea model
to yield the desired water elevations and currents in a 22 km grid. Besides of
taking predicted winds, also the observed values stemming from a careful re-analysis
of the storm situation were fed into the North-Sea model to give a "perfect forecast". The water levels obtained in this way were then compared with gauge measurements at a number of coastal stations.
Though the meteorological model simulated quite well the track and intensification of the storm cyclone the evolving pressure gradient, i.e. the geostrophic
wind at the surface, was on the whole weaker than observed. Therefore, a reasonable
correspondence with measured water elevations could only be reached by correcting
the predicted geostrophic wind with a factor of 1.55. Then the results computed
by the North-Sea model became about as good as those on the basis of observed geostrophic winds and known before they would have been a very valuable information
about the surge to be expected. It is questionable, however, whether the factor
1.55, introduced a posteriori, is valid in general. Though one knows from experience
that numerical weather predicitions tend to underestimate cyclone development, thus
justifying a correction to stronger winds, the value will certainly change from
case to case. To clarify this point too, further experiments of this kind are
planned.

1) The full article is to appear in "Deutsche Hydrographische Zeitschrift"
   Heft 1, 1979

EXTRATROPICAL STORM SURGES IN THE CHESAPEAKE BAY

DONG-PING WANG

Chesapeake Bay Institute, The Johns Hopkins University, Baltimore, MD (U.S.A.)

ABSTRACT

Two major extratropical storm (cyclone) surges in the Chesapeake Bay, in 1974-1975 are examined. The subtidal sea level was the dominant surge component, and it was induced by the local wind set-up and the nonlocal coupling with coastal sea level. The study suggests that the observational study is essential to the improvement of storm surge forecast.

INTRODUCTION

Extratropical storms (cyclones) over the U.S. Atlantic coast can cause severe damage. For example, the coastal storm of early March 1962 caused damage over $200 million. While storms causing this much damage are rare, storms of lesser damage potential do occur several times each winter. Accurate forecasts of flooding and beach erosion caused by these storms are important.

There are basically two different approachs to storm surge forecast. The empirical method relates the storm surge to meteorological data from a regression analysis. The theoretical method determines the storm surge from numerical integration of the equations of motion and continuity, with appropriate boundary conditions.

In the empirical method, physical reasoning is essential in selecting the proper predictors. The theoretical method has less uncertainty in selecting meteorological forcing. However, the numerical model is designed for limited area forecast, and therefore, the choice of model domain and boundary conditions can be critical. A better understanding of the nature of storm surge is thus vital to the improvement of forecast skill.

With the advancing of computer technology, the three-dimensional model for semi-enclosed sea, lake and estuary, has been developed (Heaps and Jones (1975), Leenderste et al. (1973), Simons (1973)). In particular, Heaps has applied the numerical model to operational surge forecast in the North Sea. In contrast, there have been few studies on the storm surge from direct observations. Lack of solid observational evidence, makes it difficult to evaluate model performance.

324

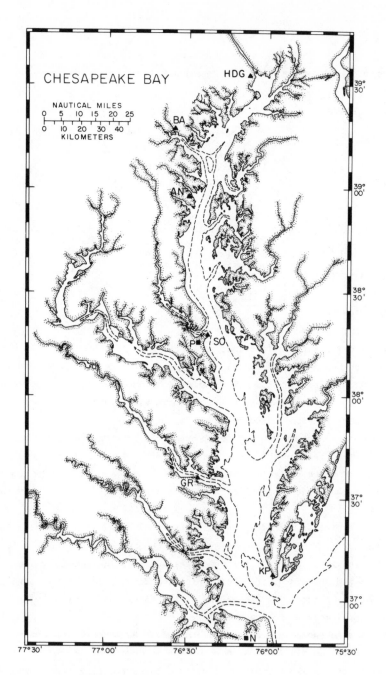

Fig. 1. Map of the Chesapeake Bay and its tributaries
(sea level and meteorological stations are marked).

Recently, Wang (1978a) has examined the subtidal sea level in Chesapeake Bay (Fig. 1) and its relations to atmospheric forcing, from a year-long record. His results indicated that the Bay water response depends on the time scale of atmospheric forcing. At time scales longer than 7 days, sea levels in the Bay were driven nonlocally by coastal sea level. Between 4 and 7 days, both coastal sea level and local forcing (particularly, lateral wind) were important. At shorter time scales (1 to 3 days), the Bay water response was local, driven by the longitudinal wind. Wang (1978a) also constructed a response model (empirical method) which accounts for over 90% of the total subtidal variance.

The success in explaining the observed sea level suggests that subtidal sea level is closely related to large-scale atmospheric forcing. In contrast, supertidal sea level was strongly affected by inhomogeneous topography, shoreline and small-scale atmospheric disturbances. It would be interesting to know if subtidal sea level is the major component of storm surge. In other words, can the storm surge be adequately determined from subtidal sea level alone, which is relatively well-understood?

We will examine the two major storm surge events in the period of our subtidal sea level study (July 1974 to June 1975). We will describe the atmospheric forcing (extratropical cyclone), the Bay water response, and the relation between subtidal sea level and storm surge.

STORM SURGE

A. Event I (December 1 to 4, 1974)

On December 1, 1974, a low pressure disturbance (cyclone) was centered around 35°N,85°W (Fig. 2a). Winds were southwestward along the Mid-Atlantic coast (Cape Cod to Cape Hatteras), which generated an onshore Ekman transport. Consequently, sea levels increased over the entire Bight. In particular, the sea level rise was about 70 cm at the mouth of Chesapeake Bay (Kiptopeake B.) (Fig. 3). Associated with coastal sea level change, sea levels also increased throughout the Bay.

The cyclone propagated to the northeast, and its center passed over the Bay area on 0600 December 2 (Fig. 2b), which resulted in a local northward wind (Fig. 3). The northward wind set-up was quite pronounced; this explains the high sea level at the Bay head (Havre de Grace).

The cyclone continued moving northeastward, and it was centered around Nova Scotia on December 3 (Fig. 2c). The intensity of the cyclone also had significantly increased; the central pressure on December 3 was 982 mb, compared to 1004 mb on December 1. Winds were northeastward along the Mid-Atlantic coast, which generated an offshore Ekman transport. Consequently, sea levels decreased

326

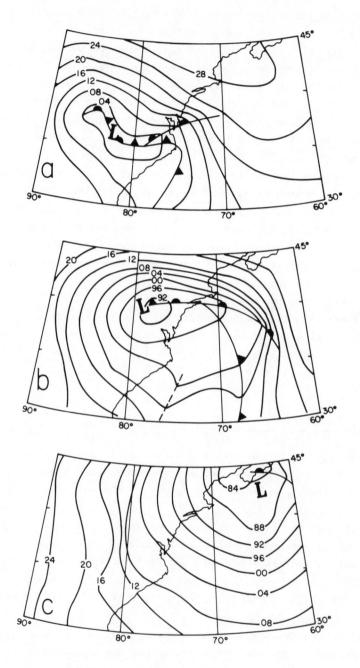

Fig. 2. Surface weather (atmospheric pressure) map on: (a) 1200 December 1, (b) 1200 December 2, and (c) 1200 December 3, 1974.

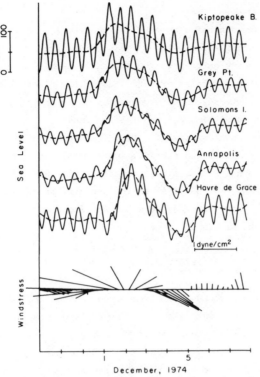

Fig. 3. The original (solid lines) and lowpass (dashed lines) sea levels, and the lowpass windstress at Patuxent.

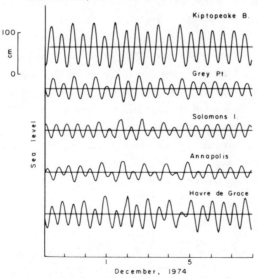

Fig. 4. The highpass sea levels.

over the entire Bight. The additional sea level drop at Havre de Grace was due
to the local wind set-down (Fig. 3).

The storm surge was dominated by subtidal sea level. In fact, the response
model (Wang, 1978a) which was developed for subtidal sea level, gives a satisfactory account of the surge event. The Bay and coastal sea levels responded to
the E-W windstress at time scales of 4 to 7 days; the rise/fall of sea level was
associated with the westward/eastward windstress. In addition, the N-S windstress
drove local set-up/down at time scales of 1 to 3 days.

The supertidal component was small. Fig. 4 shows the highpass records (difference between the original and subtidal sea levels): the semidiurnal tide was
dominant, and the diurnal tide was also clearly reflected by the "diurnal inequal-
ities." There were indications of storm influence in the upper Bay (Annapolis
and Havre de Grace). However, they were too small compared to the subtidal component, to have practical significance.

B. Event II (April 3 to 6, 1975)

On April 3, 1975, a low pressure disturbance was centered around 45°N,80°W
(Fig. 5a). Winds were westward along the New England coast, however, they were
northward over the southern Bight and Chesapeake Bay. Coastal sea levels did
not respond to the northward wind, apparently due to the lack of large-scale
(coherent) forcing. On the other hand, significant set-up in the Bay was induced
by the local wind (Fig. 6).

The cyclone propagated to the east, and it was centered around the Gulf of
Maine on April 4 (Fig. 5b), which resulted in a southeastward wind along the Mid-
Atlantic coast. As the cyclone continued moving eastward (Fig. 5c), winds became southward over the Chesapeake Bay. The local southward wind set-down was
large: the sea level difference was over 100 cm between Kiptopeake B. and Havre
de Grace (Fig. 6). Coastal sea level also dropped slightly on April 4.

The storm surge was dominated by subtidal sea level. The rise/fall of sea
level was mainly due to the northward/southward wind set-up/down. The eastward
wind was partly responsible for the sea level decrease on April 4. The supertidal component was also significant in the upper Bay (Fig.7). The regular tidal
oscillation was suppressed during the storm period.

DISCUSSION

Our analysis of two strong extratropical storm surges in the Chesapeake Bay
suggests that subtidal sea level is the dominant surge component. Our results
and Wang (1978a) also indicate that surges can be induced by local wind set-up,

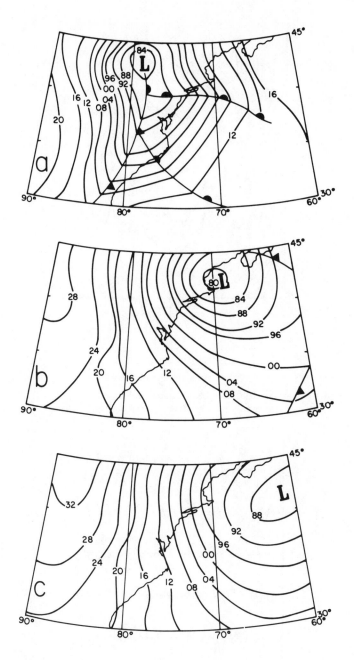

Fig. 5.   Surface weather (atmospheric pressure) map on: (a)   1200 April 3,
(b) 1200 April 4, and (c) 1200 April 5, 1975.

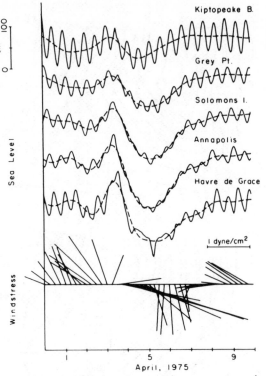

Fig. 6. The original (solid lines) and lowpass (dashed lines) sea levels, and the lowpass windstress at Patuxent.

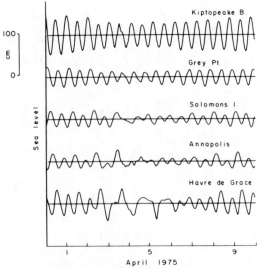

Fig. 7. The highpass sea levels.

and nonlocal coastal sea level effect. The nonlocal effect (coastal surge) can be very important under favorable large-scale forcing conditions. For example, the maximum surge height (at Havre de Grace) was comparable between the two events, despite the fact that the local longitudinal windstress was about twice the magnitude in event II. The compensation was due to the large coastal surge in event I.

The local wind set-up is well-known; Wang (1978a) found high coherence between longitudinal windstress and surface slope over a year-long period. The wind set-up can be easily adopted and calibrated in the storm surge model. The non-local effect however, is less well-known. In the estuary surge model, the coastal effect is usually modeled as "observed" surface elevations at the open ocean boundary.

Wang (1978a) indicated that the Bay and coastal water response to E-W wind forcing is coupled. Thus, it may not be appropriate to treat the two systems separately. The present modeling of "open ocean" surge is also rather poor. Wang (1978b) indicated that coastal sea levels along the Mid-Atlantic Bight are driven by: (a) the local Ekman transport, (b) the local alongshore wind set-up, and (c) the nonlocal shelf waves. The "open ocean" surge model however, mainly considers the effect of cross-shore wind set-up (Pagenkopf and Pearce, 1975). It seems unlikely that the "open ocean" surge model is applicable to extratropical storm surges.

In conclusion, our study on the storm surge in Chesapeake Bay suggests that observational study should be emphasized. Recognizing that the model validation procedure is usually rather arbitrary, governing processes must be examined from observations. Only if these processes are clearly identified, can the regional storm surge model be formulated and tested properly. A continuous feedback between model prediction and field verification is the only lead to a verified model for surge forecast.

ACKNOWLEDGEMENTS

We thank Mr. Jose Fernandez-Partagas who kindly made the weather charts available to us. This study was supported by the National Science Foundation, under Grant OCE74-08463 and OCE77-20254.

REFERENCES

Heaps, N.S. and Jones, J.E., 1975. Storm surge computations for the Irish sea using a three-dimensional numerical model. Mémoires Societe Royale des Sciences de Liége, 6e série, tome VII, 289-333.
Leenderste, J.J., Alexander, R.C. and Lin, S.K., 1973. A three-dimensional model for estuaries and coastal sea. The RAND Corp., R-1417-OWRR, 57 pp.

Pagenkopf, J.R. and Pearce, B.R., 1975. Evaluation of techniques for numerical calculation of storm surges. R.M. Parsons Laboratory, MIT, Report No. 199, 120 pp.

Simons, T.J., 1973. Development of three-dimensional numerical models of the Great Lakes. Canada Centre for Inland Waters, Scientific Series No. 12, 26 pp.

Wang, D.P., 1978a. Subtidal sea level variations in the Chesapeake Bay and relations to atmospheric forcing. To appear in J. Phys. Oceanogr.

Wang, D.P., 1978b. Low-frequency sea level variability on the Middle Atlantic Bight. Submitted to J. Mar. Res.

FIRST RESULTS OF A THREE-DIMENSIONAL MODEL ON THE DYNAMICS IN THE GERMAN BIGHT

J. BACKHAUS

Deutsches Hydrographisches Institut, Hamburg (F.R.G.)

ABSTRACT

   A three-dimensional barotropic fine mesh model of a shallow coastal sea is described. The tidal dynamics in very shallow water, e.g. wetting and drying of mud flats, are simulated by means of a movable horizontal boundary. A critical examination of the model results, especially of the vertical current structure, is carried out. In particular the influence of the wind on the horizontal and vertical current distribution is studied by simulating the extreme case of a storm surge and some idealized mean wind conditions.

INTRODUCTION

   The threat of oil spills, the increased dumping of industrial waste into the sea, and last - but not least - storm surges, are common problems in coastal oceanography. Taking these problems into account, it is essential to have detailed knowledge about the general circulation of water masses in the area under consideration, which - in the case of this study - is the German Bight.

   The spatial and temporal distribution of current and water-level data about the German Bight is rather incoherent, because a synoptic survey of the entire area has never been carried out. Therefore, knowledge about the wind and tide generated circulation in the German Bight still need improvement. The vertical distribution of residual currents in particular is rather unknown. This has given rise to the development of a three-dimensional numerical model and to extensive measuring efforts, terminating in a synoptic survey of currents and water levels taken over a period of one year within the framework of the 1979 "Year of the German Bight" experiment.

   Some locations for permanent moorings (current meters, tide gauges, meteorological buoys) to be deployed in the German Bight, were selected by means of the simple

334

model here presented.

A good way to develop a model for a particular sea area, is to improve the model stepwise; beginning with a very simple version, and always comparing the model results with measurements. In so doing, one can hope to learn a great deal about the behaviour of the model, and the physical processes in the area under consideration

In this study, the model equations and numerical techniques will be described on very briefly, more emphasis is laid upon a critical consideration of the "simulatic ability" of the model, in order to find out how it could be further improved.

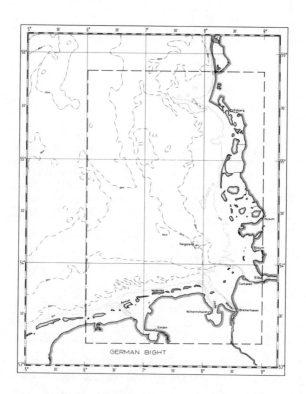

Fig. 1a. Map of German Bight, dashed line indicates area covered by the model.

THE MODEL, GENERAL DESCRIPTION

A fine, horizontal grid resolution of 3 nautical miles was chosen to approximate the German Bight's topography, which is rather complex, especially in the near shore regions. The largest system of coastal drying banks, which exists in the entire Nort Sea region, in combination with small islands, is to be found along the coast of the German Bight (Fig. 1a, 1b). Water depths vary between 45 m below mean sea level and 2 m above mean sea level (drying banks) in coastal waters. For this first three-dimensional modelling approach on the simulation of dynamics in a well-mixed shallow

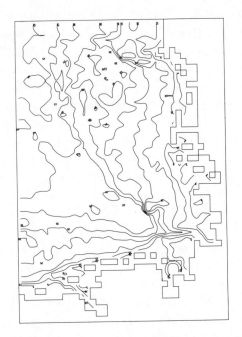

Fig. 1b. Depth (m) contours of discretisized bottom topography.

sea, a vertical equidistant discretisation of 15 m was chosen.The simulation of the wetting and drying of tidal flats is carried out in the top layer (area between two adjacent computation levels) by means of a movable model boundary (Backhaus, 1976). As the sea is considered to be well mixed, all three layers have equal homogeneous density; therefore, the model is barotropic. The assumption of well-mixed conditions is not valid during summer; however, as far as could be estimated from measurements, baroclinic effects seem to be at least one order of magnitude smaller than the effects arising from bottom turbulence and non-linear wind/tide interactions.

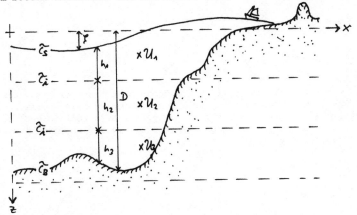

Fig. 2. Sketch of vertical configuration of the model.

336

The computation levels (Fig. 2, dashed lines) are horizontally fixed and completely permeable, so that the water can move freely in the basin. The internal she stresses τi are defined at these levels; at the surface and the bottom respectively quadratic stress laws are applied. Turbulence is parameterized by means of a consta vertical eddy viscosity coefficient Av = 40 cm^2/s and by a depth dependent horizont exchange coefficient Ah= h · 5 m/s. The model could be regarded as quasi-linear, with respect to the non-linear bottom friction. A vertically integrated flow is computed for each layer; the depth mean flow is obtained simply by integrating over the number of layers. The surface elevation is calculated from the equation of cont nuity (1), using the horizontal divergence of the depth mean flow.

In the equations of motion (2), which are given in momentum form for an arbitrar layer, the non-linear terms are omitted. As - for example - proposed by Simons (197 the layerwise vertically integrated equations of motion are coupled by the internal shear stresses and by the barotropic pressure gradient, which does not vary with depth.

No flux normal to closed boundaries may occur, slip along walls is permitted. Water levels are prescribed at open boundaries, and, for all layers, the gradient of the flux normal to the boundary is assumed to be zero. Together with the stresse given at the sea surface and bottom, this set of boundary conditions closes the pro lem for the barotropic case.

The numerical integration technique used is basing on the well-known explicit difference scheme introduced by Hansen (1956). The scheme was extended for the thir dimension in a similar manner to that proposed by Sündermann (1971). The coupled system of partial differential equations (1, 2) are solved approximatively on a temporally and spatially staggered grid.

$$\dot{H} + \bar{U}_x + \bar{V}_y = 0 \quad , \quad H = D + \zeta = \sum_L h \quad , \quad \bar{U} = \sum_L U \quad , \quad \bar{V} = \sum_L V \tag{(}$$

$$\dot{U} = f\,V - g\,h\,\zeta_x + (\,Ah\,U_x\,)_x + (\,Ah\,U_y\,)_y + (\,Av\,U_z\,)_z \tag{(}$$

$$\dot{V} = -\,f\,U - g\,h\,\zeta_y + (\,Ah\,V_x\,)_x + (\,Ah\,V_y\,)_y + (\,Av\,V_z\,)_z$$

Where U,V = horizontal transport, $\bar{U},\bar{V}$ = depth mean transport, H = actual water dept D = undisturbed water depth, ζ = surface elevation, h = layer thickness, f = corioli (constant), g = acceleration due to gravity, Ah, Av = coefficients of horizontal and vertical eddy viscosity, L = number of layers, x,y,z = coordinate system (east, nort and down respectively).

MODEL RESULTS

Before discussing the results of the model, some remarks about tidal dynamics in the German Bight are given. There is an amphidromic point (Fig. 4b) some 200 km North-West of the vertex of the right-angle shaped coastline. Therefore, tidal elevations have a wide range, varying from a few centimetres near the amphidromic point to about 1.5 m near the vertex. The tidal wave, travelling through the German Bight, shows a counter-clockwise sense of rotation, which also applies to the currents. Some examples of measured currents are shown by means of their current

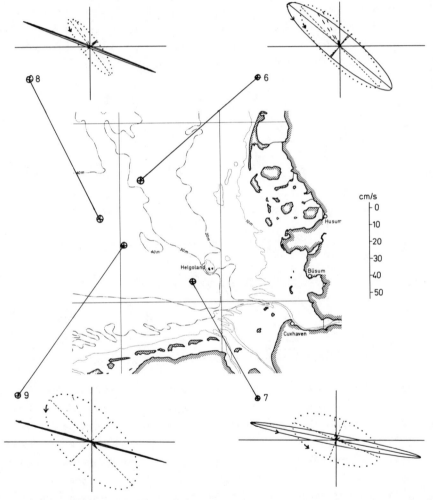

Fig. 3. Current ellipses (M_2) for near surface (full line) and near bottom (dotted line) measurements. The sense of rotation is indicated by arrows.

ellipse for the M_2 tidal constituent (Fig. 3). The 30 m depth contour in the chartlet of the Figure gives an idea of a special formation in the German Bight's bathymetry:

the remains of a post-glacial estuary of the River Elbe. From measurements, as well as from model results, it can be observed that this prehistoric estuary has a strong influence upon the vertical structure of the currents, which becomes obvious from the current ellipses shown. In the vicinity of the underwater estuary, a general narrowing of the near-surface current ellipses can be found, indicating a zone of maximum vertical shear in the German Bight.

REPRODUCTION OF THE TIDE

Since the tide is the dominant signal in the North Sea, it should be reproduced correctly in the model, and with sufficient accuracy, before that model is applied to other cases, for example, to wind and tide-induced residual currents. The propagation of the tidal wave in the German Bight is simulated for the case of the

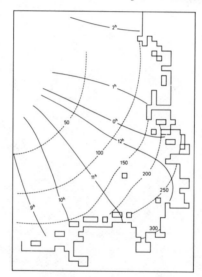

Fig. 4a. Computed co-tidal and co-range (cm) lines for M_2 tide.

dominant semi-diurnal lunar tide (M_2). The boundary values (surface elevations), prescribed at the open boundaries, were previously computed with a general two-dimensional North Sea model. The computed surface elevations of the North Sea model generally agree with observations, those of the fine-mesh German Bight model are of similar accuracy. Co-tidal and co-range lines for the M_2 tide, obtained with the German Bight model (Fig. 4a) are compared with a chart, basing upon observations (Fig. 4b). A few improvements were made, due to a better resolution of the coastal topography, especially in the vicinity of the Jade/Weser/Elbe estuaries. The horizontal resolution of the grid is far too coarse to give a correct simulation of dynamics near the coast. Here, processes of sub-grid scale are parameterised very

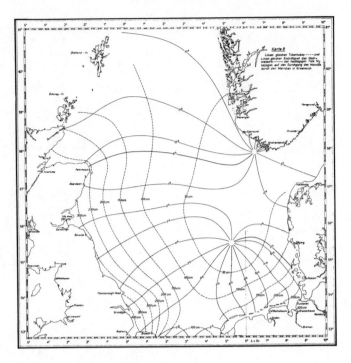

Fig. 4b. Co-tidal (related to moons transit in Greenwich) and co-range (cm) lines for M$_2$ tide. (adopted from Hansen, 1952)

roughly. However, the focus of this study is related mainly to the circulation in the deeper parts of the German Bight, and there the resolution seems to be sufficient.

For a comparison with measured tidal currents, the tidal signal had to be extracted from the current meter data by means of a bandpass filter. Filtered near-surface and near-bottom measurements for a period between spring tide and neap tide (stations 7 and 9 in Fig. 3) are compared with computed results in the corresponding layers of the model (Fig. 5, left of vertical dashed line). The agreement for the near-bottom currents is already quite close; whereas, in the near-surface regions larger deviations occur. For these regions the vertical resolution of the model seems to be insufficient, since we know, from observations, that the vertical current shear is strongest near the surface.

Comparisons were carried out for more than the two stations shown, and - in general - the same features, as described above, were observed.

Apart from the discrepancies in near-surface currents, the sense of rotation and the phase of the currents seem to have been correctly simulated. This is also valid for the amplitudes in the lower layers. Therefore, it might be justified to apply the model for cases other than the pure tide.

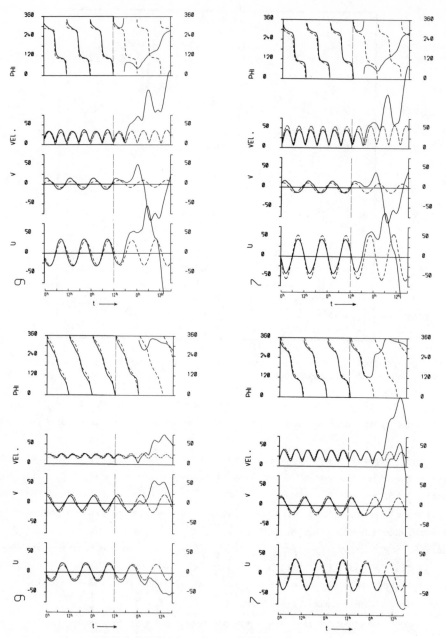

Fig. 5. Computed (full line) and observed (dashed line) currents. Direction (true north), speed, and north - and east-components (cm/s) of near surface (top) and near bottom (below) currents for station 9 (left) and station 7 (right). Vertical dashed line corresponds to January 2nd 1976, 12 noon (see Fig. 6).

SIMULATION OF A STORM SURGE

Storm surges which, from time to time, cause exceptional damage along the coast, are one of the problems in the German Bight. One question concerning modelling aspects is, whether or not a fine-mesh German Bight model - which could be regarded as a nest of the general North Sea model - will improve the accuracy of storm surge simulations in the German Bight. For that purpose, both models were run with the same set of three-hourly wind stress fields, computed from re-analysed weather maps for the storm surge of January 3rd, 1976, (Hecht, Sußebach, personal communication, 1977). The North Sea model was run first, in order to obtain a consistent set of boundary values for the German Bight model. Both models are running separately, without inter-action, because of a restricted computer memory.

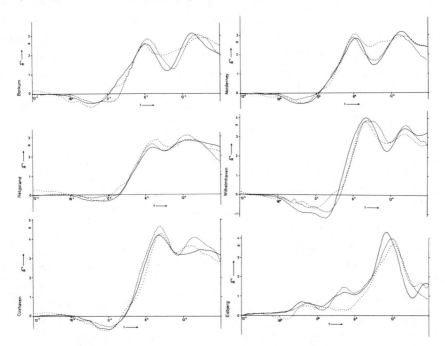

Fig. 6. Observed (dotted) and computed (dashed = 3 dim. German Bight model, full line = 2 dim. North Sea model) residuals (m) of surface elevations during storm surge of January 3rd 1976, starting at January 2nd, 12 noon.

For some coastal tide gauges, a comparison of measured and computed water level residuals is shown (Fig. 6). The residuals were obtained by subtracting the tidal surface elevations from those containing tide plus surge. When comparing the resid-uals obtained with the coarse mesh North Sea model (grid size approx. 20 km) with those of the fine mesh model no significant improvement is to be observed, in general. This could have been expected, because the boundary values, computed with the North Sea model, are the dominant forcing, besides that of the wind. Positive effects arising from a better horizontal resolution are either very small, or not present.

Obviously, a fine mesh model is not necessary, when a simulation of surface ele‹
tions only is desired. As concerns the storm surge modelling G. Fischer and other
participants at this colloquium agreed, that there are still things which are more‹
important than grid refinements. More weight should be placed upon the meteorological input data and surface stress parameterisations in combination with wave-
induced motions and surface elevations.

However, the computed vertical current structure during the storm surge (Fig.5,
(solid curves) right of vertical dashed line) show a remarkable amplification of t
current speed, especially near the surface, which is up to four times as large as
normal (see dashed curves). A similar factor is known from the very few current
measurements taken during storm surges in the German Bight. Note that the near-sur
inflow is followed almost instantaneously by an outflow near the bottom.

The circulation during the storm surge becomes clearer if current residuals are
viewed for the tidal cycle - when the maximum inflow and the peak of the surge occ‹
(Fig. 7a); and, for the subsequent cycle, when the piled-up water masses are rushi‹
back into the North Sea basin (Fig. 7b). Note the persistent outflow in the bottom
layer.

RESIDUAL CURRENTS

The residual currents in the North Sea are driven mainly by wind and tide. The
influence of the wind is much stronger (up to one order of magnitude) than
that of the tide. During spring and autumn the residual circulation is rather variable, because rapid changes in meteorological conditions occur.

The general vertically integrated mean circulation of the North Sea is known
fairly well by now (Maier-Reimer, 1977). From observations, it is known that the
circulation can vary considerably with depth, which is of extreme interest for all
marine pollution problems. Particularly in the German Bight, large differences in
speed as well as in direction between near-surface and near-bottom residual current
are observed (Mittelstaedt, personal communication, 1978).

In order to study the influence of the wind on the circulation in the German Big
some computations were carried out, using homogenous and constant wind fields of
different direction for the entire North Sea region. This rather idealized wind
forcing is far away from reality; but, nevertheless, some principal knowledge will
be gained about the processes which are causing the vertical distribution of the
residual currents observed in the German Bight.

A (moderate) wind speed of 5 m/s was chosen for all wind fields. The computation
were started from a quasi-steady state tidal cycle; again consistent sets of
boundary values for each wind situation were previously computed with the North
Sea model. A quasi-steady state was reached for all cases after at least five tidal
cycles. The model's response time on the wind field is of the order of one tidal

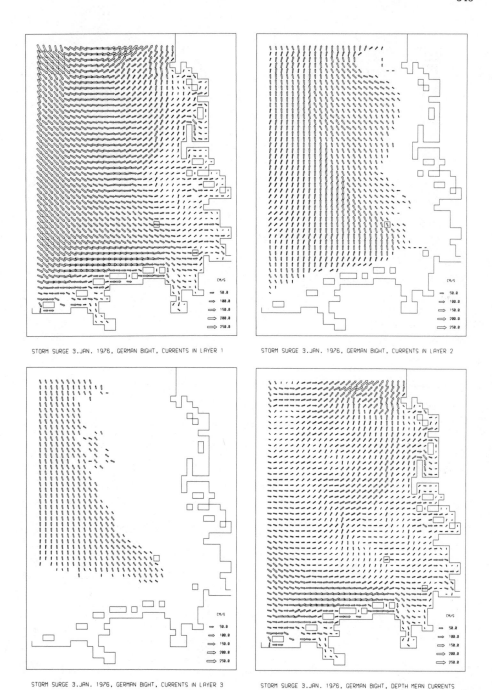

STORM SURGE 3.JAN. 1976, GERMAN BIGHT, CURRENTS IN LAYER 1

STORM SURGE 3.JAN. 1976, GERMAN BIGHT, CURRENTS IN LAYER 2

STORM SURGE 3.JAN. 1976, GERMAN BIGHT, CURRENTS IN LAYER 3

STORM SURGE 3.JAN. 1976, GERMAN BIGHT, DEPTH MEAN CURRENTS

Fig. 7a. Residual currents during storm surge of January 3rd 1976, for 'inflow period'.

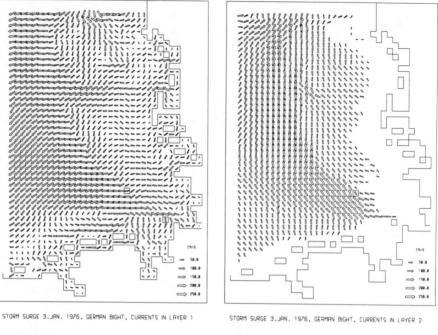

STORM SURGE 3.JAN. 1976, GERMAN BIGHT, CURRENTS IN LAYER 1

STORM SURGE 3.JAN. 1976, GERMAN BIGHT, CURRENTS IN LAYER 2

STORM SURGE 3.JAN. 1976, GERMAN BIGHT, CURRENTS IN LAYER 3

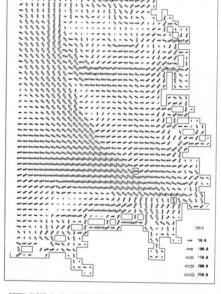

STORM SURGE 3.JAN. 1976, GERMAN BIGHT, DEPTH MEAN CURRENTS

Fig. 7b. Residual currents during storm surge of January 3rd 1976, for 'outflow period'.

cycle, which can also be observed in nature.

The residual currents $\bar{u}$, also called mean transport velocity elsewhere, were obtained by integrating over one tidal cycle, using the following formula (3):

$$\bar{u} = \int_T U \, dt \;/\; \int_T h \, dt \tag{3}$$

where T is the period of the M_2 tide.

This formula is applied for each layer, but for the surface layer only, the second integral needs to be computed, because only there the layer thickness h is dependent upon time.

For the cases of winds blowing from North-West, South-West, and South-East, the quasi-steady state residual circulation computed is shown for all three layers, and for the depth mean flow (Figs. 8, 9, 10). A considerable vertical current shear, especially in the area of the underwater estuary, can be observed in the flow patterns. Generally, these results are in good agreement with the residuals, computed and selected for certain wind situations by Mittelstaedt, using current meter data obtained in the German Bight, measured during the past 10 years.

Again, the largest discrepancies between computation and measurement occur in the near-surface region. From observations, as well as from model results, there is evidence that the vertical change from the near-surface flow to the currents in deeper regions, occurs in a rather narrow "transition zone". For comparisons between model results and measurements, it is important to know whether or not a current meter was moored above or below, or possibly right in the transition zone. It should be mentioned here, that, for technical reasons, no current meter was moored closer to the sea surface than 8 metres; all "near-surface" data has been measured in a depth range of about 8 to 12 m below the surface. However, if discrepancies between computations and measurements occur, they might be caused by both the model and/or the data. Some further and careful work is necessary here.

The variability of computed near-surface currents, in dependence of the wind direction, is much stronger than for the circulation in the deeper areas of the German Bight. There the circulation is rather persistent and significant changes only occur when the wind direction is veering from westerly winds to easterly or vice-versa. The depth circulation is mainly driven by the slope of the mean sea level. All westerly winds pile up the watermasses in the German Bight, causing a compensating outflow in near-bottom regions, focused by the underwater estuary of the River Elbe. The same mechanism causes an inflow for all easterly winds. The knowledge of these features of the vertical distribution of the residual currents is - for example - very important for the selection of dumping areas, and for the depth in which chemical waste should be dumped, in order to prevent it reaching the coast.

346

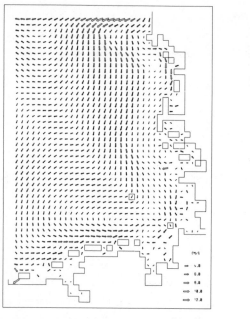

CERMAN BIGHT, RESIDUAL CURRENTS LAYER 1, WIND NW 5 M/S

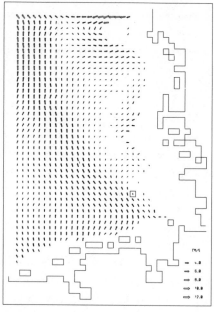

GERMAN BIGHT, RESIDUAL CURRENTS LAYER 2, WIND NW 5 M/S

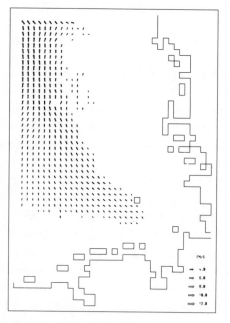

GERMAN BIGHT, RESIDUAL CURRENTS LAYER 3, WIND NW 5 M/S

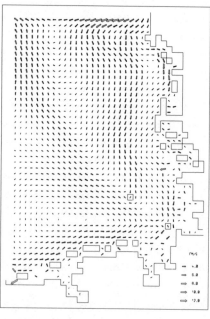

GERMAN BIGHT, DEPTH MEAN RESIDUAL CURRENTS, WIND NW 5 M/S

Fig. 8. Quasi steady state circulation for NW wind (5 m/s).

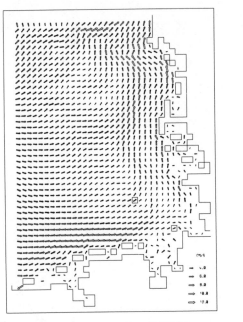

GERMAN BIGHT, RESIDUAL CURRENTS LAYER 1, WIND SW 5 M/S

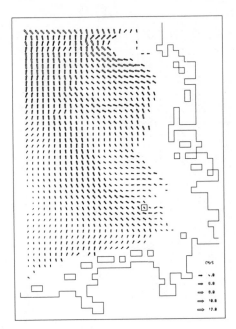

GERMAN BIGHT, RESIDUAL CURRENTS LAYER 2, WIND SW 5 M/S

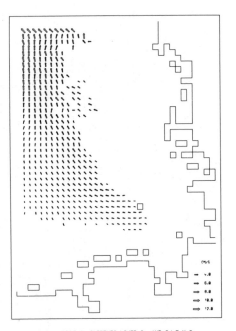

GERMAN BIGHT, RESIDUAL CURRENTS LAYER 3, WIND SW 5 M/S

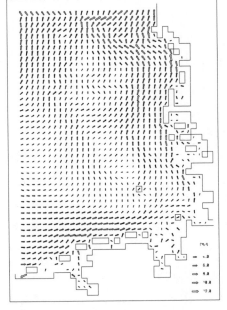

GERMAN BIGHT, DEPTH MEAN RESIDUAL CURRENTS, WIND SW 5 M/S

Fig. 9. Quasi steady state circulation for SW wind (5 m/s).

348

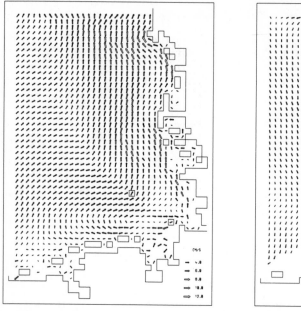

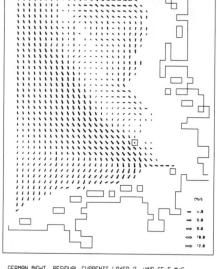

GERMAN BIGHT, RESIDUAL CURRENTS LAYER 1, WIND SE 5 M/S

GERMAN BIGHT, RESIDUAL CURRENTS LAYER 2, WIND SE 5 M/S

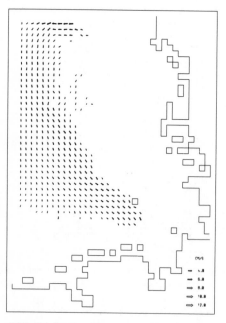

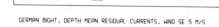

GERMAN BIGHT, RESIDUAL CURRENTS LAYER 3, WIND SE 5 M/S

GERMAN BIGHT, DEPTH MEAN RESIDUAL CURRENTS, WIND SE 5 M/S

Fig. 10. Quasi steady state circulation for SE wind (5m/s).

CONCLUDING REMARKS

Apart from insufficiencies, caused by a poor vertical resolution near the sea surface, the rather simple model version described is able to already simulate the significant features of the dynamics in the German Bight. The knowledge about the horizontal and vertical distribution of currents was improved by applying the model to some significant cases.

ACKNOWLEDGEMENTS

The author is indebted to Prof. K. Hasselmann, who encouraged him to participate at this colloquium. The assistence of Mrs. Barttels and Mrs. Petersitzke in preparing and typing the manuscript is very much appreciated. Thanks to Mr. Höntzsch for adding the final touch to the diagramms.

REFERENCES

Backhaus, J., 1976. Zur Hydrodynamik im Flachwassergebiet, ein numerisches Modell. Deutsche Hydrogr. Zeitschrift, 29:222-238.
Hansen, W., 1952. Gezeiten und Gezeitenströme der halbtägigen Hauptmondtide M2 in der Nordsee. Ergänzungsheft, Deutsche Hydrogr. Zeitschrift, Reihe A, Nr. 1.
Hansen, W., 1956. Theorie zur Errechnung des Wasserstandes und der Strömungen in Randmeeren nebst Anwendungen, Tellus No. 8.
Maier-Reimer, E., 1977. Residual circulation in the North Sea due to the M2-tide and mean annual wind stress. Deutsche Hydrogr. Zeitschrift, 30:69-80.
Neumann, H., Meier, C., 1964. Die Oberflächenströme in der Deutschen Bucht. Deutsche Hydrogr. Zeitschrift, 17:1-40.
Simons, T.J., 1973. Development of three-dimensional numerical models of the Great Lakes. Environment Canada, Scientific series no. 12.
Sündermann, J., 1971. Die hydrodynamisch-numerische Berechnung der Vertikalstruktur von Bewegungsvorgängen in Kanälen und Becken. Mitt. Inst. f. Meereskunde, XIX.
Thorade, H., 1928. Gezeitenuntersuchungen in der Deutschen Bucht. Archiv der Deutschen Seewarte, 46.

TIDAL AND RESIDUAL CIRCULATIONS IN THE ENGLISH CHANNEL

François C. RONDAY

Mécanique des Fluides Géophysiques, Université de Liège, Sart Tilman
B6, B-4000 Liège (Belgium).

Also at the Institut de Mécanique, Université de Grenoble, 38
Saint Martin d'Hères (France).

ABSTRACT

Errors introduced by various numerical schemes for hydrodynamic
models have been analysed for a real situation : the tidal circulation in the English Channel. This analysis is based on the production of harmonics of the M_2 tide. This study shows the unability of
some schemes to give a good representation of tidal harmonics. For
this reason - independently of difficulties to obtain precise boundary conditions - it is always hazardous to calculate the residual
circulation by averaging the transient circulation.

INTRODUCTION

In the English Channel, tidal harmonics are very strong and might
give a non negligible contribution to the residual flow. To carry
out this investigation different depth averaged hydrodynamic models
are used.

The first step of this study is to determine the tidal harmonics
of the main partial tide of the English Channel : the semi-diurnal
lunar (M_2) tide. The first simulation based on a simple numerical
algorithm shows an excellent agreement between calculated and observed (M_2) elevations and currents. Unfortunately this simulation
gives a poor agreement for higher harmonics. As the bottom stress
and the advection generate not only higher harmonics but also a residual component, one cannot expect to have a numerical hydrodynamic
model giving a good representation of the residual flow and a poor
representation of tidal harmonics.

In the litterature, many authors (e.g. Durance, 1974 ; Brettschneider,
1967 ; Flather, 1976) determine the residual circulation by averaging
the transient circulation without considering the generation of tidal
harmonics. Therefore it seems very interesting to verify the ability
of different numerical hydrodynamic models to reproduce the harmonics.

From this study it will be possible to show that the residual flow
calculated by averaging the transient flow is very sensitive to the
discretization of the advection.

GENERAL EQUATIONS OF DEPTH-AVERAGED TIDAL MODELS

If $\underset{\sim}{U}$ denotes the water transport vector and H the total depth
the two-dimensional (depth-integrated) hydrodynamic equations for
tides can be written (e.g. Ronday, 1976):
- in the formalism of the depth-averaged velocity

$$\frac{\partial H}{\partial t} + \bar{u}.\nabla H + H\nabla.\bar{u} = 0 \tag{1}$$

$$\frac{\partial \bar{u}}{\partial t} + \bar{u}.\nabla\bar{u} + f \wedge \bar{u} = \xi - g\nabla\zeta - \frac{D}{H}\bar{u}\|\bar{u}\| \tag{2}$$

- or in the formalism of water transport

$$\frac{\partial H}{\partial t} + \nabla.\underset{\sim}{U} = 0 \tag{3}$$

$$\frac{\partial U}{\partial t} + \nabla.(H^{-1}\underset{\sim}{U}\underset{\sim}{U}) + f \wedge \underset{\sim}{U} = H\xi - gH\nabla\zeta - \frac{D}{H^2}\underset{\sim}{U}\|\underset{\sim}{U}\| \tag{4}$$

with

$$\underset{\sim}{U} = \bar{\underset{\sim}{u}}H = \int_{-h}^{\zeta} \underset{\sim}{u}\,dz \tag{5}$$

$$H = h + \zeta \tag{6}$$

where h is the mean depth, ζ the surface elevation, f the Coriolis
rotation vector, ξ the astronomical tide-producing force per unit
mass, g the acceleration of gravity and D the drag coefficient on
the bottom.

In the English Channel (and the Dover Straits) the astronomical
tide-producing force gives only a very small contribution to the
observed M_2. Therefore, ξ can be neglected in our models, and tide
motions are induced by external forcing along open sea boundaries.

To solve these equations of motion initial and boundary conditions
must be imposed.

Initial conditions

As forced hyperbolic systems are not sensitive to initial conditions, the following initial conditions will be taken

$$\underset{\sim}{U} = 0 \quad \text{and} \quad \zeta = 0$$

or $\qquad\qquad\qquad\qquad\qquad\qquad\qquad\qquad\qquad\qquad\qquad$ (7)

$$\underset{\sim}{\overline{u}} = 0 \quad \text{and} \quad \zeta = 0$$

for all points in the English Channel and in the Dover Straits.

Boundary conditions

- along the coasts $\qquad \underset{\sim}{U}.\underset{\sim}{n} = 0$

$\qquad\qquad\qquad\qquad$ or $\underset{\sim}{\overline{u}}.\underset{\sim}{n} = 0$ $\qquad\qquad\qquad\qquad\qquad\qquad$ (8)

  where $\underset{\sim}{n}$ is the normal at the coast

- along open sea boundaries
  α) Northern open sea boundary.
  As the distance between coastal stations is not too large, a
  linear interpolation between observations at Zeebrugge and
  Foreland gives boundary conditions along the boundary.
  β) Western open sea boundary
  After different numerical simulations, (M_2, M_4, M_6) data coming
  from the physical model of Grenoble (Chabert d'Hières & Leprovost,
  1970) are used along the western boundary.

NUMERICAL METHODS FOR THE RESOLUTION OF TIDAL EQUATIONS

As described in the previous section, tidal motion can be studied
by means of two kinds of hydrodynamic models :
- the first uses the concept of depth-averaged velocity,
- the second the concept of water transport.

From a physical point of view, no differences exist between the
two sets of partial differential equations (1 to 4). However, equations (3 and 4) have a conservative form and this is extremely important in numerical analysis.

To study the propagation of long waves, hydrodynamicists have the
choice between implicit and explicit algorithms.

Implicit algorithms

Implicit algorithms are often unconditionally stable. However,
the ratio $\frac{\Delta t}{T}$ has to be taken sufficiently small to reduce the error
between the solution of the partial differential equations and that
of the finite difference equations. Leendertse (1967) and Nihoul &
Ronday (1976) have shown that the time step (Δt) must remain small

354

when a small phase deformation is imposed. Moreover, implicit algorithms require the resolution of algebraic equations at each time step.

Since the advantage of unconditionally stable schemes cannot be exploited for coastal seas, implicit algorithms are not considered in this study.

Explicit algorithms

All explicit algorithms have a stability condition. The critical time step is a function of the maximum depth, of the maximum velocity, and of the spatial step. For the English Channel, the critical time step is approximatively :

$\Delta t \sim 200$ sec

with $\Delta x = 10$ km.

Only explicit algorithms will be considered in this study.

NUMERICAL MODELS USED TO STUDY THE GENERATION OF TIDAL HARMONICS

To carry out the present investigation, three numerical models based on typical numerical algorithms are used. These models have several characteristics in common :
- the same geographical area,
- the external forces
- the empirical coefficients
- the numerical staggered grid (e.g. Ronday, 1976).

These models differ by the discretization of the equations (1 to 4). The quality of the numerical solution is a function of
- the accuracy of the algorithm
- the conservative or non conservative form of the equations.

Model 1 is based on the concept of the depth averaged velocity, and has been described by many authors (e.g. Hansen, 1966 ; Ramming, 1976 and Ronday, 1976).

The algorithm of resolution is explicit and its accuracy is only $O(\Delta t, \Delta x)$ due to a simple discretization of the advection terms : forward or backward derivatives according to the direction of the current. There arises from this discretization a numerical viscosity $\left(\nu_n \sim \frac{\bar{u} \Delta x}{2} \right)$.

Model 2 is based on the concept of the water transport, and has been used by Fisher (1959) and Ronday (1972).

The algorithm of resolution is explicit and its accuracy is $O(\Delta t, \Delta x^2)$. The centered discretization of the advection terms induces a weak instability. To eliminate this instability, an artificial viscosity $(\nu_n \sim 10^3 \ m^2/s)$ and viscous terms $(\nu_n \Delta \underset{\sim}{U})$ are introduced. An order of magnitude analysis shows that the artificial viscous terms are small compared to the pressure or Coriolis terms.

Model 3 also uses the water transport formalism. The long wave propagation is studied by means of an explicit predictor-corrector procedure.

First, a dissipative procedure (the advection terms are calculated with forward or backward derivatives) gives an estimate of the solution. Secondly, a "weakly" instable procedure corrects this first estimate (the advection terms are calculated with centered derivatives). The accuracy of this two steps procedure is approximatively equal to $O(\Delta t^2, \Delta x^2)$.

ANALYSIS OF RESULTS

Comparison between observed and calculated elevations

Fig. (1 to 18) show the amplitudes and phases of M_2, M_4 and M_6 tides calculated with models 1, 2, and 3. Tables (1 to 3) give the comparison between the observations and the numerical results for some coastal stations. Data are taken from the Deutsches Hydrographisches Institut - Hamburg (1962) and the Bureau Hydrographique de Monaco (1966). The statistical analysis of the elevations is based on eighteen stations (Fig. 19).

α) M_2 tide

i) Fig. (1 to 6) and Table 1 show that the differences existing between the results of the three simulations are very small. The standard deviations are :

for the phases $\sigma_\phi \sim 2°$ (or 4 minutes)

for the amplitudes $\sigma_A \sim 0.13$ m

ii) The agreement between the in situ observations (Table 1) and the numerical results {Fig. (1 to 6)} is excellent. The different models give the following standard deviations :

. for the phases

    model 1 $\sigma_\phi \sim 5.4°$ (or 11 minutes)

    model 2 $\sigma_\phi \sim 6.5°$ (or 13 minutes)

    model 3 $\sigma_\phi \sim 5.4°$ (or 11 minutes)

. and for the amplitudes

    model 1 $\sigma_A \sim 0.12$ m

    model 2 $\sigma_A \sim 0.08$ m

    model 3 $\sigma_A \sim 0.09$ m

TABLE 1

M_2 tide

Comparison between the observations and the numerical results for some coastal stations (amplitude in meters ; phases in degrees)

STATIONS	OBSERVATIONS	MODEL 1	MODEL 2	MODEL 3
St. Servan	3.84/180°	3.74/173°	3.86/173°	3.71/172°
Cherbourg	1.91/230°	1.87/226°	1.90/225°	1.92/227°
Le Havre	2.68/285°	2.56/280°	2.66/279°	2.72/281°
Dieppe	3.11/312°	3.01/307°	3.11/305°	3.18/307°
Hastings	2.47/323°	2.48/321°	2.55/318°	2.63/320°
New Haven	2.27/322°	2.05/313°	2.15/311°	2.24/312°
Nab Tower	1.47/317°	1.41/313°	1.48/310°	1.57/313°
Lyme Regis	1.11/178°	1.28/170°	1.26/171°	1.22/171°
Salcombe	1.48/159°	1.62/156°	1.62/155°	1.62/157°

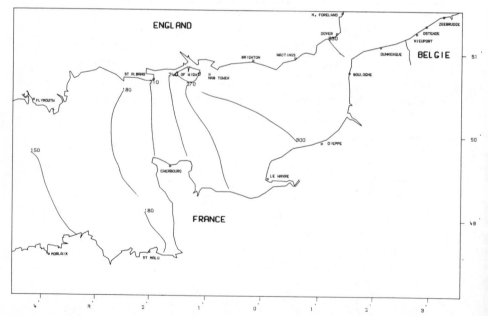

Fig. 1. Lines of equal phases for the M_2 tide calculated with model (in degrees).

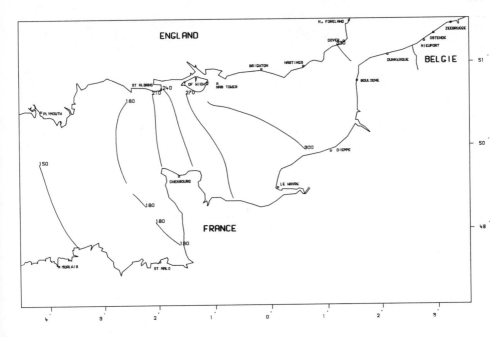

Fig. 2. Lines of equal phases for the M_2 tide calculated with model 2.
(in degrees)

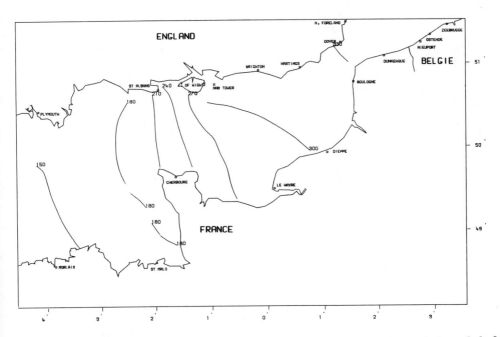

Fig. 3. Lines of equal phases for the M_2 tide calculated with model 3.
(in degrees)

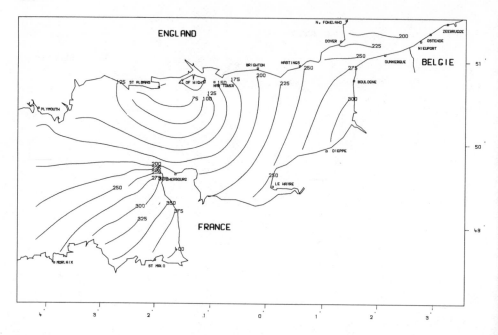

Fig. 4. Lines of equal amplitudes for the M_2 tide calculated with model 1. (in centimeters)

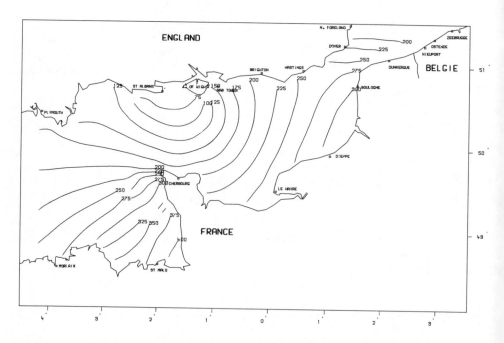

Fig. 5. Lines of equal amplitudes for the M_2 tide calculated with model 2. (in centimeters)

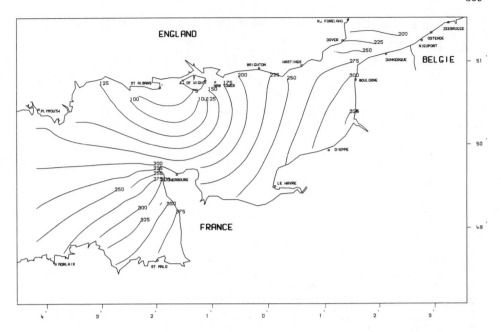

Fig. 6. Lines of equal amplitudes for the M_2 tide calculated with
model 3. (in centimeters).

β) M_4 tide

Results from the three models are presented in Fig. (7 to 12)
and Table 2 gives the comparison between the numerical results and
the observations at different stations. The features which distinguish the respective solutions are as follows :

i) The calculated phases are in general in good agreement with
the observations :

model 1 $\sigma_\phi \sim 23°$ (or 24 minutes)

model 2 $\sigma_\phi \sim 21°$ (or 22 minutes)

model 3 $\sigma_\phi \sim 20°$ (or 20 minutes)

Differences between these simulations also remain small :

$\Delta\phi_{max}$ = 62° (or 64 minutes)

σ_ϕ $\sim 27°$ (or 28 minutes)

ii) Fig. (10 to 12) show the spatial distributions of the M_4 amplitudes in the English Channel. Shapes of these lines are similar, but there are large differences in intensity between the
different simulations :

.<u>Model 1</u> overestimates the M_4 tide (Fig. 10 and Table 2) :
$\sigma_A \sim 0.06$ m

The error is amplified with increasing distance from Cherbourg.
For example, at The Havre the calculated M_4 is 0.34 m instead
of the 0.25 m observed.

Chabert d'Hières and Le Provost (1970) have shown that the M_4
tide is mostly generated near the "Cap de la Hague" and the
"Cap de Barfleur" where the advection is very strong. As the
accuracy of the scheme is poor for the advection terms $0(\Delta t, \Delta x)$
one can expect a radiation of errors from these capes.

TABLE 2

M_4 tide

Comparison between the observations and the numerical results
for some coastal stations (amplitudes in meters ; phases in degrees)

STATIONS	OBSERVATIONS	MODEL 1	MODEL 2	MODEL 3
St. Servan	0.28/286°	0.26/242°	0.30/304°	0.28/293°
Cherbourg	0.14/359°	0.19/ 14°	0.09/338°	0.14/350°
Le Havre	0.25/ 77°	0.34/ 89°	0.20/ 80°	0.25/ 76°
Dieppe	0.27/187°	0.32/174°	0.20/164°	0.27/172°
Hastings	0.22/228°	0.24/212°	0.14/208°	0.23/207°
New Haven	0.09/245°	0.06/205°	0.035/209°	0.065/214°
Nab Tower	0.16/354°	0.16/ 4°	0.08/330°	0.09/333°
Lyme Regis	0.10/ 75°	0.23/ 60°	0.10/ 53°	0.14/ 41°
Salcombe	0.10/132°	0.11/112°	0.09/117°	0.09/110°

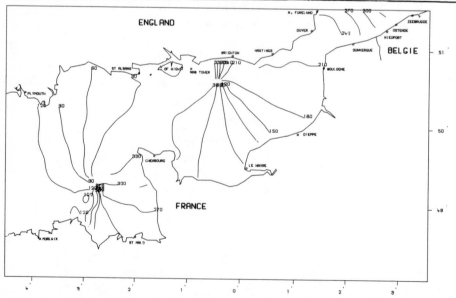

Fig. 7. Lines of equal phases for the M_4 tide calculated with model 1
(in degrees).

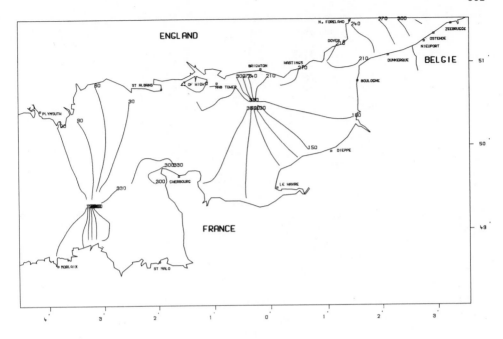

Fig. 8. Lines of equal phases for the M_4 tide calculated with model 2.
(in degrees)

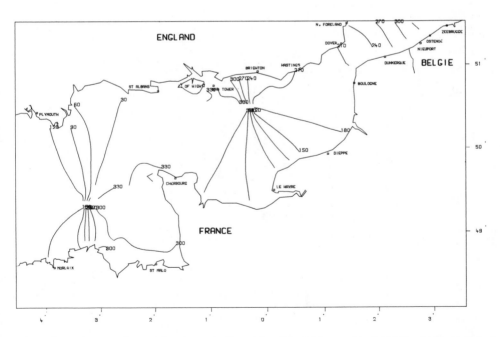

Fig. 9. Lines of equal phases for the M_4 tide calculated with model 3.
(in degrees)

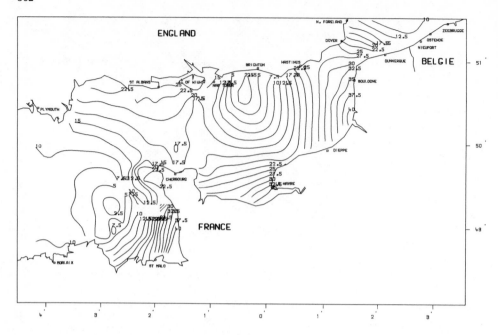

Fig. 10. Lines of equal amplitudes for the M_4 tide calculated with model 1. (in centimeters).

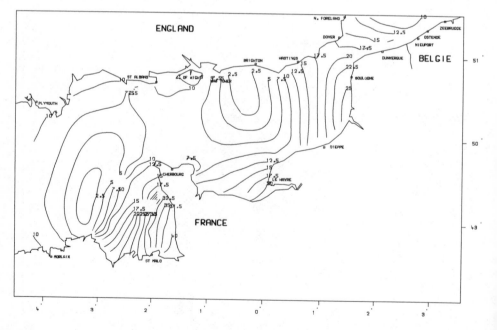

Fig. 11. Lines of equal amplitudes for the M_4 tide calculated with model 2. (in centimeters).

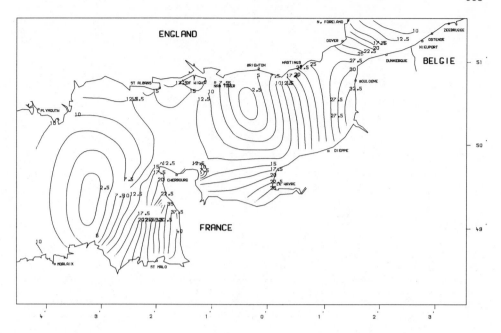

Fig. 12. Lines of equal amplitudes for the M_4 tide calculated with
model 3. (in centimeters).

.Model 2 underestimates the M_4 tide (fig. 11 and Table 2) :
$\sigma_A \sim 0.06$ m
The damping of M_4 comes from the discretization of the advection terms (To maintain a stable procedure with centered derivatives, artificial viscous terms have been introduced).
As the advection is very strong near Cherbourg, the additional
viscosity must be high ($\nu_n \sim 10^3$ m²/sec) to keep a stable
scheme. Therefore this numerical viscosity induces a too
large damping of the solution elsewhere. It is possible to
improve the solution a little by increasing the drag coefficient ($D \sim 2.5 \; 10^{-3}$) and reducing the viscosity 10^2 m²/sec
(Pingree & Maddock, 1977).

.Model 3 gives the best reproduction of the M_4 tide (Fig. 12
and Table 2) :
$\sigma_A \sim 0.03$ m
Now, there is no difference between calculated and observed
tide at The Havre. However, there remains an error (0.07 m)
at Nab Tower. This might be due to the spatial discretization
of this area : the narrow and shallow channel between Nab Tower
and Southampton is not taken into account.

γ) M_6 tide

The analysis of Fig. (13 to 18) and Table 3 leads to the following conclusions :

i) There are few differences between the three simulations : the
shape and the intensity of the iso-lines are similar.

Phases. The concentration of cotidal lines and the lack of resolution near St. Malo explain why the standard deviation seems
large : $\sigma_\phi \sim 31°$ (or 22 minutes).

Amplitudes. The standard deviation $\sigma_A \sim 0.021$ m is small, but
the intensity of the M_6 is also small (of the order of 0.05 m)

TABLE 3

M_6 tide

Comparison between the observations and the numerical results for
some coastal stations (amplitude in meters ; phases in degrees).

STATIONS	OBSERVATIONS	MODEL 1	MODEL 2	MODEL 3
St. Servan	0.02/352°	0.03/289°	0.01/283°	0.01/320°
Cherbourg	0.03/101°	0.04/100°	0.05/ 87°	0.04/ 98°
Le Havre	0.16/286°	0.26/288°	0.28/264°	0.25/269°
Dieppe	0.02/298°	0.04/307°	0.03/300°	0.02/289°
Hastings	0.04/173°	0.05/ 95°	0.06/ 78°	0.06/ 90°
New Haven	0.024/160°	0.04/156°	0.03/137°	0.025/142°
Nab Tower	0.04/119°	0.07/146°	0.06/ 86	0.05/ 87°
Lyme Regis	0.05/103°	0.07/ 97°	0.11/ 53°	0.09/ 73°
Salcombe	0.03/172°	0.03/151°	0.02/128°	0.02/149°

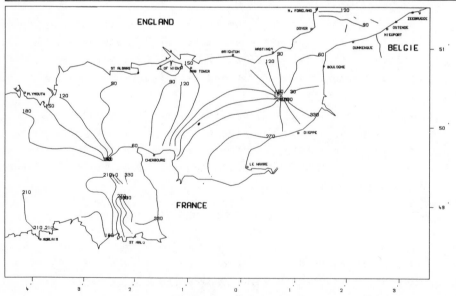

Fig. 13. Lines of equal phases for the M_6 tide calculated with model
(in degrees).

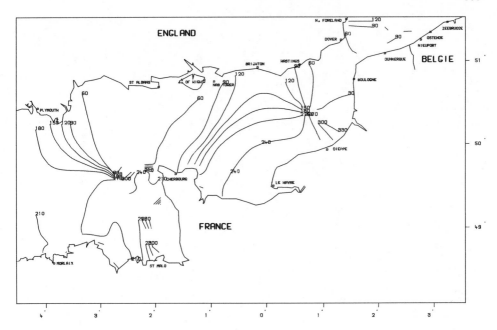

Fig. 14. Lines of equal phases for the M_6 tide calculated with model 2. (in degrees)

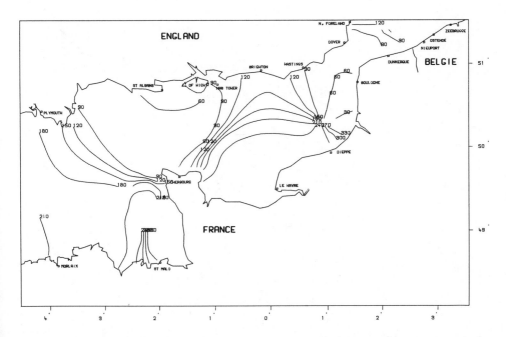

Fig. 15. Lines of equal phases for the M_6 tide calculated with model 3. (in degrees).

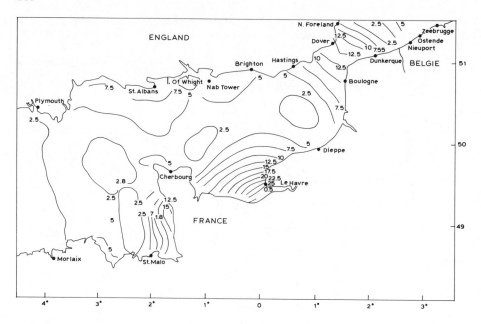

Fig. 16. Lines of equal amplitudes for the M_6 tide calculated with model 1 (in centimeters).

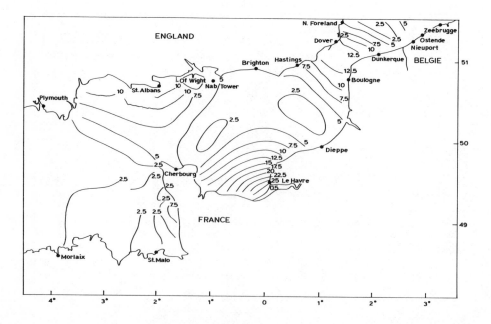

Fig. 17. Lines of equal amplitudes for the M_6 tide calculated with model 2 (in centimeters).

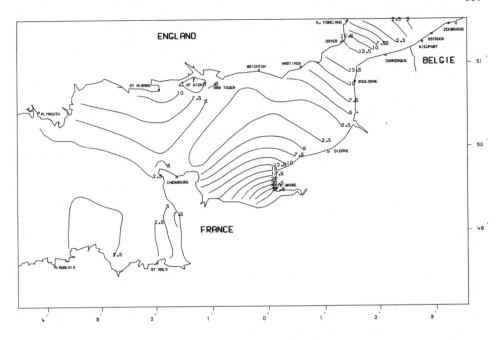

Fig. 18. Lines of equal amplitudes for the M_6 tide calculated with
model 3 (in centimeters).

ii) The agreement between the observations and the numerical results
   is satisfactory for the phases, and poor for the amplitudes if
   one considers the intensity of M_6 in the English Channel :
   for model 1 : σ_ϕ ∿ 26° (or 17 minutes)
   σ_A ∿ 0.034 m
   for model 2 : σ_ϕ ∿ 38° (or 26 minutes)
   σ_A ∿ 0.044 m
   for model 3 : σ_ϕ ∿ 20° (or 14 minutes)
   σ_A ∿ 0.032 m

iii) As no serious improvement exists from one model to another,
   the origin of discrepancies between the observations and the
   numerical results has to be found elsewhere.
   It is well known that a good reproduction of the S_2 tide is
   impossible without the combination of S_2 and M_2 tides. Moreover, the M_6 tide generated by friction depends not only on
   M_2, but also on S_2, N_2,... For a station located between the
   "Cap de la Hague" and Guernesey, Le Provost (1976) showed that
   the $3\omega_{M_2}$ component of the friction term (a source of M_6) is

overestimated (about 20 % at spring tides) if the S_2 tide is
not taken into account.

A spectral analysis of the friction term for the three simulations locates the main source of M_6 near the "Cap de la Hague".
Therefore, the radiation of an error, estimated at about 20 %
near the "Cap de la Hague", can produce much larger errors near
Lyme Regis and The Havre. This error is not affecting regions
located near the open sea boundaries where correct M_6 elevation
are prescribed.

In conclusion, a good reproduction of M_6 is impossible with M_2
only.

COMPARISON BETWEEN CALCULATED AND OBSERVED TIDAL CURRENTS

It is always difficult to compare calculated currents to the observations : currents rapidly vary from point to point, and they are
often reduced to their surface values by means of empirical formulas
in atlas of currents (Sager, 1975).

α) In order to visualize the differences existing between the three
models, fifteen stations (fig. 19) are chosen.

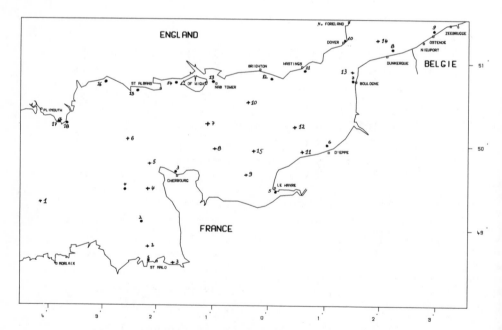

Fig. 19. Stations of comparison for
  • vertical tides
  + horizontal tides

Fig. (20.1 to 20.15) give the amplitude (in cm/s), the direction (in degrees), and the tidal ellipse of currents calculated with the models :

model 1 ——————
model 2 - - - - -
model 3 · · · · ·

The analysis of the figures leads to the following conclusions :

i) Maxima of tidal currents. The three numerical simulations approximately give the same results ($\sigma \sim 0.15$ m/s). The coherence between models 2 and 3 is higher.

ii) Minima of tidal currents. Here the differences are smaller ($\sigma \sim 0.1$ m/s). Currents calculated with models 2 and 3 are more similar.

iii) Phases of tidal currents. Models 2 and 3 approximatively give the same results. Differences with model 1 are of the order of 30 minutes.

iv) Direction of tidal currents. If one excepts the time of tide reversal, the direction of currents is not very sensitive to the scheme used in the model.

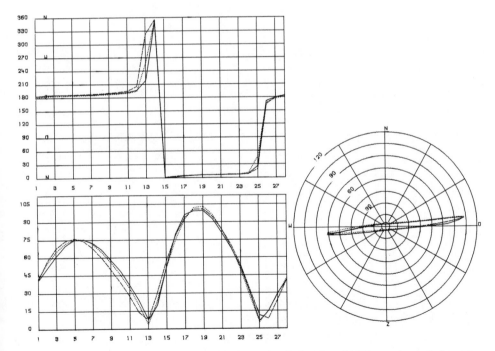

Fig. 20.1. Tidal currents calculated with the different models at station 1. (in cm/s).

370

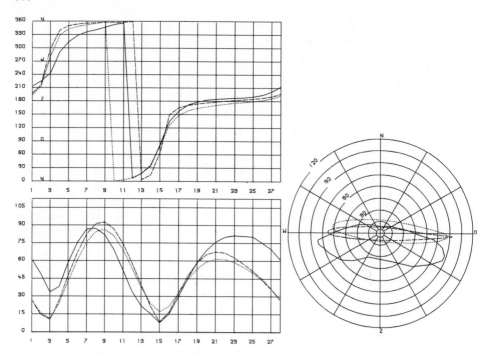

Fig. 20.2. Tidal currents calculated with the different models at
station 2 (in cm/s).

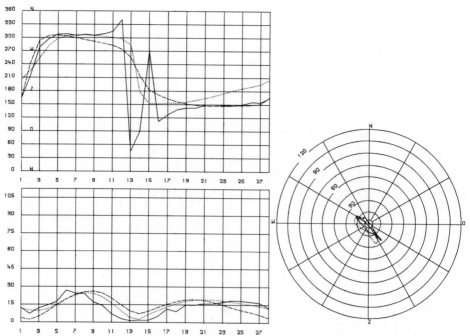

Fig. 20.3. Tidal currents calculated with the different models at
station 3 (in cm/s).

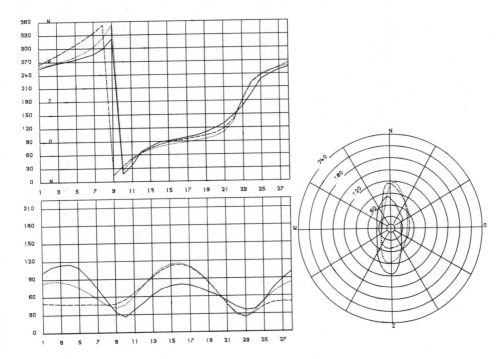

Fig. 20.4. Tidal currents calculated with the different models at
station 4 (in cm/s).

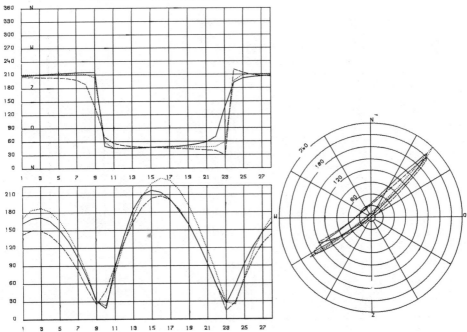

Fig. 20.5. Tidal currents calculated with the different models at
station 5 (in cm/s).

372

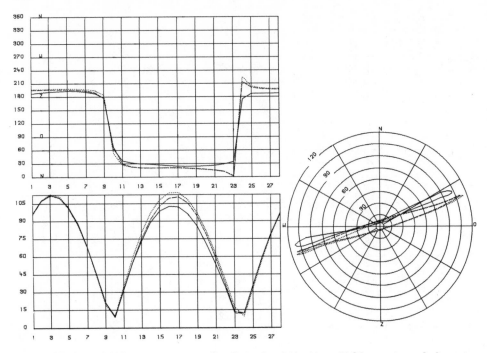

Fig. 20.6. Tidal currents calculated with the different models at station 6 (in cm/s).

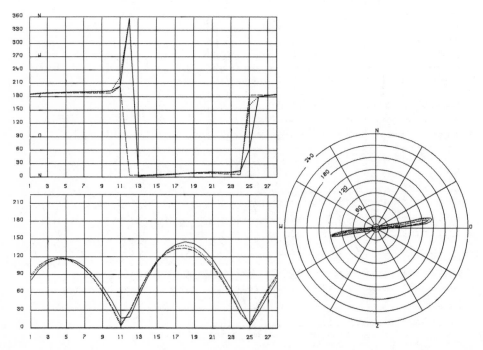

Fig. 20.7. Tidal currents calculated with the different models at station 7 (in cm/s).

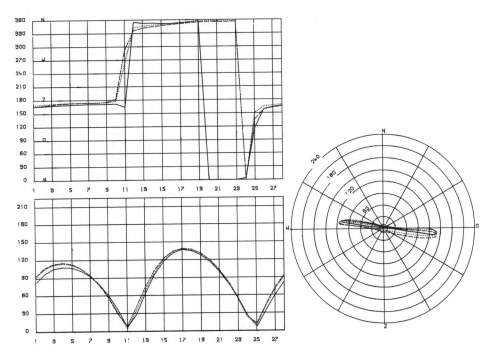

Fig. 20.8. Tidal currents calculated with the different models at station 8 (in cm/s).

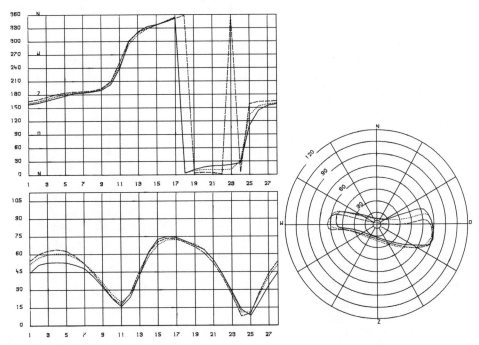

Fig. 20.9. Tidal currents calculated with the different models at station 9 (in cm/s).

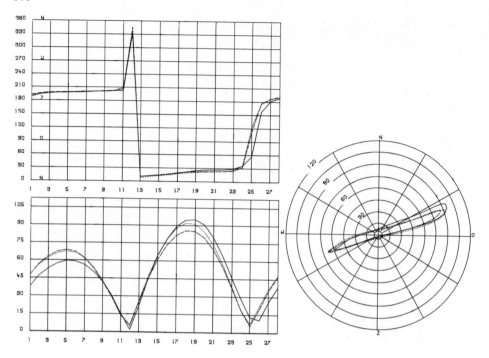

Fig. 20.10. Tidal currents calculated with the different models at station 10 (in cm/s).

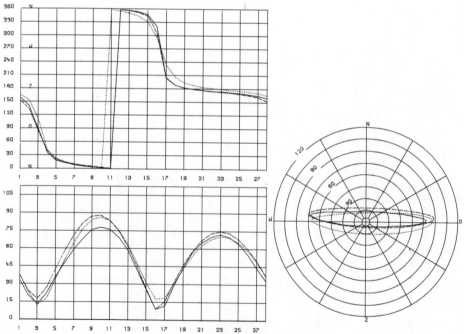

Fig. 20.11. Tidal currents calculated with the different models at station 11 (in cm/s).

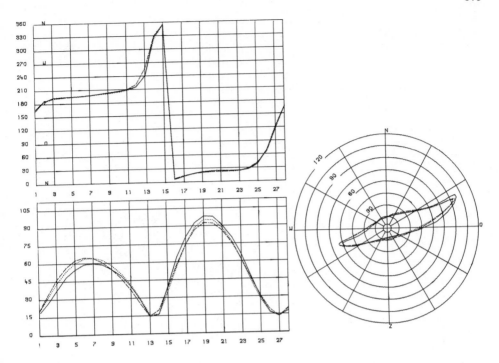

Fig. 20.12. Tidal currents calculated with the different models at
station 12 (in cm/s).

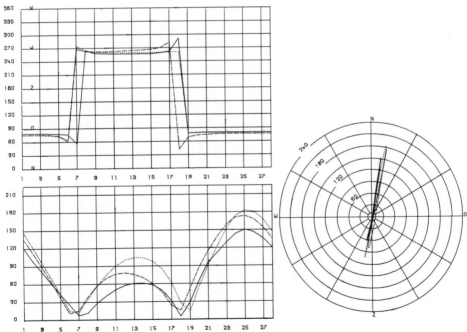

Fig. 20.13. Tidal currents calculated with the different models at
station 13 (in cm/s).

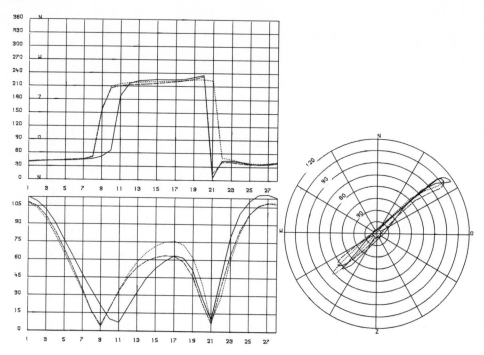

Fig. 20.14. Tidal currents calculated with the different models at
station 14 (in cm/s).

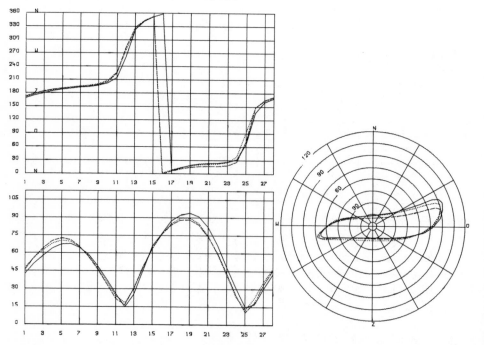

Fig. 20.15. Tidal currents calculated with the different models at
station 15 (in cm/s).

β) As a purpose of this study is to show that the residual flow calculated by averaging the transient flow is very sensitive to the
discretization of the advection, a Fourier analysis of currents is
made.   To clarify ideas, characteristic stations are chosen (Fig. 21).

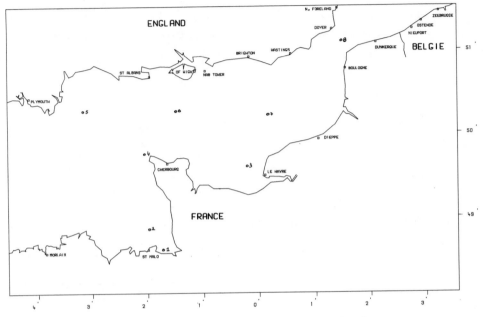

Fig. 21. Stations of comparison for Fourier Analysis.

- for the M_2 currents

TABLE 4

Amplitude of M_2 currents

STATION	Amplitude of the eastern (u) and northern(v) currents	Model 1 (m/s)	Model 2 (m/s)	Model 3 (m/s)
3	u	0.53	0.54	0.56
	v	0.20	0.17	0.21
4	u	1.22	1.42	1.61
	v	1.07	1.08	1.37
6	u	1.32	1.35	1.41
	v	0.05	0.06	0.06
8	u	0.61	0.62	0.60
	v	0.97	0.95	1.03

The analysis of Table 4 shows that the three models give similar results. However, Model 1 has a weak tendency to underestimate currents if one considers Model 3 as the best (higher numerical accurac

- for the M_4 and M_0 (residual) currents

M_4 and M_0 tides are both generated for a large part by the advective terms. For this reason, it seems interesting to compare simultaneously the differences between the models. Since Model 3 gave th best reproduction of M_4 elevations, results of Model 3 are taken as reference values.

Table 5 shows the amplitudes of the eastern and western component of M_4 and Table 6 the eastern and western components of the residual currents.

TABLE 5

M_4 currents

STATION	Amplitude of the eastern (u) and northern (v) currents	Model 1 (m/s)	Model 2 (m/s)	Model 3 (m/s)
3	u	0.05	0.02	0.03
	v	0.09	0.05	0.07
4	u	0.06	0.02	0.03
	v	0.12	0.12	0.12
6	u	0.04	0.03	0.03
	v	0.02	0.00	0
8	u	0.10	0.05	0.07
	v	0.16	0.09	0.12

TABLE 6

Residual currents

STATION	Eastern (u) and northern (v) currents	Model 1 (m/s)	Model 2 (m/s)	Model 3 (m/s)
3	u	0.02	0.00	0.02
	v	0.01	-0.04	0.00
4	u	-0.05	0.04	0.01
	v	0.22	0.30	0.25
6	u	0.01	0.00	0.04
	v	-0.06	-0.03	0.00
8	u	0.26	0.10	0.10
	v	0.40	0.24	0.21

The features which distinguish the respective solutions are :

- the intensity of M_4 and M_0 currents are similar
- model 1 has the tendency to overestimate the currents and model 2 to underestimate them
- According to the results of Table 6 the residual currents are very sensitive to the discretization of the advective terms. At station 3, M_0 current goes north-east with model 1, south with model 2, east with model 3. Near the "Cap de la Hague", the u component of the current is negative with model 1 and positive with models 2 and 3.

γ) Calculated currents have to be compared with the observations. The quality of current measurements is not sufficient to decide the ability (or unability) of models to reproduce harmonics of M_2 currents. Nevertheless, one might expect the same conclusions for M_4 (and M_0) currents than those for M_4 elevations. For this reason, only M_2 currents will be considered in this section.

An important parameter for the comparison is the intensity of the largest M_2 current. The analysis of Fig. (22 to 25) shows a good agreement between the observations and the three simulations.

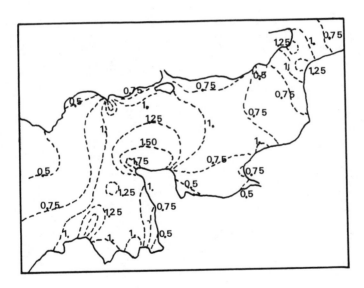

Fig. 22. Largest M_2 currents deduced from the observations (in m/s) (Sager, 1975).

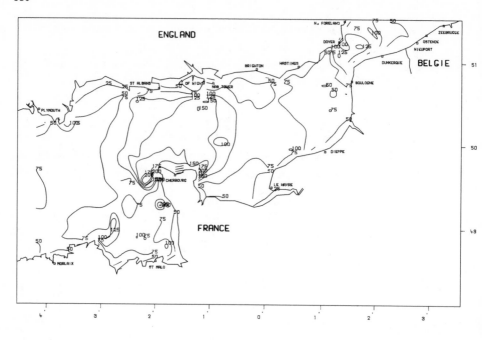

Fig. 23. Largest M_2 currents calculated with model 1 (in cm/s).

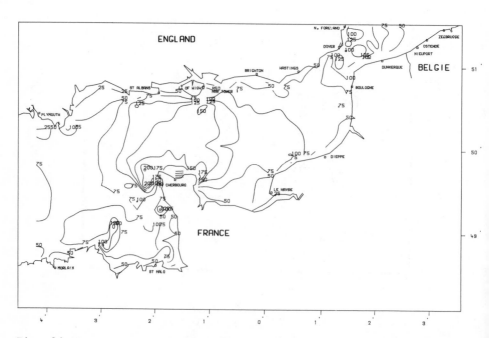

Fig. 24. Largest M_2 currents calculated with model 2 (in cm/s).

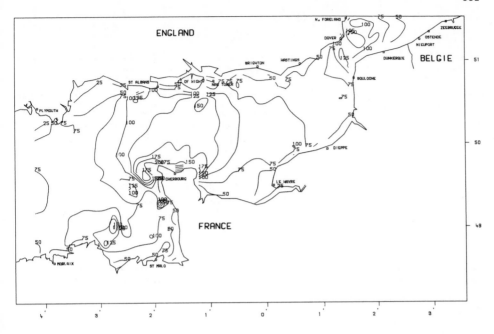

Fig. 25. Largest M$_2$ currents calculated with model 3 (in cm/s).

Tables (7 to 10) show that the differences between the three models and the observations are reasonable (errors less than 20 %). However, the parameter R - ratio between the small and the great axis of the M$_2$ tidal ellipse - is much larger than that observed at stations 3 (near The Havre). That might be due to the closure of the Seine's estuary.

TABLE 7

Amplitude of the M$_2$ current along the great axis of the tidal ellipse (in m/s).

STATION	Observation	Model 1	Model 2	Model 3
3	0.51	0.54	0.54	0.57
4	1.71	1.62	1.79	1.93
6	1.20	1.33	1.36	1.42
8	1.20	1.16	1.14	1.19

382

TABLE 8

Direction of the great axis of the M_2 tidal ellipse (relative to the North)

STATION	Observation	Model 1	Model 2	Model 3
3	270°	282°	282°	260°
4	230°	239°	243°	220°
6	255°	278°	278°	258°
8	215°	222°	223°	200°

TABLE 9

Ratio between the small and the great axis of the M_2 tidal ellipse

STATION	Observation	Model 1	Model 2	Model 3
3	0.15	0.39	0.32	0.38
4	0.08	0.07	0.07	0.01
6	0.08	0.01	0.03	0.01
8	0.00	0.02	0.00	0.03

TABLE 10

Delay (in hours) between the time of maximum of current and the passage of the moon at the Greenwich meridian

STATION	Observation	Model 1	Model 2	Model 3
3	1h. 30 min	1h. 18 min	1h. 18 min	1h. 34 min
4	1h.	0h. 42 min	0h. 42 min	0h. 52 min
6	1h. 25 min	1h. 18 min	1h. 24 min	1h. 22 min
8	5h. 40 min	5h. 54 min	5h. 36 min	5h. 46 min

CONCLUSION

Even if a model yields a good representation of M_2 elevations and currents, its ability to give correct harmonics and subharmonics (especially residual currents) of M_2, strongly depends on the quality of the discretization of the advection terms.

To overcome this difficulty, one must (Nihoul & Ronday, 1976)

i) solve the transient motions by means of a simple model (model 1 or 2) ;

ii)average the transient equations (1-2 or 3-4) over T and solve the steady state resulting equations for the residual flow.

In the averaged equations, the transient motions still appear in the non-linear terms producing the equivalent of an additional stress on the mean motion. This stress can be calculated explicitly using the results of the preliminary long wave equations, and the question of numerical stability is obviously ignored in the calculation of this stress.

ACKNOWLEDGEMENTS

The author is indebted to Dr. Ch. Le Provost for his valuable advice during the course of this work. He also wishes to express his appreciation to Mr. G. Chabert d'Hieres for his constant encouragement and most appreciated support so vital to a project of this nature. Thanks are also due to Prof. J.C.J. Nihoul for computer time facilities. Support for this research has been provided by the Centre National de la Recherche Scientifique - A.T.P. Internationale 1976-1977, N°1563.

REFERENCES

Brettschneider, G., 1967. Anwendung des Hydrodynamisch-numerischen Verfahrens zur Ermittlung der M_2-Mitschwingungsgezeit der Nordsee. Mittl. Inst. Meereskunde. Univ. Hamburg, 7:1-65.

Bureau Hydrographique International, 1966. Marées - Constantes harmoniques. Monaco, Publication spéciale, 26.

Chabert d'Hières, G., & Le Provost, Ch., 1970. Etude des phénomènes non linéaires dérivés de l'onde lunaire moyenne M_2 dans la Manche. Cahiers Océanographiques, 22:543-570.

Durance, A., 1975. A mathematical model of the residual circulation of the Southern North Sea. Sixth Liège Coll. On Ocean Hydrodynamics, Mem. Soc. R. Sci., Liège, pp. 261-272.

Fisher, G., 1959. Ein numerisches Verfahrens zur Errechnung von Windstau und Gezeiten in Randmeeren. Tellus, 9:60-76.

Flather, R.A., 1976. A tidal model of the North-West european continental shelf. Seventh Liège Coll. on Ocean Hydrodynamics, Mem. Soc. R. Sci. Liège, pp. 141-164.

Hansen, W., 1966. The reproduction of the motion in the sea means of hydrodynamical - Numerical methods. NATO Subcommittee on Oceanographics Research, Tech. Rep. 25:1-57.

Hyacinthe, J.-L., & Kravtchenko, J., 1967. Modèle mathématique des marées littorales. Calcul numérique sur l'exemple de la Manche.

384

La Houille Blanche, 6:639-650.

Leendertsee, J.J., 1967. Aspects of a computational model for long period water-wave propagation. Ph. D. Dissertation, Technische Hogeschool Delft, 165 pp.

Leprovost, Ch., 1976. Technical analysis of the structure of the tidal wave's spectrum in shallow water areas. Seventh Liège Coll. on Ocean Hydrodynamics, Mem. Soc. R. Sci. Liège, pp. 97-112.

Nihoul, J.C.J. & Ronday, F.C., 1976. Hydrodynamic models of the North Sea. Seventh Liège Coll. on Ocean Hydrodynamics, Mem. Soc. R. Sci. Liège, pp. 61-96.

Pingree, R.D., & Maddock, L., 1977. Tidal residual in the English Channel. J. Mar. Biol. Ass. U.K., 57:339-354.

Ramming, H.G., 1976. A nested North Sea model with fine resolution in shallow coastal areas. Seventh Liège Coll. On Ocean Hydrodynamics, Mem. Soc. R. Sci. Liège, pp.9-26.

Ronday, F.C., 1972. Modèle mathématique pour l'étude de la circulatio de marées en Mer du Nord. Marine Sciences Branch, Manscp. Rep. Ser. Ottawa, 29:1-42.

Ronday, F.C., 1976. Modèles hydrodynamiques de la Mer du Nord. Ph. D. Dissertation, Université de Liège, 269 pp.

Sager, G., 1975. Die Gezeitenströme im Englischen und Bristol-Kanal. Seewirtschaft, 7:247-248.

RECENT RESULTS FROM A STORM SURGE PREDICTION SCHEME FOR THE NORTH SEA

R.A. FLATHER

Institute of Oceanographic Sciences, Bidston Observatory, Birkenhead, U.K.

ABSTRACT

During the last four years a new system for the prediction of storm surges in the North Sea has been under development at IOS Bidston. The scheme is based on the use of dynamical finite-difference models of the atmosphere and of the sea. The atmospheric model, the Bushby-Timpson 10-level model on a fine mesh, used in operational weather prediction at the British Meteorological Office, provides the essential forecasts of meteorological data which are then used in sea model calculations to compute the associated storm surge. The basic sea model, having a coarse mesh, covers the whole of the North West European Continental Shelf. Additional models of the North Sea and its Southern Bight, the eastern English Channel and the Thames Estuary with improved resolution are also under development. First real-time predictions were carried out early in 1978.
This paper outlines the prediction scheme and presents some recent results.

INTRODUCTION

This paper deals with some aspects of the implementation of the storm surge prediction scheme, based on the use of dynamical finite difference models, proposed by Flather and Davies (1976). The essence of the scheme is to take data from numerical weather predictions carried out by the British Meteorological Office using a 10-level model of the atmosphere (Benwell, Gadd, Keers, Timpson and White, 1971), then to process the data in order to derive, in advance, the changing distribution of wind stress and gradients of atmospheric pressure over the sea surface. Subsequently a numerical sea model taking the processed data as input is used to compute the associated storm surge.

The original scheme has undergone considerable development and improvement as a result of a series of experiments carried out in the last four years. The basic linear sea model covering the continental shelf has been replaced by a much-improved non-linear version capable of reproducing the tidal distribution with good accuracy (Flather, 1976a). Tide and surge can now be calculated together taking account

of the important effects associated with their interaction. A second component
consisting of a North Sea model with finer spatial resolution has also been established and incorporated in the scheme (Davies and Flather, 1977). A test of the
scheme with both sea models covering a continuous period of 44 days in November
and December 1973 is perhaps one of the longest successful surge simulations yet
carried out (Davies and Flather, 1978). Other experiments of a practical nature
in which the predictability of surges was examined led to the design of a first
operational scheme giving predictions up to about 30 hours in advance (Flather,
1976b). The procedure described here is based on this scheme.

The question of how to derive the best possible estimate of the meteorological
forces on the sea from limited atmospheric information is of fundamental importance
for surge prediction. Many alternative procedures exist with varying degrees of
dynamical and empirical content (see for example Duun-Christensen (1975),
Timmerman (1975)). Some of the alternatives were compared for the storm surge
of 2nd to 4th January 1976 (Flather and Davies, 1978). Since then the Meteorologic
Office has been able to provide atmospheric pressure, surface wind and near-surface
air temperature instead of the basic dependent variables (the height of the 1000 mb
pressure surface, the 1000 mb wind in components, and the thickness of the 1000-
900 mb layer) from the 10-level model. These require modified procedures for
deriving the meteorological forces, which are described here.

The plan of the paper is as follows. First, two sections give an outline of
the sea model and the meteorological data with alternative methods for processing
it into the required form. These two ingredients make up the prediction scheme
as described in the section which then follows. First real-time predictions using
the scheme were carried out from 13th to 17th February 1978, with a second sequence
of forecasts from 7th to 15th March covering a period of spring tides. These tests
are described in the fourth section. Since no substantial surges occurred during
the period of real-time running, the accuracy of the prediction scheme is illustrated for the case of the storm surge of 11th and 12th January 1978: the most
recent severe surge on the east coast of England. Four separate predictions were
made and are compared with observations. Finally, a first comparison between
predictions for the high tide on the night of 11th and 12th January obtained from
the present dynamical method and corresponding predictions obtained from a statistical procedure are presented. Further comparisons of this kind will be possible
during the 1978-79 storm surge season when the new dynamical scheme is to be operat
on a routine basis at the Meteorological Office, Bracknell (U.K.).

THE SEA MODEL

The hydrodynamical equations which constitute the basis of the sea model are

$$\frac{\delta\zeta}{\delta t} + \frac{1}{R\cos\phi}\left\{\frac{\delta\,(Du)}{\delta\chi} + \frac{\delta\,(Dv\cos\phi)}{\delta\phi}\right\} = 0, \tag{1}$$

$$\frac{\delta u}{\delta t} + \frac{u}{R\cos\phi}\frac{\delta u}{\delta\chi} + \frac{v}{R\cos\phi}\frac{\delta\,(v\cos\phi)}{\delta\phi} - 2\omega\sin\phi v$$

$$= -\frac{g}{R\cos\phi}\frac{\delta\zeta}{\delta\chi} - \frac{1}{\rho R\cos\phi}\frac{\delta p_a}{\delta\chi} + \frac{1}{\rho D}\,(F^{(S)} - F^{(B)}) \tag{2}$$

$$\frac{\delta v}{\delta t} + \frac{u}{R\cos\phi}\frac{\delta v}{\delta\chi} + \frac{v}{R}\frac{\delta v}{\delta\phi} + \frac{u^2\tan\phi}{R} + 2\omega\sin\phi u$$

$$= -\frac{g}{R}\frac{\delta\zeta}{\delta\phi} - \frac{1}{\rho R}\frac{\delta p_a}{\delta\phi} + \frac{1}{\rho D}\,(G^{(S)} - G^{(B)}) \tag{3}$$

where the notation is:

χ,ϕ	east-longitude and latitude, respectively
t	time
ζ	elevation of the sea surface
u,v	components of the depth mean current
$F^{(S)}, G^{(S)}$	components of the wind stress $\underline{\tau}^{(S)}$ on the sea surface
$F^{(B)}, G^{(B)}$	components of the bottom stress $\underline{\tau}^{(B)}$
p_a	atmospheric pressure on the sea surface
D	total depth of water ($=h+\zeta$)
h	undisturbed water depth
ρ	density of sea water, assumed uniform
R	radius of the Earth
g	acceleration due to gravity
ω	angular speed of rotation of the Earth

(1) - (3) are depth-averaged equations written in spherical polar co-ordinates.
The component directions are those of increasing χ,ϕ; i.e. to the east and to
the north respectively.

Adopting a quadratic law, relating bottom stress to depth mean current, we write

$$F^{(B)} = k\rho u(u^2 + v^2)^{\frac{1}{2}}, \quad G^{(B)} = k\rho v(u^2 + v^2)^{\frac{1}{2}}, \tag{4}$$

where k, the coefficient of bottom stress, takes the value 0.0025. The components
of wind stress, $F^{(S)}$, $G^{(S)}$, and gradients of atmospheric pressure supply the meteorological forcing which drives the surge motion. They are specified externally as
described in the next section.

Equations (1)-(4) are to be solved starting from a prescribed initial distribution
of elevation and motion

$$\zeta = \zeta(\chi,\phi,t), \quad u = u(\chi,\phi,t), \quad v = v(\chi,\phi,t) \text{ at } t = t_o. \tag{5}$$

At a coastline the component of the depth-mean current along the outward directed
normal to the boundary, q, is required to vanish. On open sea boundaries a
generalised radiation condition is employed taking the form

$$q = \hat{q}_M + \sum_i \hat{q}_i + \frac{c}{h}\,(\zeta - \hat{\zeta}_M - \sum_i \hat{\zeta}_i) \tag{6}$$

where $c = (gh)^{\frac{1}{2}}$; $\hat{\zeta}_M$ and $\hat{q}_M$ are contributions, associated with the meteorologically
driven motion, to the total elevation and current ζ and q; $\hat{\zeta}_i$ and $\hat{q}_i$ are similar

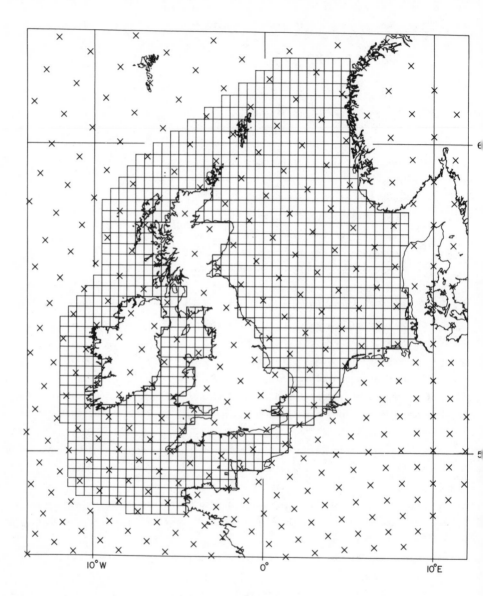

Figure 1. Finite difference mesh of the continental shelf sea model with grid points (×) of the 10-level model of the atmosphere.

contributions associated with the ith constituent of the tide. With ζ_M, $\hat{q}_M$, $\hat{\zeta}_i$ and $\hat{q}_i$ given, equation (6) relates ζ and q, seeking to prevent the artificial reflection at the open boundaries of disturbances generated within the model. The surge input on the boundary, ζ_M, $\hat{q}_M$, is specified externally as described in the next section. The tidal input, $\hat{\zeta}_i$, $\hat{q}_i$, is derived from offshore measurements (Cartwright, 1976) and model experiments (Flather, 1976a), as indicated by Davies and Flather (1977). Only the two largest tidal constituents, M_2 and S_2, are included.

Equations (1)-(4) are now replaced by finite difference approximations which allow the numerical solution, defined on a staggered spatial grid, to be developed from the prescribed initial state (5) by means of a simple one-step explicit time integration. The open boundary condition (6) is used to determine the normal com-ponent of current q from prescribed values of $\hat{q}_M$, $\hat{q}_i$, $\hat{\zeta}_M$, $\hat{\zeta}_i$ employing the total elevation ζ derived from the equation of continuity. A complete description of the scheme is given in Davies and Flather (1978).

The basic sea model used in the present work covers the whole of the North West European continental shelf. The spatial grid, shown as an array of elemental boxes, appears in Figure 1. For each box element, u is defined at the mid-point of longitudinal sides, v at the mid-point of latitudinal sides and ζ at the box centre. The grid spacing is $\frac{1}{2}^{\circ}$ in longitude by $\frac{1^{\circ}}{3}$ in latitude and the time increment used in the integration is 180 seconds. Additional sea models covering the North Sea with improved resolution of shallow coastal waters (Davies and Flather, 1977) and the Southern Bight of the North Sea with the tidal part of the River Thames (Prandle, 1975) are also being developed for use in the surge prediction procedure.

THE METEOROLOGICAL DATA

In this section we consider the derivation of wind stress and gradients of atmos-pheric pressure together with the surge input along the open boundaries of the sea model. These data are deduced from the output of the fine mesh version of the Bushby-Timpson 10-level numerical model of the atmosphere (Benwell et al., 1971) used in routine weather prediction at the British Meteorological Office.

In earlier work the available atmospheric data consisted of hourly values of the height of the 1000 mb pressure surface, in some cases with the addition of 6-hourly values of the thickness of the layer between the 1000 mb and 900 mb pressure surfaces and the components of the 1000 mb wind, all taken directly from and defined at grid points of the 10-level model shown superimposed on the shelf model mesh in Figure 1. Different methods used to calculate the required input to the sea model from these variables are described and compared in Flather and Davies (1978) using the storm of 2nd-4th January 1976. More recently the Meteorological Office have, themselves, made estimates of hourly grid point values of p_a, $\underset{\sim}{w}$: the surface

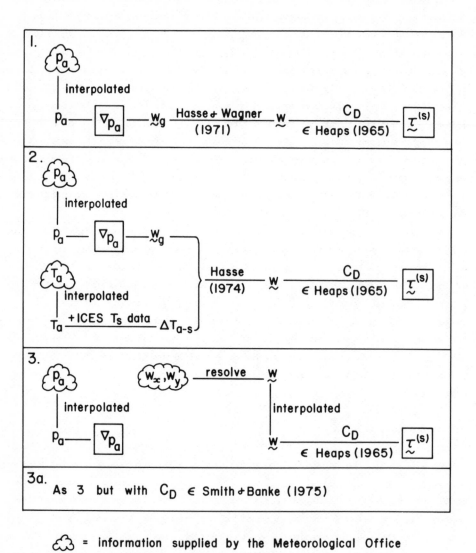

Figure 2. Alternative methods used to derive input data for the dynamical storm surge prediction model.

wind, and T_a: the air temperature close to the sea surface, for use in operational forecasts of waves and swell (Golding, 1977). The methods used are based on the surface and 900 mb wind relationships due to Findlater, Harrower, Howkins and Wright (1966) and employ information from the two lowest layers of the atmospheric model (see Gadd and Golding (1977) for details). The new surface data has been available for our surge experiments since late in 1976 and is used in the present real time predictions.

The grid points of the atmospheric model comprise a rectangular array on a stereo-graphic map projection. The relationship between the non-dimensional Cartesian co-ordinates (x,y) of the atmospheric model and latitude and longitude, the co-ordinates of the sea model, is

$$\left. \begin{array}{l} x = (2R/s) \tan (\pi/4 - \phi/2) \sin (\chi + 35^\circ) \\ y = -(2R/s) \tan (\pi/4 - \phi/2) \cos (\chi + 35^\circ) \end{array} \right\} \tag{7}$$

where s, the grid length at the pole is approximately 100 km. The atmospheric data are defined at intersections of the lines x = 10(1)33, y = -44(1)-19.

Alternative methods used to derive ∇p_a and $\underset{\sim}{\tau}^{(s)}$ from the meteorological data are outlined in Figure 2. Method 1 assumes a knowledge of atmospheric pressure only. The data are first interpolated from the 10-level model onto the sea model grid where the east and north components of ∇p_a are calculated using simple centred finite difference approximations. The geostrophic wind, $\underset{\sim}{w}_g$, is then derived in components in the usual way. To estimate the surface wind a simple empirical law

$$w = 0.56 w_g + 2.4 \text{ m/s} \tag{8}$$

is used. Obtained by Hasse and Wagner (1971) from measurements taken in the German Bight, this relationship requires no additional information about the atmospheric boundary layer. A constant angle of backing, $\delta = 20^\circ$, between the directions of $\underset{\sim}{w}$ and $\underset{\sim}{w}_g$ is assumed to approximate the frictional influence. The wind stress is calculated from a quadratic law

$$\underset{\sim}{\tau}^{(s)} = C_D \rho_a |\underset{\sim}{w}| \underset{\sim}{w} \tag{9}$$

where the drag coefficient, C_D, varies with wind speed according to the equation (Heaps, 1965)

$$C_D \times 10^3 = \left. \begin{array}{ll} 0.565 & \text{for } w \leqslant 5 \text{ m/s} \\ -0.12 + 0.137w & \text{for } 5 < w \leqslant 19.22 \\ 2.513 & \text{for } w > 19.22 \text{ m/s} \end{array} \right\} \tag{10}$$

The east and north components of ∇p_a and $\underset{\sim}{\tau}^{(s)}$ constitute the required forcing terms in the sea model equations.

Method 2 assumes a knowledge of both atmospheric pressure and near surface air temperature, which makes it possible to take into account the important influence of stability on the geostrophic to surface wind relationship. The derivation of ∇p_a and $\underset{\sim}{w}_g$ follows exactly Method 1. In addition T_a values are interpolated onto the sea model grid where sea surface temperatures, T_s, digitised from the chart showing their climatological distribution for the appropriate month in the year

(ICES, 1962) are also defined. Subtracting gives $\Delta T_{a-s} = T_a - T_s$. Then using the results of a later investigation by Hasse (1974), we take

$$\omega = a\omega_g + b \qquad (11)$$

$$\left.\begin{array}{l} \text{with } a = 0.54 - 0.012\Delta T_{a-s} \\ b = 1.68 - 0.105\Delta T_{a-s} \text{ in m/s.} \end{array}\right\}$$

The angle of backing is defined as a numerical function of ΔT_{a-s} (from Hasse (1974) Fig. 3). The derivation of wind stress then follows Method 1. Since Method 2 takes account of the effect of stability explicitly, it might be expected to give a better estimate of the surface wind and hence the wind stress than does Method 1.

Method 3 uses surface wind as well as atmospheric pressure provided by the Meteorological Office. The derivation of ∇p_a follows Method 1. The surface winds, defined as x and y components at 10-level model grid points are first resolved into east and north components, using relationships deduced from Equations (7), which are then interpolated on to the sea model grid. The wind stress then follows from the quadratic stress law (9) with drag coefficient (10). In a variant, Method 3a, the drag coefficient obtained by Smith and Banke (1975) from measurements on Sable Island, Nova Scotia, namely

$$C_D \times 10^3 = 0.63 + 0.066\omega \qquad (12)$$

replaced the one defined in (10). Method 3 differs from 1 and 2 in that the surfac winds are derived from 900 mb rather than geostrophic winds. The 900 mb wind distribution is calculated directly by the 10-level model and so includes dynamical effects associated with the curvature of isobars, the motion and development of depressions, friction and all other physical processes represented in the atmos-pheric model. It should therefore approximate the actual wind at the top of the boundary layer more closely than does the geostrophic wind. Also the 900 mb and surface wind relationships of Findlater et al. (1966) are defined numerically to take account of the influence of stability (or lapse rate) as does (11). Method 3 might therefore be expected to give, overall, the best estimate of surface wind and hence wind stress.

In all cases the surge input elevation at open sea boundary points is derived from the hydrostatic law in the form

$$\zeta_M = (\bar{p}_a - p_a)/\rho g, \qquad (13)$$

where $\bar{p}_a$ is the mean atmospheric pressure taken to be 1012 mb. In the absence of any other information, the associated current, $\hat{q}_M$, is assumed to be zero.

THE PREDICTION SCHEME

It remains to describe how the meteorological data and the sea model are used to provide storm surge forecasts in real-time on a day-to-day operational basis. A preliminary scheme was described by Flather (1976b). This has subsequently been modified and the revised version is outlined here. The scheme is illustrated

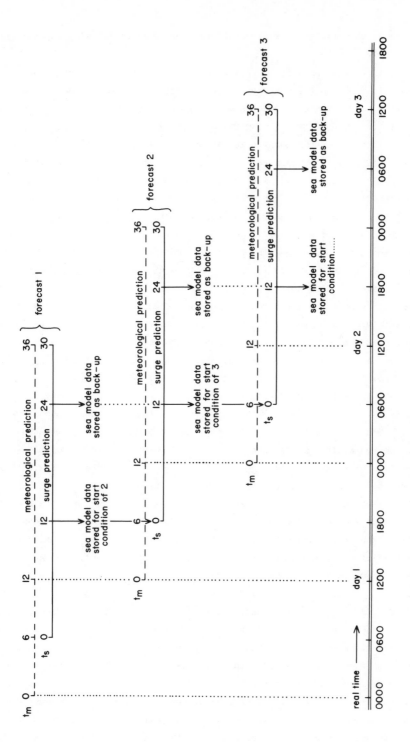

Figure 3.   Scheme for operational surge forecasting.

in Figure 3, to which the reader should refer in conjunction with the description which follows.

Numerical weather predictions using the 10-level model are carried out by the Meteorological Office twice a day. Each prediction run covers the period $0 < t_m < 36$ hours, where t_m denotes meteorological model time with $t_m = 0$ corresponding to either 0000 GMT or 1200 GMT on the day. Collection of observations and initialisation of the model takes about $2\frac{1}{2}$ hours of real time, so that the model calculation, which requires about 14 minutes of central processor time on an IBM 360/195 computer, begins at about 0230 GMT or 1430 GMT. Data required for wave prediction and for the surge calculation are derived from the model output and stored on magnetic disc. The data comprise values of p_a, w_x, w_y, (the x and y components of $\underset{\sim}{w}$) and T_a from a 24×26 rectangular subset of grid points of the atmospheric model covering north-west Europe (see Figure 1). The data are provided at hourly intervals from $t_m = 6$ to $t_m = 36$ hours, and become available at $t_m = 3$, i.e. at 0300 GMT or 1500 GMT.

The surge prediction itself can now begin. Each prediction consists of a number of separate steps carried out in turn. First the data are processed using one of the methods described in the previous section to give appropriate values of the components of ∇p_a and $\underset{\sim}{\tau}^{(s)}$ at internal grid points and the values of ζ_M at open boundary points of the sea model at hourly intervals, $t_m = 6(1)36$.

Because of the limited description of tide in the sea model, confined to only the M_2 and S_2 constituents, it is not intended that predictions of total water level be obtained from the model directly. Rather, the sea model is used to forecast surge residuals, taking into account the important influence of tide-surge interaction. The final estimate of total water level at ports of interest is then obtained by combining these residuals with the predicted tide calculated, using 60 constituents or more, by the harmonic method. This procedure requires that two separate sea model computations are carried out for each forecast. In the first, the meteorological forcing terms and surge input are set to zero, so that a prediction of the model tide is obtained; in the second, tide and surge are calculated together. Subtracting the two solutions gives the required residual.

Each sea model run covers a period $0 < t_s < 30$ hours, where t_s denotes sea model time and $t_s = 0$ corresponds to $t_m = 6$, being either 0600 GMT or 1800 GMT. The initial state of the sea is taken directly from fields, representing the distribution of tide or tide + surge as appropriate, computed in the previous forecast. Thus, referring to Figure 3, the state of elevation and motion at time $t_s = 0$ in forecast 3, say, is identified with that at time $t_s = 12$ in forecast 2, stored in the preceding run. If, because of computer failure or some other cause, forecast 2 had not been carried out, then fields from $t_s = 24$ in forecast 1 would be used as initial conditions for forecast 3. Thus, by storing complete ζ, u, v data twice, at t_s = 12 and t_s = 24, during each sea model run, a degree of robustness sufficient to survive the loss of one meteorological forecast is embodied in the procedure. If more than one successive forecast is lost then the sequence can continue without restarting only if meteorological data from another source is available. Within the tide + surge run constant components of ∇p_a and $\underline{\tau}^{(s)}$ are applied in the hourly interval $t_i - \tfrac{1}{2} \leqslant t_s < t_i + \tfrac{1}{2}$ centred on the data time t_i. Surge input ζ_M is interpolated in time linearly between the hourly specified values in order to prevent the introduction of sudden elevation changes on the boundary.

Having obtained the tide only and tide + surge solutions, the former is subtracted from the latter to give the residual. Further steps in the procedure produce output in graphical form of time series of tide, tide + surge and residual at selected ports and contours of residual elevation over the shelf at suitable time intervals, showing the temporal and spatial development of the surge. Some of the results are archived for subsequent analysis. In particular, hourly elevation and current arrays are retained for use in extracting statistical extremes. The complete surge calculation starting from the hourly atmospheric data and including the two sea model runs and all associated steps requires about 5 minutes of central processor time on an IBM 370/165 computer at the Science Research Council's Daresbury Nuclear Physics Laboratory.

REAL-TIME PREDICTIONS

During 1977 it was agreed with the Meteorological Office that experimental real-time predictions using the scheme described above would be undertaken during the winter of 1977-78. The meteorological data was to be transmitted to Bidston in real time and the sea model calculations performed on the computer at the Daresbury Laboratory using existing communication facilities with Bidston. The co-operation of computer experts in both the Institute of Oceanographic Sciences (IOS) and the Meteorological Office was required to establish the necessary data link between Bracknell and Bidston, a link which uses a dial-up connection along telephone lines within the public telephone network. The link became operational at the end of January 1978 and the experiments commenced soon after.

First real-time predictions were carried out from Monday 13th to Friday 17th February 1978. Thus ten successive forecasts were run with initial times 0600 GMT and 1800 GMT on each of the five days. The procedure described in the last section was carried out with additional steps required to transmit the essential atmospheric data from the Meteorological Office to IOS Bidston and then on to the computer at Daresbury. To minimise the time taken for data transmission along the telephone lines, Method 1 using only the single variable p_a was employed. Results were produced in the form of line printer output at Bidston. No special priority was granted for the experiments so that all computer work was carried out as though

by an ordinary computer user. Consequently the time taken to complete each forecast depended on how heavily the computers were being used. The morning forecasts,
carried out outside normal working hours, took between 40 and 60 minutes with the
exception of the 0600 GMT forecast on Monday 13th February, which was delayed by
engineering work on the Daresbury computer and ended at 1307 GMT. In the after-
noons general computer usage was higher and consequently delays were longer so
that the 1800 GMT forecasts required from 3 to $4\frac{1}{4}$ hours, ending between 1800 GMT
and 1920 GMT.

In a subsequent experiment the scheme was operated again from Tuesday 7th to
Wednesday 15th March 1978 to provide predictions over the period of spring tides
on the 10th and 11th. The aim here was to extend the system to return some of the
output to the Meteorological Office, where it could be examined by members of the
Storm Tide Warning Service (STWS) - the organisation responsible for the provision
of surge warnings in Britain. In the event of any interesting surges occurring,
comparisons between predictions from the present dynamical scheme and from the
statistical procedure, on which surge warnings are currently based, would have been
possible. Fourteen forecasts were carried out in all, though the sequence was not
completed without interruption. Two of the scheduled forecasts were lost. On
Wednesday afternoon 8th March the Daresbury computer broke down during the sea
model run and the forecast had to be abandoned. Similarly the IBM 360/195 at
the Meteorological Office suffered a power failure on Sunday afternoon 12th March.
In this eventuality essential work is transferred to a second computer, an IBM 370/1
on which the time requirement is much greater and consequently non-essential jobs
are not processed. Again the forecast had to be abandoned. In both instances,
the prediction following the one lost was run successfully using the back-up initial
data for the sea model stored by the system. In practice, mainly because of the
additional time requirement for data transmission, results were returned to the
Meteorological Office on only a limited number of occasions. These results were
received satisfactorily by STWS.

The only storm surge of note occurring on the east coast of England during the
periods covered was on 8th March, giving a maximum residual of 0.51 m at 1200 GMT
at North Shields increasing to 0.96 m at 2200 GMT at Southend. Unfortunately, the
surge occurred within about a day of the start of the sequence of predictions with
the 1800 GMT forecast on 7th March. Switching on the meteorological forcing
suddenly at this time generated a substantial artificial negative surge, -0.82 m
at 0900 GMT on 8th March at Southend compared with the true residual of -0.24 m.
The influence of the start-up can be expected to persist for perhaps two or three
days, reducing the accuracy of predictions during this period. As a result the
surge peak was badly underestimated. It seems unlikely that the 1800 GMT forecast
on 8th March, lost because of computer breakdown, would have significantly improved
on these results from earlier forecasts. Some moderately substantial surges were

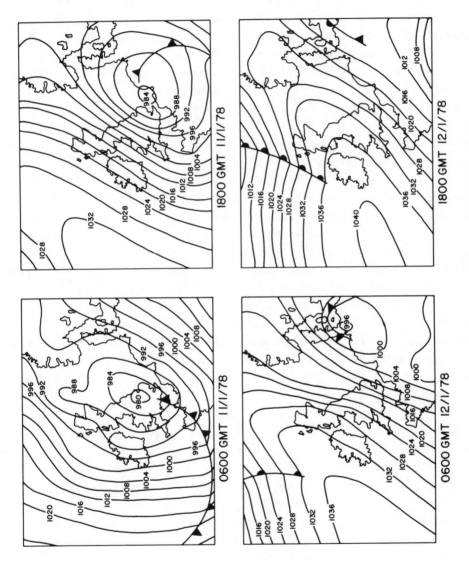

Figure 4. Weather charts for 11th and 12th January 1978.

predicted in other areas towards the end of the period as a complex depression
moved east from Scotland over the North Sea. Maximum residuals of 0.87 m at
Heysham at 2300 GMT 13th March; 0.96 m at Avonmouth at 2100 GMT on 14th; 1.24 m
at Dieppe at 2300 GMT on 14th; 0.79 m at Hilbre Island at 0100 GMT on 15th; and
1.36 m at Büsum in the German Bight at 1300 GMT on 15th were predicted but have
not been checked against observations.

The main outcome of these experiments was to demonstrate that surge prediction
using the scheme is practicable. Clearly, the need to transmit atmospheric data
along telephone lines from one computer to another, returning results to their
ultimate destination in the same way, increases the complication and the possibility of faults causing failures of the system. Ideally all the calculations should
be carried out on the computer used to make the meteorological predictions, so that
data transmission delays can be eliminated. Then, assuming no delay in obtaining
access to this computer, the surge forecast would be completed within a very short
time after the atmospheric data becomes available. It was not the aim of these
experiments to examine the accuracy of the predictions. There is no reason why
the requirement for real-time running as opposed to running in a hindcast mode
should in any way influence the results achieved by the scheme given identical
atmospheric model data. It is more interesting, then, to choose a period of large
surges, extract the atmospheric data from the 10-level model predictions archived
at the Meteorological Office, and compute the associated storm surges at any convenient time. The results will be exactly the same as would have been obtained
in real time. This procedure has been followed for the storm surge of 11th and
12th January 1978 as described in the next section.

THE STORM SURGE OF 11-12 JANUARY 1978

In this section computations of the storm surge of 11th and 12th January 1978
are described. These computations were carried out in order to compare the accuracy
of the results obtained using the alternative methods 1, 2 and 3 (outlined earlier)
to calculate the meteorological forces on the sea. The results were obtained from
computer programs written to perform real time forecasts - used in the experiments
described in the last section. The atmospheric data were extracted from the actual
operational 10-level model predictions. The present results, therefore, are identical to those which would have been obtained by running the scheme in real time
to predict the storm surge.

The surge was associated with a depression which developed to the west of the
British Isles on 10th January. The depression moved east and intensified reaching
the Wash with a central pressure of about 978 mb on the morning of the following
day (see Figure 4). It continued east south east across the Southern Bight of the
North Sea before turning north of east into Holland and Germany, slowly filling

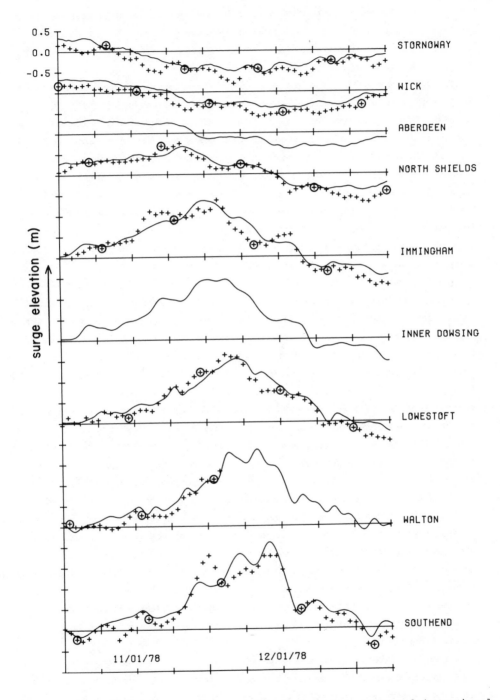

Figure 5. Comparison between surge elevations computed in solution 3 (————) and observations (+++++). ⊕ indicates the residual closest to the time of tidal high water.

as it went. Pressures rose quickly as a following ridge of high pressure extended eastward into Ireland, bringing a strong pressure gradient and north to north-easterly gales over the western part of the North Sea from midday on the 11th to mid-morning on 12th January.

The resulting storm surge coincided with a spring tide to produce levels on the north-east coast of England in excess of those occurring during the major surges of 31st January 1953 and 3rd January 1976. Further south the levels were less extreme, although those experienced on 10th December 1965, when flooding last occurred in London, were equalled in the Thames Estuary. On this occasion the Thames levels came within 60 cm of the top of the defences in central London and within 30 cm of the top in other parts of the river. At Vlissingen on the Dutch coast the predicted high water early on 12th January was exceeded by more than one metre, but further north little effect was felt. On the west side of the British Isles, in the Irish Sea and in the English Channel, a fairly substantial negative surge was produced by the north-easterly winds.

Four separate model calculations have been carried out, using the alternative methods of deriving meteorological input data illustrated in Figure 2. Each calcu-lation started from a state of rest with no elevation disturbance at 0000 GMT on 7th January, and a 54 hour run-in period with tide only established the tidal distr bution at 0600 GMT on the 9th when the atmospheric forcing was introduced. There-after eight complete forecasts were carried out starting at 0600 GMT and 1800 GMT on each of the four days 9th to 12th January. For the purposes of comparison with observations, time series covering the period 0000 GMT on the 11th to 0600 GMT on 13th January were constructed by taking appropriate calculated residuals from the first 12 hours of these forecasts. Figure 5 shows a plot of the resulting time series at British ports from solution 3 with available hourly values derived from observations. The relationship of tide and surge is indicated by circles, which identify the observed residual occurring closest to the time of predicted high water. It can be seen that the observed surge profile is reproduced reasonably well at most ports. Perhaps the most obvious deficiency occurs at Southend, where the surge peak at 0000 GMT on 12th January on the rising tide is omitted. Since no equivalent feature appears at other ports, the peak is probably associated with tide-surge interaction or local wind effect within the Thames Estuary. However, the residual at high water some two hours later is predicted quite well. The behaviour of the remaining three solutions was generally similar to that shown in the figure.

Numerical comparisons have been carried out by calculating root mean square (RMS) errors based on differences between observed hourly residuals and equivalent model predictions. The results are shown in Table 1. The typical RMS error of about 20 cm is rather better than has been achieved in some previous experiments (see Davies and Flather, 1978, Flather and Davies, 1978). The improvement might

TABLE 1

Root mean square errors (cm) for the storm surge period 0000 GMT 11/1/78 to
0600 GMT 13/1/78 based on comparisons between observed hourly residuals and
values computed in different solutions based on methods 1, 2, 3 and 3a respectively.

Port	1	2	3	3a
Stornoway	23.3	21.5	17.4	19.1
Wick	21.8	20.7	16.7	17.9
North Shields	16.3	16.4	13.3	14.1
Immingham	19.8	19.3	18.8	18.8
Lowestoft	14.5	15.9	17.3	17.9
Walton*	19.0	18.3	15.2	15.2
Southend	23.3	21.1	23.7	22.8
All ports	20.1	19.2	17.9	18.4

*0000 GMT 11/1/78 to 0200 GMT 12/1/78 only.

TABLE 2

Regression coefficients C_1, C_o (cm) where $\zeta_{observed} = C_1 \zeta_{computed} + C_o$ based on
hourly residuals in the period 0000 GMT 11/1/78 to 0600 GMT 13/1/78.

Port	1		2		3		3a	
	C_1	C_o	C_1	C_o	C_1	C_o	C_1	C_o
Stornoway	0.94	−21.3	0.91	−19.9	0.88	−15.9	0.94	−16.7
Wick	1.01	−20.7	0.96	−19.6	0.96	−15.4	1.03	−16.1
North Shields	1.11	−12.0	1.07	−11.7	1.16	− 9.8	1.29	− 9.0
Immingham	1.02	−12.9	1.02	−13.7	1.05	− 9.1	1.24	− 9.9
Lowestoft	1.11	−10.0	1.21	−12.8	1.10	−13.0	1.29	−14.2
Walton*	0.86	− 1.5	1.03	− 9.0	1.05	−10.6	1.16	−11.3
Southend	0.93	− 4.2	1.02	− 8.6	1.00	−10.5	1.18	−11.3
Mean C_1	1.00		1.03		1.03		1.16	

*0000 GMT 11/1/78 to 0200 GMT 12/1/78 only.

402

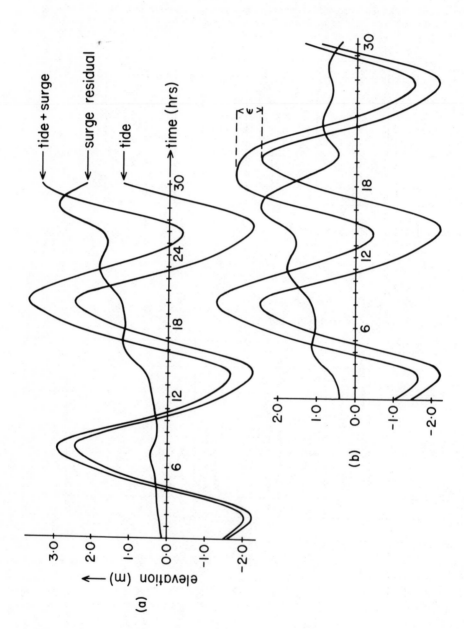

Figure 6. Successive model forecasts for Southend from solution 3 starting at (a) 0600 GMT and (b) 1800 GMT on 11th January 1978. ε denotes the exceedance of high water.

be attributed to the new form of meteorological data used in the present case.
Overall solution 3 gives the best results with RMS errors between 13.3 cm at North
Shields and 23.7 cm at Southend. Using the same time series, linear regression
analyses were carried out for each port seeking coefficients C_1 and C_0 in the
relationship

$$\zeta_{observed} = C_1 \zeta_{computed} + C_0. \tag{14}$$

The values obtained are given in Table 2. A tendency for surge variations to be
overestimated at Stornoway, where C_1 ranges from 0.88 to 0.94, and underestimated
at North Shields and Lowestoft, with C_1 between 1.07 and 1.29, is evident. In
every case C_0 turns out to be negative indicating that calculated residuals are
consistently higher than observed. The effect is particularly marked at Stornoway
and Wick and probably contributes to the relatively large RMS errors at these ports.
This kind of consistent error in level may be introduced through incorrect surge
elevation input along the open boundaries of the model. Solution 3a, with an
average C_1 value of 1.16, underestimates surge variations as compared with the
other solutions, reflecting the fact that the drag coefficient from Smith and
Banke (1975), Equation (12), gives a lower estimate of wind stress than Equation (10)
due to Heaps (1965) in the important range of wind speeds from 10.6 to 28.5 m/s.

   The foregoing discussion concerns results strictly from hours 0 to 12 only of
each sequence of 30 hour sea model predictions. Figure 6 shows two complete forecasts for Southend, beginning at 0600 GMT and 1800 GMT, respectively, on 11th
January. Successive forecasts overlap, the first giving, in this instance, an
early warning more than 20 hours before the high water at 0200 GMT on 12th January
most affected by the surge. The second forecast, available for issue 12 hours
later, updates this prediction. Because observations are used to redefine the
initial state of the atmosphere before each 10-level model run, consecutive
meteorological predictions, and hence also consecutive surge predictions, differ
in the period of overlap, the latest forecast for a given instant being in general
most accurate. Thus, although the surge profiles in Figure 6 look very similar,
significant differences occur in the predicted residuals at high water (1.26 m in
(a) and 1.13 m in (b) compared with the observed value of 1.04 m). In addition,
the maximum predicted residuals are 2.94 m in (a) and 2.48 m in (b), later reduced
to 2.10 m in the forecast beginning at 0600 GMT on 12th January, still 0.34 m
higher than actually occurred. These differences are also apparent in plots of
the spatial distribution of surge. Figures 7a and 7b show two such predicted distributions for 0300 GMT on 12th January when the surge peak was between Immingham
and Lowestoft. Levels in the north-eastern Irish Sea and in the English Channel
are about 25 cm lower in Figure 7b (hour 9 of the 1800 GMT 11/1/78 forecast) than
in Figure 7a (hour 21 of the 0600 GMT 11/1/78 forecast). Similar differences appear
in other areas but the general features remain the same. The sequence of contour
plots, produced at 3 hourly intervals throughout each prediction, should therefore

404

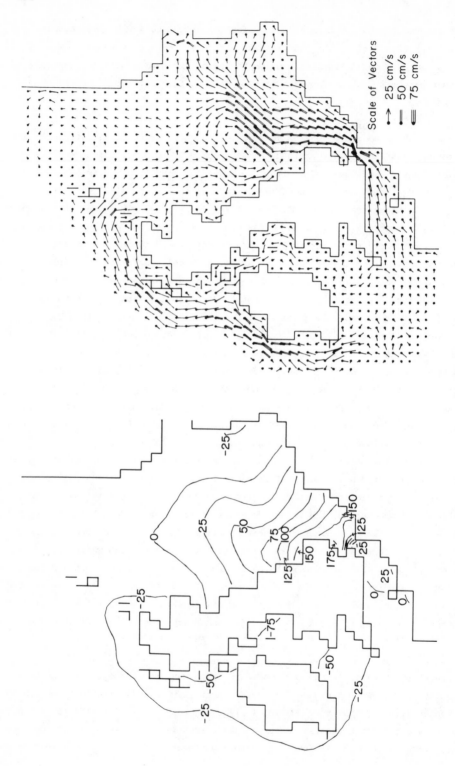

Figure 7a. Spatial distribution of surge elevation (cm) and current at 0300 GMT 12/1/78 from hour 21 of the forecast starting at 0600 GMT 11/1/78 (solution 3).

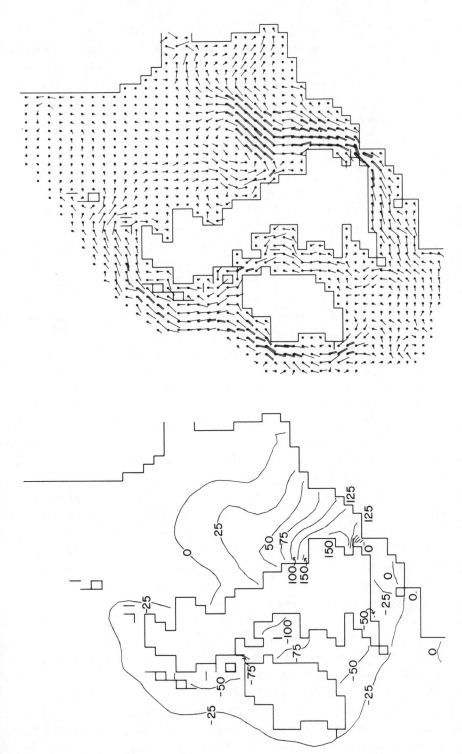

Figure 7b. Spatial distribution of surge elevation (cm) and current at 0300 GMT 12/1/78 from hour 9 of the forecast starting at 1800 GMT 11/1/78 (solution 3).

give to the surge forecaster a useful picture of the developing situation.

Finally in this section we present a first comparison between predictions from the model and from the established statistical procedure operated by the Storm Tide Warning Service (Townsend, 1975). STWS make predictions of surge residuals at the time of harmonically predicted high water, which are then used to estimate maximum levels at ports along the east coast of England. When a maximum level in excess of the prescribed danger level for the port is predicted, appropriate flood warnings are issued.

The procedure used to derive comparable information from the model output was as follows. First, the difference ε between the maximum calculated tide plus surge elevation and the associated maximum calculated tidal elevation gives the expected amount by which the actual tidal high water should be exceeded (see Figure 6b). This quantity seems more appropriate for flood prediction than the residual at a particular instant since tidal high water and actual high water will not, generally coincide in time. The appropriate danger level for a particular model tide is assumed to be the same height above high water of the model tide as the real danger level is above the high water level of the actual predicted tide. The maximum heigh above this effective danger level, the times at which danger level is reached and at which the level falls below danger are then read from tabulated tide plus surge levels produced by the model. Finally, the critical times are corrected for the error in the time of occurrence of the model tidal high water as compared with the actual predicted tidal high water, thus $t_{corrected} = t + t_{actual\ HW} - t_{model\ HW}$.

The results for the night tide of 11th to 12th January at east coast ports are given in Table 3. It is assumed that the model predictions, obtained from solution 3, would be available one hour before the indicated forecast data time. STWS predictions are those actually produced and issued at the time of the surge. Also included are values derived by STWS from observations. Thus, referring to the section for Southend in Table 3, the model forecast starting at 0600 GMT on 11th Januar predicted that the tidal high water (HW) level at 0200 GMT on the following day would be exceeded by 1.26 m and that levels would be above the danger mark from 0049 GMT until 0349 GMT on 12th January. This information would have been available at about 0500 GMT on the 11th - about 21 hours before the predicted time of high tide. The statistical forecast about 7½ hours later was marginally less accurate with ε = 1.32 m. The next model prediction from the 1800 GMT forecast on 11th January gave ε = 1.13 m, slightly better than the short term STWS forecast (1.19 m) and only 9 cm above the observed value. In general, the earliest model predictions 24 to 30 hours before the event are not very accurate but are perhaps good enough to be useful as early warnings. Later forecasts show considerable improvement and seem to compare well with the statistical predictions. The model also appears to show some skill in predicting the critical times, especially at Immingham. Clearly, many more comparisons of this kind will be required to assess the usefulness of the model in surge forecasting.

TABLE 3

Comparison of various predictions of high water for the night tide of 11-12 January 1978. Model predictions are identified by the forecast initial data time and assume that the results would be available one hour before this. Storm Tide Warning Service (STWS) predictions are those actually obtained from the established statistical procedure and issued at the times indicated.

Port and observed HW time	Source and time of forecast	Exceedence of HW (m)	Height above danger level (m)	Time danger reached GMT	Time below danger GMT	Length of warning (hours)
North Shields 1701 GMT 11/1/78	1800 10/1/78 model 0600 11/1/78 model 0625 11/1/78 STWS 1315 11/1/78 STWS observed	0.34 0.45 0.75 0.31 0.69	0.05 0.16 0.46 0.02 0.40	1627 1612 – – 1600	1712 1737 – – 1800	24 12 10½ 4 0
Immingham 1942 GMT 11/1/78	1800 10/1/78 model 0600 11/1/78 model 0625 11/1/78 STWS 1318 11/1/78 STWS 1800 11/1/78 model observed	0.66 0.85 1.15 0.84 0.88 0.96	0.18 0.37 0.67 0.36 0.40 0.48	1848 1833 – – 1833 1825	2003 2028 – – 2033 2035	26½ 14½ 13½ 6½ 2½ 0
Lowestoft 2259 GMT 11/1/78	1800 10/1/78 model 0600 11/1/78 model 1150 11/1/78 STWS 1800 11/1/78 model 1810 11/1/78 STWS observed	0.74 1.00 1.42 0.94 1.18 1.21	-0.02 0.24 0.66 0.18 0.42 0.45	not reached 2208 – 2218 – 2125	 0118 – 0103 – 0025	30 18 11 6 5 0
Walton ~0130 GMT 12/1/78	0600 11/1/78 model 1320 11/1/78 STWS 1800 11/1/78 model *2048 11/1/78 STWS observed	1.18 1.01 1.07 1.21 *	0.68 0.51 0.57 0.71 *	2357 – 0007 – 0020	0322 – 0312 – *	20½ 12 8½ 4½ 0
Southend 0200 GMT 12/1/78	0600 11/1/78 model 1320 11/1/78 STWS 1800 11/1/78 model 2048 11/1/78 STWS observed	1.26 1.32 1.13 1.19 1.04	0.72 0.78 0.59 0.65 0.50	0049 – 0104 – 0050	0349 – 0344 – 0320	21 12½ 9 5 0

*tide gauge broken.

408

CONCLUDING REMARKS

Recent developments in the establishment of a storm surge prediction scheme for British waters based on the use of dynamical finite difference models of the atmosphere and of the sea have been described. In particular, first real-time prediction carried out during February and March 1978 have demonstrated the practicability of the method.

Using as a test case the storm surge of 11th and 12th January 1978, a severe surge which caused flooding and considerable damage on the English coast, four alternative methods of deriving the meteorological forces acting on the sea from atmospheric data have been compared. As expected, the method (3) using atmospheric pressure and surface winds derived from 900 mb winds in the 10-level model by the Meteorological Office gave the best results when used with the drag coefficient from Heaps (1965). The alternative drag coefficient, due to Smith and Banke (1975) gave a lower estimate of the wind stress from the same surface winds and an inferior prediction of the storm surge.

A first comparison has been made between predictions obtained by the statistical procedure operated by the Storm Tide Warning Service and those using the present dynamical prediction scheme for ports on the east coast of England. Generally the model predictions emerge quite well from the comparison. The accuracy of the two methods for a given length of warning is similar and the model may also give a reasonably good indication of the times at which danger levels are passed. In addition, the new scheme can give predictions more than 30 hours in advance whereas the statistical method, depending on observations along the east coast is limited by the propagation time of the tide and travelling component of the surge to a maximum of about 12 hours warning. The spatial coverage of the model scheme is an important advantage offering the possibility of offshore predictions.

It is evident that further comparisons of this kind will be required to gauge the value of the new scheme. To this end it is hoped that sea model predictions will be carried out on a routine basis at the Meteorological Office during the winter of 1978-79. The results will be passed directly to the Storm Tide Warning Service. By the end of the season, comparisons of dynamical and statistical predictions in a wide variety of surge situations will have been carried out. Hopefully the results will justify the continued use of the model-based scheme. Further development and improvements can then be anticipated in the light of the valuable experience gained.

ACKNOWLEDGEMENTS

The author is indebted to the Meteorological Office for their continuing co-operation in this project and to Lt. Cdr. J. Townsend, Officer in Charge, Storm

Tide Warning Service for observations and permission to include statistical predictions. Thanks are also due to Dr N.S. Heaps for valuable comments and criticism, to Mr R.A. Smith, who prepared the diagrams and to Mrs Young, who typed the manuscript.

The work described in this paper was funded by a consortium consisting of the Natural Environment Research Council, the Ministry of Agriculture, Fisheries and Food, and the Departments of Energy, the Environment and Industry.

REFERENCES

Benwell, G.R.R., Gadd, A.J., Keers, J.F., Timpson, M.S. and White, P.W., 1971. The Bushby-Timpson 10-level model on a fine mesh. Sci. Pap. Met. Off., London, 32: 23 pp.
Cartwright, D.E., 1976. Shelf boundary tidal measurements between Ireland and Norway. Mém. Soc. R. Sci. Liège, Ser. 6, 10: 133-140.
Davies, A.M. and Flather, R.A., 1977. Computation of the storm surge of 1 to 6 April 1973 using numerical models of the north west European continental shelf and the North Sea. Dt. hydrogr. Z., 30: 139-162.
Davies, A.M. and Flather, R.A., 1978. Application of numerical models of the north west European continental shelf and the North Sea to the computation of the storm surges of November-December 1973. Dt. hydrogr. Z. Erg.-H. A, Nr. 14.
Duun-Christensen, J.T., 1975. The representation of the surface pressure field in a two-dimensional hydrodynamic numeric model for the North Sea, the Skaggerak and the Kattegat. Dt. hydrogr. Z., 28: 97-116.
Findlater, J., Harrower, T.N.S., Howkins, G.A. and Wright, H.L., 1966. Surface and 900 mb wind relationships. Sci. Pap. Met. Off., London, 23: 41 pp.
Flather, R.A., 1976a. A tidal model of the north west European continental shelf. Mém. Soc. R. Sci. Liège, Ser. 6, 10: 141-164.
Flather, R.A., 1976b. Practical aspects of the use of numerical models for surge prediction. Institute of Oceanographic Sciences Report No. 30, 18 pp.
Flather, R.A. and Davies, A.M., 1976. Note on a preliminary scheme for storm surge prediction using numerical models. Quart. J. R. met. Soc., 102: 123-132.
Flather, R.A. and Davies, A.M., 1978. On the specification of meteorological forcing in numerical models for North Sea storm surge prediction, with application to the surge of 2-4 January 1976. Dt. hydrogr. Z. Erg.-H. A, Nr. 15.
Gadd, A.J. and Golding, B.W., 1977. Temperature and wind modelling for the 10-level model. Meteorological Office Met. O 2b Technical Note No. 13. (Unpublished manuscript).
Golding, B.W., 1977. Waves and swell can be forecast. Offshore Services 10, No. 10: 100-102.
Hasse, L., 1974. On the surface to geostrophic wind relationship at sea and the stability dependence of the resistance law. Beitr. Phys. Atmosph., 47: 45-55.
Hasse, L. and Wagner, V., 1971. On the relationship between geostrophic and surface wind at sea. Mon. Weath. Rev., Wash., 99: 255-260.
Heaps, N.S., 1965. Storm surges on a continental shelf. Phil. Trans. R. Soc. (A) 257: 351-383.
ICES, 1962. Mean monthly temperature and salinity of the surface layer of the North Sea and adjacent waters from 1905 to 1954. International Council for the Exploration of the Sea, Copenhagen.
Prandle, D., 1975. Storm surges in the southern North Sea and River Thames. Phil. Trans. R. Soc. (A) 344: 509-539.
Smith, S.D. and Banke, E.G., 1975. Variation of the sea surface drag coefficient with wind speed. Quart. J. R. met. Soc., 101: 665-673.
Timmerman, H., 1975. On the importance of atmospheric pressure gradients for the generation of external surges in the North Sea. Dt. hydrogr. Z., 28: 62-71.
Townsend, J., 1975. Forecasting 'negative' storm surges in the southern North Sea. The Marine Observer, XLV: 27-35.

BELGIAN REAL-TIME SYSTEM FOR THE FORECASTING OF CURRENTS AND ELEVATIONS IN THE
NORTH SEA

Y. ADAM
Unité de Gestion du Modèle Mathématique de la Mer du Nord et de l'Estuaire de
l'Escaut. Ministère de la Santé Publique et de l'Environnement.

ABSTRACT

   A real time system of storm surge forecasting based on a mathematical model
of the North Sea is presented.
   Problems concerning the reliability of results and time lags are dealt with,
and sources of errors are analysed.

PRINCIPLES

   The aim of this work is to build a system which has the ability to forecast
the response of currents and surface elevations in the North Sea to the influence
of the tide and the meteorological forces. The system should help warning civil
authorities of possible floods along the Belgian coast and in the Scheldt estuary,
and it should help forecasting strong coastal currents that could cause damage
to harbour constructions or other man-built structures.

   In this preliminary stage, however, we are mainly (if not only) concerned with
elevations, and will restrict our attention to the specific problem of surges
(positive or negative) induced by stormy weather.

   As we want to design a simulation system for forecasting purposes, we restrict
ourselves to using only data that can be forecast at least several hours in advance, and we prohibit the use of real measurements which could not be used in
a true real-time system. Therefore, as the results of our simulations depend
on unascertained data, they will probably be less accurate than results obtained
by models using real inputs.

   As data we use only the following

- tidal data (amplitudes and phases of tidal harmonics) for boundary conditions,
  and computation of pure tide-driven circulation

- forecasts of wind and atmospheric pressure at the sea surface, for boundary conditions and external forces computation.

The results of real time simulations are compared, after a while, to computed and measured elevations in several seaports along the English and Belgian coasts.

METHODS

1. Model

The method used to compute the occurence of storm surges, is the numerical integration of a mathematical model of water circulation in the North Sea, using as input tidal and meteorological forecasts.

The mathematical model was developed during the "Programme National R-D sur l'Environnement-Projet Mer" by Nihoul and Ronday (1976).

It is integrated using a numerical algorithm on a finite difference grid covering not only the North Sea, limited by the Dover Strait, as in the original model, but also the English Channel in order to be freed from a boundary condition too difficult to deal with in real time. Equations and numerical algorithms are fully described in Nihoul and Ronday (1976).

Fig. 1 shows the portion of ocean which is being modelled, and the limits of the computational grid.

2. Boundary conditions

The boundary condition easiest to impose is the elevation along the northern and western boundaries across the Channel.

The total elevation ζ_B is the sum of a tidal elevation ζ_M, and an elevation ζ_p generated by a variation of the atmospheric pressure

$$\zeta_B = \zeta_M + \zeta_p \tag{1}$$

where

$$\zeta_p = - \frac{1}{\rho_w g} (p - p_m) \tag{2}$$

$$\zeta_M = \sum_{i=1}^{n} f_i \, a_i \, \cos(\omega_i t + \varphi_i) \tag{3}$$

where ρ_a is the air specific weight, ρ_w the water specific weight, p the atmospheric pressure, p_m the mean atmospheric pressure, n the number of tidal harmonics (Darwin theory), a_i the main amplitude of the ith harmonic at a given point, ω_i its wave number, φ_i its initial phase (slowly varying with time), and f_i a correction factor, also very slowly varying with time.

For the northern boundary, the values of a_i and φ_i are computed from Cartwright

(1976) at a series of points plotted on Fig. 1 (with labels A-M) and translated
to the boundary line using interpolation, extrapolation, and phase correction.

For the western boundary, amplitudes and phases are assumed to be constant
across the Channel, and equal to these quantities at Devonport. Eq. 3 is also
used to compute a theorical elevation at a series of locations along the coast.

3. Air-sea interactions

The influence of the atmospheric conditions on the circulation is modelled by
2 parameters : air pressure gradient and wind stress at the sea surface.
In the equations describing the evolution of the depth-averaged velocity, $\bar{u}$,
the pressure gradient acts as a force

$$\underline{F}_p = - \frac{1}{\rho w} \nabla p$$

and the wind stress as a force

$$\underline{F}_w = (1 + m) \frac{\rho_a}{\rho_w} \frac{C_d}{H} \underline{w} \| \underline{w} \|$$

where $\underline{w}$ is the wind velocity, m a parameter (m ~ 0.1), and C the surface drag
coefficient depending upon the wind speed. In a preliminary stage, we have chosen the following wind stress model :

$$C_d = \begin{array}{ll} 1.26 \ 10^{-3} & \text{for } \| \underline{w} \| \leqslant 10 \text{ m/sec} \\ 2.4 \ 10^{-3} & \text{for } \| \underline{w} \| > 10 \text{ m/sec} \end{array}$$

but the following formulation has also been used, with little sensible improvement
(Flather and Davies, 1975)

$$C_d = \begin{array}{ll} 0.565 & \text{for } \| \underline{w} \| < 5 \text{ m/sec} \\ - 0.12 + 0.137 \| \underline{w} \| & \text{for } 5 < \| \underline{w} \| \leqslant 19.22 \text{ m/sec} \\ 2.513 & \text{for } \| \underline{w} \| > \quad 19.22 \text{ m/sec} \end{array}$$

4. Available data

Input data are :
- amplitudes and phases of tidal harmonics, from Cartwright (1976) and tables
  of harmonic constants (Bureau Hydrographique International, 1953)
- pressure and wind velocities at the nodes of a grid covering the North Sea, the
  British Isles and Western Europe (see Fig. 2). These parameters are the results
  of numerical weather forecasts by the Meteorological Office, Bracknell, which

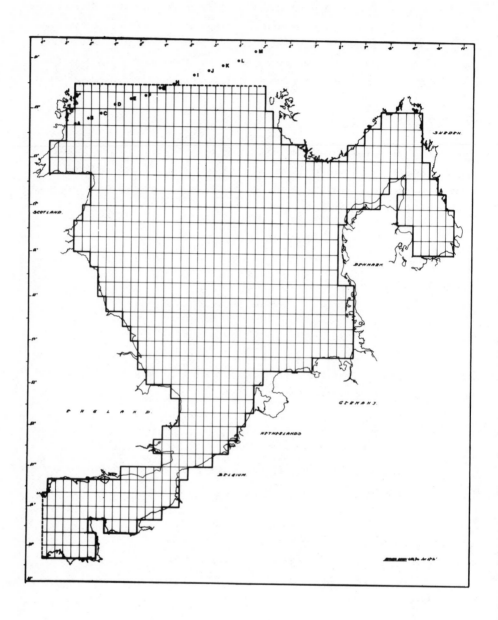

Fig. 1. The finite difference grid used for the computations with the mathematical model.

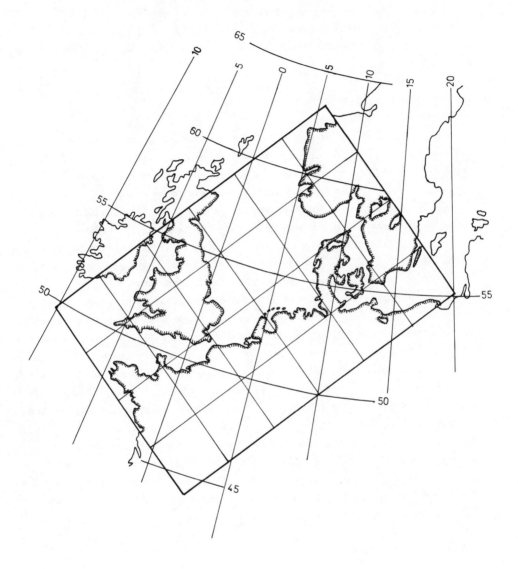

Fig. 2. The meteorological grid covering the North Sea.

are sent to Liege through the Régie des Voies Aériennes by telex.

Time interval between 2 data sets is 6 hours; by interpolation, meteorological parameters are transfered from the very coarse atmospheric grid to the finer hydrodynamical grid.

GENERAL SCHEME OF SURGE WARNING

The first purpose of this system is to help giving an early warning of surge and flood danger. As in the present stage of computational costs and facilities it is not possible to run the simulation programs throughout the whole winter period, some kind of warning and alarm system has to be designed.

The proposed scheme is given on Fig. 3 ; it is not fully operational yet, in the sense that not all parts have been tested in true stormy conditions.

A first warning of storm danger comes from the Institut Royal Météorologique. If the risk of an important storm surge exists (mainly, strong winds from N, NW, or very deep depression) the simulation system is set into action.

First, an initial condition is calculated (distribution of $\underline{u}$ and ζ) over the North Sea and the Channel, using the numerical model with no atmospheric input. In the meantime, the file of meteorological information is updated with the last available forecasts. When the pure tidal regime is almost reached, at a time corresponding to the available meteorological data, a first run of the storm surge computational procedure is done; this run gives a forecast of the circulation in the North Sea for more than 12 hours ahead. In the meantime, another program has computed the pure tidal circulation. By subtracting two distributions of variables (at the same instant), one can see whether there is a surge forming somewhere. If there is, a next simulation is run as soon as another weather forecast becomes available. If not, one checks whether the weather evolution is favourable or not. If it isn't, the simulation goes on with up-to-date data; if it is, the simulation is terminated.

SCHEME OF A SIMULATION RUN

As our system is designed to work in real time, we must now take care of data acquisition, data-files updating, transfer of data from the coarse meteorological grid to the hydrodynamical grid, delays, stability, safety in case of lack of data.

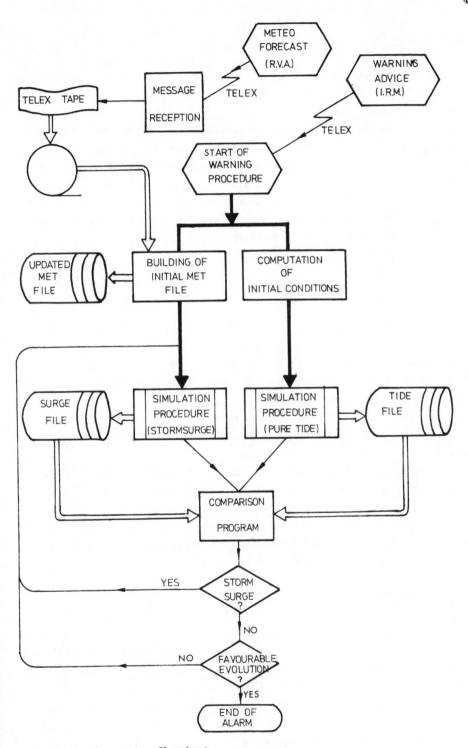

Fig. 3. Warning system flowchart.

1. Data acquisition

Meteorological data come from the R.V.A. through telex. The messages are sent twice a day (at 8 GMT and 20 GMT), contain each date and time of transmission, and two sets of useful data:
- time of forecasting (the first message provides forecasts for GMT 18 and GMT 0 the following day, the second forecasts for GMT 6 and GMT 12, the following day)
- for each node of the coarse grid: pressure, wind speed, wind direction.

Every day, one receives thus four sets of meteorological parameters.
Each telex tape is written onto a magnetic tape; telexcode is then translated to EBCDIC code in a disk file; the latter is then read, and data are converted into usable data and written on a third file; this file is used to update the formerly-built meteorological data file.

2. File updating

The resulting file contains:
- 2 data sets which are of no use, except for safety purposes (see later)
- 4 data sets, at six hours intervals, covering thus a time period of 18 hours
- 19 data sets, at one hour intervals, covering the same period, derived from the former data by cubic interpolations.

Let us note that the updating program can also:
- fill lacking data by time interpolation using, if needed, the first two data sets
- fill lacking data by time extrapolation using, if needed, the first two data sets, under automatic programmed procedure or programmer request.

The latter case happens when longer-term forecast is wanted

These 19 data sets are then used to provide, by spatial interpolation, 19 sets of data, i.e. pressure, and two wind velocity components at each node of the hydrodynamical grid, which are stored on a "grid file".

3. Hydrodynamical computations

Using as initial condition the state of the sea at the time of the first data set of the grid file, and as input the meteorological data in that file, the numerical program computes the evolution of the hydrodynamical variables, and stores their distribution at every hour on a hydrodynamical file.

The computation simulates 18 hours of real time; the sea state corresponding to the 12th hour of simulation will be used as initial condition for the following run. The whole procedure is summarized in Fig. 4.

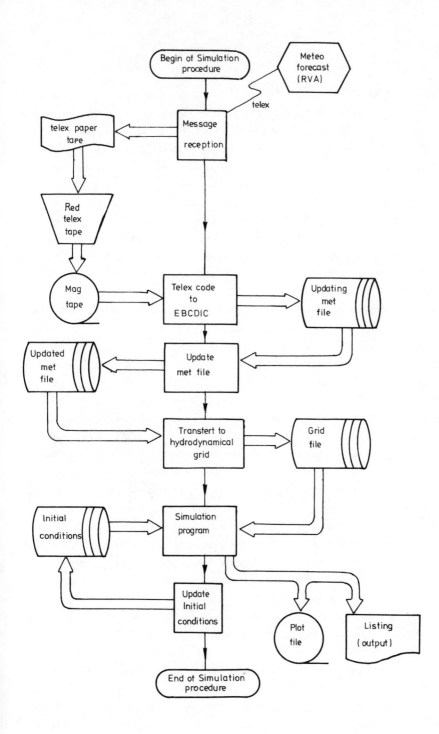

Fig. 4. Real time system flowchart.

4. Delays

The problem of time lags in the forecasting and the availability of results
is of course crucial in surge forecast. Let us estimate how much in advance one
can forecast the elevation of the sea surface, and a critical high tide.

At 8 GMT, a telex message is received. It takes about 1 hour to transfer the
telex tape onto magnetic tape; this operation is performed off-line and requires
a lot of manipulations. The next stages can be executed completely automatically
by the operating system. In emergency priority, but taking into account that the
computer can not be totally dedicated to this work (existence of a time sharing
system), it takes another hour to get the results on listing, and a last one to
get the plots, which are drawn off-line; thus, around 11 GMT, a forecast is available, which runs to 24 GMT, i.e. 13 hours in advance; during this period there
is one high tide.

We come later to the case where high tide (without meteorological influence),
takes place in the first seven hours. If high tide occurs in the last six hours of
the day, then, its actual height can be predicted at least 7 hours in advance,
and a warning can be given. If the high tide occurs before 18 GMT, it is too
near for a forecast to be useful.

However, as was said under Section 3 of this chapter, meteorological data can
be extrapolated. In the case here mentioned (high tide before 18 GMT) it would
have been wise to extrapolate the data received in the evening of the previous
day and to run a simulation with these extrapolated data; at 23 GMT (the day
before), one can, in so doing, get an idea of the high tide at least 12 hours
in advance. Of course, extrapolated parameters are probably less reliable than
forecast parameters, but one has to take into account the inertia of the atmosphere and, above all, of the sea. Systematic tests have been run to check the
reliability of the results yielded by such a procedure. A provisional conclusion can be drawn from the tests: hydrodynamical computations using extrapolated
data are almost as good as results computed with true forecasts, as soon as successive extrapolations are distant in time of at least 48 hours. This relatively
good fit might also be due to the poor resolution of the true forecasts.

The delays estimated here hold when one wants to forecast the situation at
any point, using the numerical output of the model at that point. However, surge
waves travel at a finite speed, and the computation of the surge at one point
could be used to have a good idea (through regression analysis of typical cases)
of what will happen, some time later, at another place which is further on the
way of the surge wave. For instance, the surges most dangerous for the Belgian
coast come from the North West. Such surge waves travel from North Shields to
Ostend in about 12 hours. Therefore, by using a numerical forecast of the ele-

vation at North Shields (for instance), to infer a forecast of the elevation
at Ostend, one can predict a surge 18 to 24 hours in advance.

ANALYSIS OF SOME RESULTS

It has taken relatively long time to develop the whole system (decoding of
telex messages, file structure, updating and interpolating programs, being taken
into account the number of safety procedures such programs include) and to extend
the model to the English Channel.

Moreover, to check the results, true measurements are needed, which arrive with
some delays. However, we have simulated, in pseudo-realtime, a bad weather period
in November 1977, and compared elevations computed by the model to elevations
measured in situ (more thorough tests will be run during the winter 78-79).
The comparison done was mainly between computed and measured surelevations (surge
residuals). However, such a procedure is rather questionable: indeed, what is a
surelevation ? It is the difference between something real (measured), and something that does not exist (tidal elevation computed by harmonic series) and is
never observed, even during fine weather periods. A better method could be to compare total elevations (measured and simulated), but it is then very hard to make
a discrimination between good and bad results, because variations of the amplitudes become very small compared to the total amplitudes. Figs. 5 to 9 show some
results of comparison for some places along the British Coast. At first sight
the model appears to give very bad results at Wick (Fig. 5): the residuals are
positive following the model (depression near the Northern boundary), while the
measurements show negative residuals, probably due to a negative external surge
which is, of course, completely ignored by the model. Positive residuals are
due to the boundary condition

$$\zeta_p = \frac{1}{\rho_w g} (p - p_m)$$

Results are much better for North Shields (Fig. 6), where the general features
of the surges are well simulated by the model. However, positive surges are
generally overestimated, while negative surges are underestimated; this general
trend could be due to the influence of a rather bad boundary condition as stated
above. The same pattern shows up at Immingham (Fig.7) , where a small negative
surge is completely missed. However, confidence in measurements in Immingham
should not be too high, because of the particular configuration of the coastline in this region. At Lowestoft (Fig. 8), the general trend of the surge is
well simulated, but the value of the surelevation is strongly overestimated.
Apart from the influence of the boundary condition, one could suggest, as res-

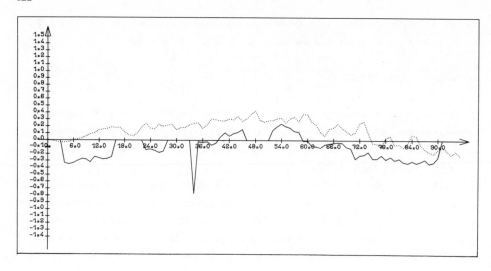

Fig. 5. Computed and observed surge at Wick.

    ——— : measured residual elevation
    ..... : computed residual elevation

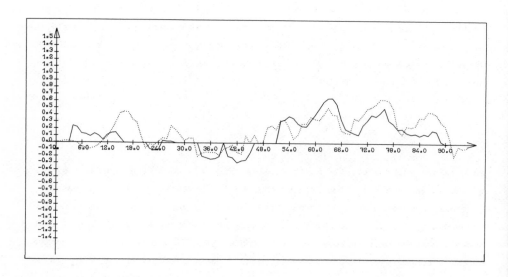

Fig. 6. Computed and observed surge at North Shields.

    ——— : measured residual elevation
    ..... : computed residual elevation

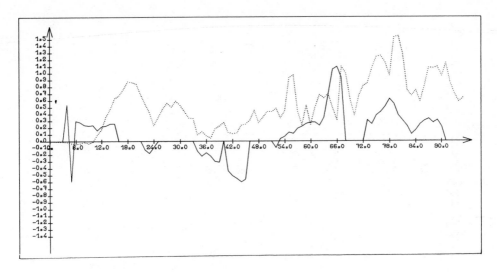

Fig. 7. Computed and observed surge at Immingham.

——— : measured residual elevation

..... : computed residual elevation

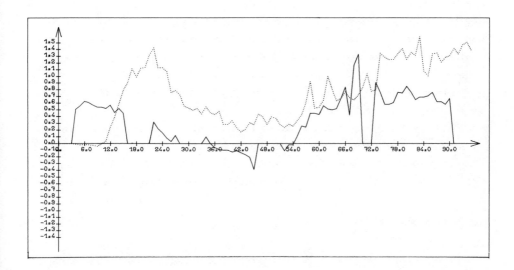

Fig. 8. Computed and observed surge at Lowestoft.

——— : measured residual elevation

..... : computed residual elevation

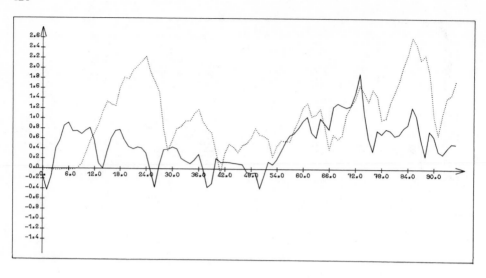

Fig. 9. Computed and observed surge at Ostend.

    ——— : measured residual elevation

    ····· : computed residual elevation

ponsible for this, the accumulation (in the model) of water masses pushed by the wind in the Southern Bight due to insufficient flow through the Dover Strait. However, since the original model (Nihoul and Ronday, 1976) gave rather better results with true meteorological fields, one can also suspect, to some extent, the poor resolution of the meteorological forecasts.

   At present, too few numerical experiments are available to distinguish between the following sources of error:

- the hydrodynamical-mathematical model itself (which, for instance, would not reproduce a good water flow through the Dover Strait)
- the meteorological input (for instance, overestimation of surface winds)
- the mathematical model of air sea interaction (overestimating the wind stress ?).

FUTURE IMPROVEMENTS

   In its present state the real-time forecast model of the North Sea cannot give satisfactory results in all situations. One can distinguish several kinds of improvements that could be included in the general scheme to deal more correct-ly with a majority of possible meteorological situations; short term and mid-long term modifications are suggested.

1. Improvements to the mathematical model

- introducing a better air-sea interaction formula; however, results with Heap's model (Flather and Davies, 1975) are only slightly better
- introducing a better condition at the northern boundary, if this makes sense without an external surge model
- coupling the existing model to
    - an external surge model, capable of giving better real time elevations at the boundaries. This model should cover the whole continental shelf and, possibly, a part of the open ocean. This model will be useless if the meteorological input does not cover a more extended region than it currently does
    - a finer mesh model of the Southern Bight-Dover Strait region, to enhance the computation of the flow in this region. This is necessary for the reason stated above, and if one also wants to compute the storm currents in the region.

2. Improvements to the meteorological input

- extension of the coarse meteorological grid to cover the whole continental shelf and a part of the adjacent open ocean. This is necessary if one wants to develop an external surge model
- refinement of the mesh of the meteorological model, together with higher frequency of meteorological forecasts.

3. Improvements to the system

The above-mentioned improvements are suggested to enhance the accuracy of the forecasts. One will also try to get the forecasts faster, by performing the whole computation on-line.

The only limitation to such improvements is the cost, which is likely to be worthwile when the system is completely operational.

REFERENCES

Bureau Hydrographique International, 1953. Constantes harmoniques. Publication Spéciale n° 26
Cartwright, D.E. , 1976. Shelf boundary tidal measurements between Ireland and Norway. Proc. 7th Liege Colloquium on Ocean Hydrodynamics, Mem. Soc. Roy. Sci. Liège, 10 : 133-139
Flather, R.A. and Davies, A.M., 1975. The application of numerical models to storm surge prediction, I.O.S. report n° 16
Nihoul, J.C.J. and Ronday, F. , 1976. Modèles hydrodynamiques. In : Projet Mer, Rapport final, Vol. 3. Programmation de la Politique Scientifique, Brussels, 270 pp.

CYCLOGENESIS AND FORECAST OF DRAMATIC WATER ELEVATIONS IN VENICE

A. TOMASIN[1] and R. FRASSETTO[1]

[1]Laboratorio per lo Studio della Dinamica delle Grandi Masse, C.N.R.,Venezia (Italy)

ABSTRACT

The oceanographic and meteorological aspects of the Adriatic surges are investigated. Since Venice floods are only due to such anomalous tides, their forecast is obviously requested for practical purposes. Short term warning (about
six hours ahead) is now possible with good accuracy, thanks to present understanding of the surge generation. The study of the cyclogenesis in the atmosphere over
the western Alps can extend the time freeboard for alarm.

INTRODUCTION AND HISTORICAL REMARKS

In the history of Venice, floods are reported as a recurrent calamity for the
city (Unesco, 1969, pp. 34-66). This is not surprising when considering that
streets are normally less than one meter over mean water level and the tidal
range is comparable to this figure. Troubles can clearly arise even with a surge
half a meter high, which is a rather common occurrence. Figure 1 shows an example.

Even though Venice is located in a lagoon, and not directly exposed to the
sea, from the point of view of tides and surges no difference is observed, except
for a delay of about an hour from the open sea to the town. The situation in
this century is certainly worse than in the past, as far as the frequency of floods
is concerned. This is due to various reasons: the town has sunk about 20 cm
since the last century (which is comparable, presumably, to the total sinking in
the previous millennium). Also, the communication between the sea and the lagoon
is larger now than 200 years ago, due to extensive dredging in the lagoon inlets.
In those days, the peak levels of the external sea were presumably "cut" by the
narrow openings of the lagoon, except for surges of long duration (and the most
tremendous flood, in 1966, was of this kind).

428

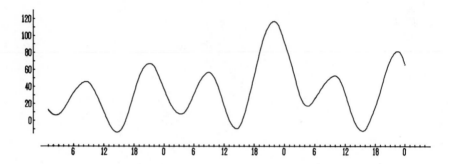

Fig. 1. Recorded water level in Venice, starting April 20, 1967. The anomalous
rise at the end of the second day is evident (Time is GMT + 1).

ADRIATIC DYNAMICS

Only the Adriatic phenomena (tides and surges) will be considered now, since,
as it was stated above, there is no significant change (other than a time lag)
in the transition from the open sea to Venice (Goldmann et al., 1975). An important feature is the independence of tide and surge in this area. They seem not
to have any interaction, and simply build up. At a first glance, this can be
surprising, since the northern "half" of the sea is a plateau slowly declining
southwards, up to 100 m depth (see Fig. 2).

In such a shallow area, oceanographers would expect substantial nonlinearities
in the tidal constituents and related interactions between tide and surge. Indeed, the ratio of the depth to the tidal range is still very large, and this accounts for the observed linearity.

Since tide and surge are independent, one can easily imagine the various mix-
tures that can appear. There can be a coincidence of flood tide and surge, so
that the level rises to dangerous values, or they can otherwise cancel each other.
What is important for numerical analysis is that, given the tidal record, by subtracting the ordinary astronomical tide one obtains a kind of meteorological tide
to be studied separately.

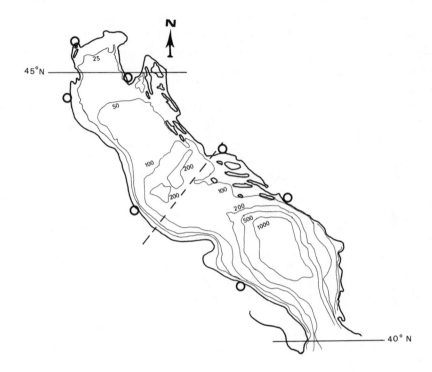

Fig. 2.   The Adriatic Sea.   Depths are in meters.

TIDES AND SEICHES

The range of the astronomical tide increases from the SE opening of the Adriatic through the NW end where Venice is: the spring range $2(M_2 + S_2)$ rises from 25 to 80 cm (about) (Sterneck, 1919). This is consistent with the classical picture according to which the tide, as observed in the Adriatic, is forced by the oscillation of the Mediterranean by the Otranto Channel, with little relevance of the direct lunisolar attraction. It should be remarked that recent calculations tend to change this picture (Tomasin, 1976). At least for the diurnal part of the tide, a contribution not less to 30% of the observed range should be attributed to the direct local forcing of the astronomical effects. Indeed, this should have only theoretical interest. Much more important, in practice, is another remark: the dominant tidal periods, 12 and 24 hours, are close to the resonant periods of this sea. The study of its eigenmodes is crucial, since free oscillations are observed frequently and with relevant amplitude. This way a common mechanism is found for a regular phenomenon, the tide, constant through the centuries, and for the random bursts of energy stimulating the sea in the storms.

The fundamental mode of oscillation ("seiche") of the sea has a period of approximately 22 hours (Kesslitz, 1910). Ordinary storms have a much shorter duration, being sharp atmospheric fronts that cross the sea moving eastwards. The sea reacts like a pendulum hit by a quick pulse and starts oscillating. The rather peculiar shape of the Adriatic makes it as an organ pipe, well tuned on its tone and with little dissipation (Robinson et al., 1973). Indeed, one can believe that there is little internal friction for the same reason of the linearity of the tide. There is also little external radiation of energy since, after the gate, the Mediterranean immediately offers a wide and deep basin, very close to an ideal infinite ocean where no energy can be radiated (Tomasin, 1971). A very slow damping of the free oscillations is observed, if there is no further perturbation. The Q-value (the figure of merit) is larger than 100, and the oscillations persist for many days (see Fig. 3).

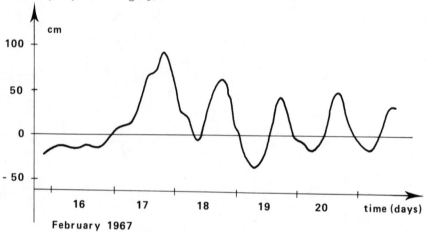

Fig. 3. An example of "seiche". The ordinary tide was subtracted from the record.

A similar behaviour is shown by the first harmonic, 11-hour oscillation, usually (but not always) smaller in amplitude. For this frequency an amphidromy is observed in the northern Adriatic, both in tides and seiches (Artegiani et al., 1972).

It is now clear that a surge has usually a peculiar shape, since, due to seiches, it is followed by many replicas of substantial height, spaced less than one day. It may occur that the first arrival of the surge does not coincide with the flood tide, while the second one does: people forget the storm and the flood comes, as was the case in February 1972. As an interesting curiosity, cases can be found of "negative" surges, caused by reverse meteorological conditions, regularly followed by a seiche sequence (e.g. June 30, 1975). Storm surges, the real core of the problem, can be faced at this point.

ORIGIN OF SURGES

It turns out from the records that the seasonal plot of the floods has a peak
in November and a certain relevance in December and January. Since there is
little difference in the tidal pattern through the seasons, special meteorological conditions must favour the surges. By analyzing case after case, one sees
that pulses of SE wind along the Adriatic are the true cause.

A more detailed analysis of the meteorological dynamics giving origin to floods
shows that the Ligurian Sea, facing the northwestern regions of Italy, is an important reference point, since sometimes the storm is born there. More frequently,
moderate perturbations from the Atlantic reaching the Ligurian area become much
stronger, as if a local mechanism were triggered. One can say that this place is
subject to cyclogenesis, or formation of depressions, that eventually migrate
eastward. There are important studies concerning this weather feature (Speranza,
1975; Buzzi et al., 1978). It can be mentioned here that the surge sketched in
figure 1 was selected to illustrate a specific case of cyclogenesis analyzed in
the literature, whose weather maps are shown in fig. 4. The effort to forecast
the floods will obviously take advantage of this knowledge.

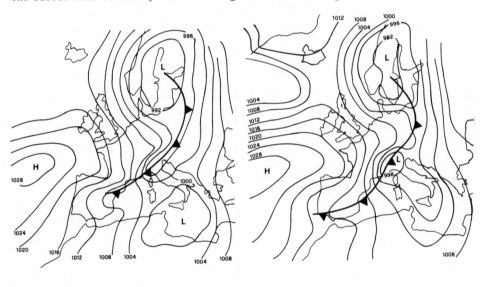

April 21, 1967, 1800 GMT   April 22, 1967, 0000 GMT

Fig. 4. A typical case of cyclogenesis. A depression is formed at the western
end of the Alps.

PREDICTION BY STATISTICAL METHODS

A big effort for the forecast of the floods was made in the last ten years.

The purpose is warning the city in order to reduce the damages of the flood. Also, if the planned protections for the lagoon will be built (like sluices at the inlets, to be closed only when a surge is coming), a warning system will be vital.

A variety of predictive schemes was developed, using either statistical or deterministic methods - or both. A distinction could also be made between simple low accuracy schemes (frequently useful for a long term forecast) and more sophisticated and precise methods.

An idea for the former ones comes from the above considerations about seiches. They persist for many days, and this means that the Adriatic has a good memory. Troubles coming from the surge "returns" can be predicted by simply observing the tide of the last hours. On these assumptions, a simple scheme was developed (Tomasin, 1972) where the future level of the sea is estimated by a predictive linear filter applied to the observed tide. The filter weights were statistically obtained from the recordings of the past.

In mathematical symbols, the estimate of the sea level in Venice, at time t, s_t^* , is obtained as the inner product of two vectors built by proper filter weights ($\underline{f} \equiv f_0 , f_1 , \ldots , f_n$) and sea levels ($\underline{s} \equiv s_{t-T} , s_{t-T-1} , \ldots , s_{t-T-n}$) observed up to T hours before the predicted time. Then

$$s_t^* = \underline{s} \cdot \underline{f}$$

The length of the filter, $n + 1$, was found to be convenient at about 60 hours. Statistics offer another simple idea for an approximate expectation: since the weather behaviour is somehow determined in the cases of interest, the simplest local atmospheric parameter, the pressure, could give some indication. A numerical filter was built, like the one described above, but now using also pressure figures of the last hours.

Now $s_t^* = \underline{s} \cdot \underline{f}_1 + \underline{p} \cdot \underline{f}_2$, where $\underline{p}$ is the vector of the observed pressure values. The length of the filter was now reduced to 24 hours. In spite of its simplicity (any interested Venetian could use it by himself), the advantage is remarkable. Figure 5 shows, for the case already considered (April 21, 1967) how the simple scheme would have predicted the flood six hours in advance. (Obviously, the weights of the filter were obtained from a sample which did not include this case).

Further advances for the latter method (which is still in progress) are expected by relaxing the linearity of the method.

As stated above, more complicate methods are required for better reliability, and the precise dynamics of the sea must be studied. What really gives origin to surges is the SE wind and, with smaller effect, the low pressure or more precisely a certain pressure gradient along the Adriatic. The good coverage of weather stations along the shores is obviously a favourable feature. Their reports can be correlated to sea level either via statistics or using storm surge

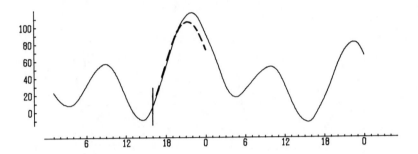

Fig. 5. An example of the prediction of a flood using a simple statistical regression. Predictors are available for the scheme up to 16.00, April 21, 1967.

equations.

The statistical approach (Sguazzero et al., 1972) extends the method described above to the whole Adriatic. There are more independent variables: not the simple atmospheric pressure in Venice, but estimates of the pressure gradient and wind stress in certain ideal points in the middle of the Adriatic. In formula

$$s_t^* = \underline{s} \cdot \underline{f} + \underline{P}_1 \cdot \underline{F}_1 + \underline{P}_2 \cdot \underline{F}_2 + \ldots + \underline{W}_1 \cdot \underline{G}_1 + \underline{W}_2 \cdot \underline{G}_2 + \ldots$$

Here $\underline{s}$ is again the vector of the sea levels in Venice $s_{t-T}, s_{t-T-1}, \ldots; P_i$ is the vector $(P_{i,t-T}, P_{i,t-T-1}, \ldots)$ of the pressure values in the i-th standard point; $\underline{W}_j$ is the vector of the wind stress figures in the j-th standard point. $\underline{F}_k, \underline{G}_k$ and $\underline{f}$ are obvious "filter" vectors. More specifically, this model used five points for gradient and stress estimates (equally spaced along the axis of the Adriatic) and this field reconstruction was made using seven coastal stations. Data were the ordinary synoptic three-hourly measurements. The "past" data entering the formula referred to two days for level and one day for weather. It should be remarked that gradient and stress were taken only in the component along the axis of the Adriatic. This is a point of interest, since it emphasizes that the Adriatic dynamics is almost completely one-dimensional for surges.

In the range of forecast of six hours (T = 6 in the previous formula) floods were predicted within a few centimeters in the dozens of cases when the model was tested (many of them in real time). Fig. 6 gives an example.

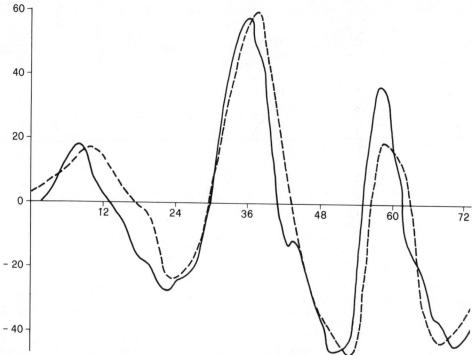

Fig. 6. An example of prediction using the statistical method by Sguazzero et al. The surge (i.e. the deviation from ordinary tide) is shown by the solid line. Dashed curve is the forecast, each point being given six hours in advance. (starting point February 12, 1972, 0300 GMT).

One can conclude that using only the weather reports from the Adriatic coast and the observed tidal levels, the prediction is possible for about six hours a-head. (Further forecasting is progressively unreliable, due to the arrival of other atmospheric phenomena in the meantime).

PREDICTION BY DETERMINISTIC MODELS

Better satisfaction comes from a detailed inspection of what happens in the Adriatic when the atmosphere forces it. More or less the same information (i.e. weather reports from coastal stations) can be used for the integration of the storm surge equations where the driving forces are, again, the wind stress and the pressure gradients. It is obviously not only a matter of satisfaction but also a better knowledge of the physics of the surge and a more complete informa-tion for the various places. The storm surge equations can be written (Proudman, 1954)

$$\frac{1}{a \cos \phi} \frac{\partial}{\partial \phi} (U \cos \psi) + \frac{\partial V}{\partial \chi} + \frac{\partial \zeta}{\partial t} = 0,$$

$$\frac{\partial U}{\partial t} + 2\omega \sin \phi \, V = -\frac{gh}{a} \frac{\partial \zeta}{\partial \phi} - \frac{h}{\rho a} \frac{\partial P_a}{\partial \phi} + \frac{1}{\rho}(F_s - F_B),$$

$$\frac{\partial V}{\partial t} - 2\omega \sin \phi \, U = -\frac{gh}{a \cos \phi} \frac{\partial \zeta}{\partial \chi} - \frac{h}{\rho a \cos \phi} \frac{\partial P_a}{\partial \chi} + \frac{1}{\rho}(G_s - G_B)$$

where ϕ and χ are latitude and east-longitude, t is time, ζ the elevation of the
sea surface, U and V the components of the total stream, i.e. velocity times
depth, F_s and G_s the components of the friction of the wind on the sea surface,
F_B and G_B the components of the water friction on the sea bottom, p_a the atmospheric pressure on the sea, h the water depth, ρ the density of the water (assumed uniform), a the mean radius of the earth, g the acceleration of gravity,
ω the angular speed of the earth's rotation.

These equations are linearized and integrated with respect to depth. Linearity
can be safely assumed after the above remarks. This accounts for the suppression
of all non linear terms in deriving these equations and allows the assumption of
a linearized bottom friction, as it was done in practice.

A variety of works were developed (Tomasin, 1971; Stravisi, 1972; Accerboni et al.,
1973) using the above basic equations, correctly reproducing tides, seiches and
surges. Integration techniques were essentially already known from the literature of hydrodynamical numerical models (Hansen, 1956; Heaps, 1969). In particular, reduction to one dimension was possible due to the shape of the Adriatic.
And this was really the scheme more extensively tested in practice (Finizio, 1970;
Tomasin, 1973). In formula

$$\frac{1}{w} \frac{\partial Q}{\partial t} = -gh \frac{\partial \zeta}{\partial x} - \frac{h}{\rho} \frac{\partial P_a}{\partial x} + \frac{1}{\rho}(F_s - F_B)$$

$$\frac{\partial \zeta}{\partial t} = -\frac{1}{w} \frac{\partial Q}{\partial x}$$

where the x axis is directed along the sea, w is the width of the cross section
and Q the discharge across it, or, say, the average velocity times the area of
the section.

The equations are clearly simplified also because the earth curvature is dis-
regarded: needless to say, all these simplifications would not be possible for
other areas, like for example the Mediterranean, whose modelling is mentioned below.

In forecasting, the one dimensional scheme was used on the field, using synoptic meteorological data and integrating the equations by an explicit finite-differences scheme. The weather field is known up to the time when the forecast is issued: the situation is supposed constant for the following hours and the equations give future sea level under this hypothesis. Consistent with the statistical approach, it turns out that at any time there is enough information to predict safely the sea level for about six hours. As expected, the above method turned out to be satisfactory as a predictor.

FURTHER DEVELOPMENTS

Improvements of the prediction are expected by widening the limits in which research has been confined so far. It will be remarked that the use of weather predictions was avoided. Over the Mediterranean they are made difficult by the influence of orography which complicates the flow of air masses and generates local effects: cyclogenesis on the western lee of the Alps was mentioned above. The Global Atmospheric Research Program (GARP) has created a subprogram on Air Flow Over and Around Mountains and the first experiment around the Alpine barrier (Alpex) is proposed for 1981. This effort should help in understanding the dynamics of perturbations in the Mediterranean, with the final objective of an accurate prediction 12 to 24 hours ahead: Venice, Italy and several countries will substantially benefit.

For the specific warning of floods, a wind forecasting model is being developed (Palmieri at al., 1976). Statistical methods also are being implemented in Venice for the prediction of the pressure field in the central Mediterranean.

From the hydrodynamical point of view, it is noticeable that no attempt was made to forecast external surges, i.e. phenomena originating out of the Adriatic, but giving significant effect in it. From experience, they seem not to exist; so far the low frequency changes in the level of the Mediterranean have been monitored close to the Otranto Channel and transmitted to Venice for a direct use in the predictive models (Mazzoldi et al., 1973). For a better insight in these large scale phenomena affecting the Mediterranean, a specific numerical model is being developed, to fill up the theoretical gap that was so far overcome by direct inspection.

Time is ripe for wider analysis, where the Mediterranean is fully considered for its hydrodynamical contributions and for the meteorological patterns that develop over it. The goal of a warning for Venice floods with an advance of twelve hours or one day seems not to be chimerical.

ACKNOWLEDGEMENTS

The help of R. Dazzi, G. Aldighieri, and Mrs. J. Zanin, of the CNR laboratory, has been of utter importance for this work. Precious support was given by the Centro Scientifico IBM of Venice and the CRIS-ENEL branch of Venice-Mestre.

REFERENCES

Accerboni, E. and Manca, B., 1973. Storm surges forecasting in the Adriatic Sea by means of a two-dimensional hydrodynamical numerical model. Boll. Geofis. Teor. Appl., 15:3-22.

Artegiani, A., Tomasin, A. and Goldmann, A., 1972. Sur la dynamique de la mer Adriatique due aux excitations météorologiques. Rapp. Comm. Int. Mer Médit., 21:181-183.

Buzzi, A. and Tibaldi, S., 1978. Cyclogenesis in the lee of the Alps: A case study. Quart. J. R. Met. Soc., 104:271-287.

Finizio, C., Palmieri, S. and Riccucci, A., 1970. A numerical model of the Adriatic Sea for the study and prediction of sea tides at Venice. Ist. Fis. Atmosf., STR 12.

Goldmann, A., Rabagliati, R. and Sguazzero, P., 1975. Propagazione della marea nella laguna di Venezia: analisi dei dati rilevati dalla rete mareografica lagunare negli anni 1972-73. Riv. Ital. Geofis., 2:119-124.

Hansen, W., 1956. Theorie zur Errechnung des Wasserstandes und der Strömungen in Randmeeren nebst Anwendungen. Tellus, 8:287-300.

Heaps, N.S., 1969. A two-dimensional numerical sea model. Phyl. Trans. A 265: 93-137.

Kesslitz, W.V., 1910. Das Gezeitenphänomen in Hafen von Pola. Mittl. Geb. Seewesens, 38:H. V-VI.

Mazzoldi, A., Dallaporta, G., Gaspari, A., Curiotto, A. and Peruzzo, G., 1973. Sistema di telemisure da stazioni automatiche fisse. CNR-LSDGM, Venezia, Rapporto Tecnico n. 37.

Palmieri, S., Finizio, C. and Cozzi, R., 1976. The contribution of meteorology to the study and prediction of high tides in the Adriatic. Boll. Geofis. Teor. Appl., 19:191-198.

Proudman, J., 1954. Note on the dynamics of storm-surges. Mon. Not. R. astr. Soc. Geophys. Suppl., 7:44-48.

Robinson, A.R., Tomasin, A. and Artegiani, A., 1973. Flooding of Venice: Phenomenology and prediction of the Adriatic storm surge. Quart. J. R. Met. Soc., 99:688-692.

Sguazzero, P., Giommoni, A. and Goldmann, A., 1972. An empirical model for the prediction of the sea level in Venice. IBM Italia Tech. Rep. CSV006.

Speranza, A., 1975. The formation of baric depressions near the Alps. Ann. di Geofis., 28:177-217.

Sterneck, R.V., 1919. Die Gezeitenerscheinungen in der Adria. Denschr. Akad. Wiss. Wien, 96:277-324.

Stravisi, F., 1972. A numerical experiment on wind effects in the Adriatic Sea. Acc. Naz. Lincei, Rend. Sc. Fis., mat., nat., 52:187-196.

Tomasin, A., 1971. Application of the hydrodynamical numerical integration method to the Adriatic Sea. Physics of the Sea, Trieste, 13-16 Oct., Accad. Naz. Lincei, Quad. n. 206, 119-121.

Tomasin, A., 1972. Autoregressive Prediction of Sea Level in the Northern Adriatic. Riv. Ital. Geofis., 21:211-214.

Tomasin, A., 1973. A computer simulation of the Adriatic Sea for the study of its dynamics and for the forecasting of floods in the town of Venice. Comp. Phys. Comm., 5:51-55.

438

Tomasin, A., 1976. The dynamics of the diurnal tide in the Adriatic Sea: preliminary result of a revisiting analysis. Rapp. Comm. Int. Mer Médit., 23:49.

Unesco, 1969. Rapporto su Venezia. Mondadori, Milano, 348 pp.

THE RESPONSE OF THE COASTAL WATERS OF N.W. ITALY

ALAN J. ELLIOTT

SACLANT ASW Research Centre, Viale San Bartolomeo 400, 19026 La Spezia, Italy

ABSTRACT

Two-month long low pass records of the coastal currents and winds have been analysed for two locations off the N.W. coast of Italy. Most of the energy was found to be in the long period (>20 day) motions and there was low coherence between the currents and wind except for time scales around 5 days. This suggests that either the wind was exciting a rotational mode of the entire Western Mediterranean or else that the weather systems were more coherent spatially at the 5 day time scale. A depth-integrated hydrodynamic model is being used to resolve the time scales and the effects of bottom topography.

The coastal currents may make a significant contribution to the day-to-day variability in sound speed, especially near frontal zones, due to the alongshore advection of water of differing acoustic properties. Consequently, accurate prediction of the sound speed at a fixed location may not be obtained until the coastal dynamics are clearly understood.

INTRODUCTION

During recent years there has been an increasing interest in problems related to shallow water acoustics. As a result, two distinct problem areas have arisen which need to be addressed by the research. The first of these involves the acoustic propagation itself, the second is more oceanographic in nature and concerns the variability of the acoustic properties of the water near a coast. In the deep ocean, for many acoustic purposes, the water can be considered as being well-mixed in the horizontal plane and only the vertical variations need to be considered. Thus the majority of sonar models do not incorporate range-dependent temperature fields and variable bottom topography but, instead, use a single vertical temperature profile and a flat bottom to characterise a region of interest. For many purposes this is an acceptable approximation, but it is not generally valid in shallow water. In the coastal zone, as well as the complicating effect of a sloping bottom, there can be significant variations in the salinity, temperature and sound speed characteristics of the water over relatively short horizontal distances. Among the mechanisms which can cause the variations are the fresh water input by rivers, upwelling induced by the coastal winds, and the enhanced vertical mixing due to the strong tidal and storm generated currents. As a result the coastal zone is usually a region of high acoustic variability. The problem

440

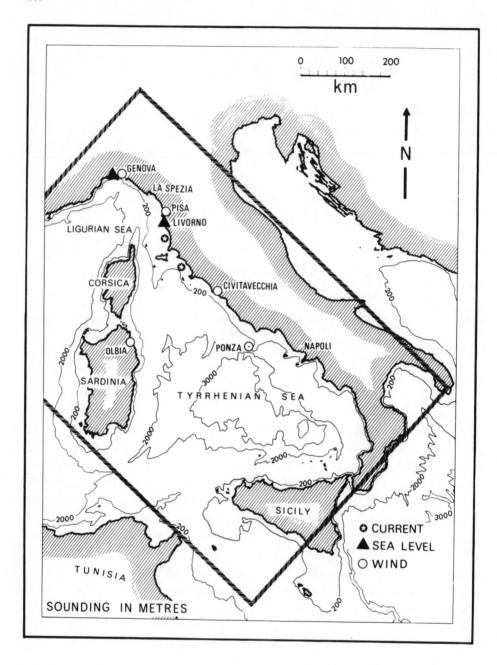

Fig. 1. The Ligurian and Tyrrhenian Seas showing the mooring locations. The dashed line represents the boundary of the numerical model.

is complicated further since the coastal waters are not static but are constantly being moved by the coastal currents under the influence of the winds. Consequently, measurements made at a selected location on one day may not be valid for the following day due to the combined effect of the high spatial variability in the sound speed and the advective effects of the currents. For the oceanographer there are two distinct problems to be resolved: first, can we obtain insight into the mechanisms which lead to the high variability in the coastal waters, e.g. the processes which generate fronts; second, given that high spatial variability exists in the coastal waters, what are the time scales associated with the coastal currents which would contribute to the temporal acoustic variability at a fixed location, and to what extent can we succeed in modelling the dynamics of the coastal response?

In order to answer some of these questions, and to provide an oceanographic input into what is essentially an acoustic problem area, a series of field measurements were made in the coastal waters of N.W. Italy and this has been combined with a numerical study of the region. The purpose of this paper is to present some of the observational results and show that, in general, there was a lack of coherence between the coastal currents measured at different locations. The second part of the paper describes how a depth-integrated numerical model is being used to resolve the role that might be played by the variations in bottom topography: one of the factors which may have contributed to the low coherence in the measurements.

THE OBSERVATIONS

During April and May of 1977 current measurements were made at two locations off the N.W. coast of Italy in water approximately 100 m deep and 15 km offshore. The two moorings, which were 100 km apart, were located near Elba on opposite sides of the shallow water which extends between the Italian mainland and the island of Corsica (Fig. 1). Three oceanographic cruises were made at approximately monthly intervals to survey the temperature/salinity properties of the coastal waters near the mooring positions, and meteorological and sea level data were obtained from established coastal recording stations. Each mooring supported two current meters: one at a depth of 20 m and the other at 80 m. The data series were filtered with a low pass filter to remove the fluctuations with periods less than two days; the data were then resampled at 6-hour intervals.

The hydrography

There was a uniform warming of the coastal waters during the months of April and May, the surface temperature increasing from about 14°C to 18°C. Near-bottom temperatures remained constant and the warming was confined to the upper layers of the water column. Since there were no significant horizontal temperature gradients

Fig. 2. Infrared satellite image of the Tyrrhenian Sea, April 23, 1977; the warm water masses show up as darker patches, parts of Sicily and Calabria are cloud covered (courtesy of the Univ. of Dundee).

along the coast and there was no evidence of a warm water mass to the south, it appears that advective effects were not important and that the warming was due to solar heating. This view was supported by a satellite image of the Tyrrhenian Sea taken on April 23 (Fig. 2) which showed insignificant large-scale horizontal temperature gradients. Therefore, there appeared to have been a uniform heating of the coastal waters, due to seasonal changes, throughout the period of the study. Miller et al. (1970) measured water temperature along a section parallel to the coast and extending from Elba to Calabria. Their data show an alongshore temperature change of about 0.5°C over a distance of about 600 km (the warmer water lying to the south). If we assume that the mean flow was to the N.W., i.e. up the coast, at 5 cm s^{-1} and we write

$$\frac{dT}{dt} = \frac{\partial T}{\partial t} + u \frac{\partial T}{\partial x}$$

then $\frac{dT}{dt}$ represents the total temperature change and can be estimated from the data. Consequently, $\frac{\partial T}{\partial t}$ which represents the effect of local heating can be computed. The data showed that:

$$7.70 \times 10^{-7} = 7.28 \times 10^{-7} + 0.42 \times 10^{-7}$$

Total change (100%)	Local heating (95%)	Advection (5%)

which supports the conclusion that the increase in temperature was mainly due to a seasonal warming and not due to the advection of warm water from the south.

The increase in temperature caused a reduction in the density of the surface water and at both locations the water column changed from being well-mixed during March to a two-layered system, with a surface mixed-layer about 10 m deep during May. At both positions, the current meters were moored beneath the depth of the surface mixed-layer.

The coastal winds

There was significant spatial variability in both the strength and the direction of the low pass coastal winds. Whereas the mean stress at the three southern stations (Civitavecchia, Olbia, and Ponza) was directed eastwards with a strength of about 0.5 dynes cm^{-2}, the mean stress at Genova and Pisa was about an order of magnitude weaker; at Pisa the mean stress vector was directed towards the N.E. while at Genova it was towards the N.W. The anti-clockwise rotation of the mean wind may have been caused by cyclogenesis (Cantu, 1977), or it may have been due to orographic effects. To resolve the structure of the wind, empirical orthogonal function analysis (Wallace and Dickinson, 1972) was applied to the five components of E-W wind stress. Of the five modes isolated (since there were five input series)

ALONGSHORE COMPONENTS

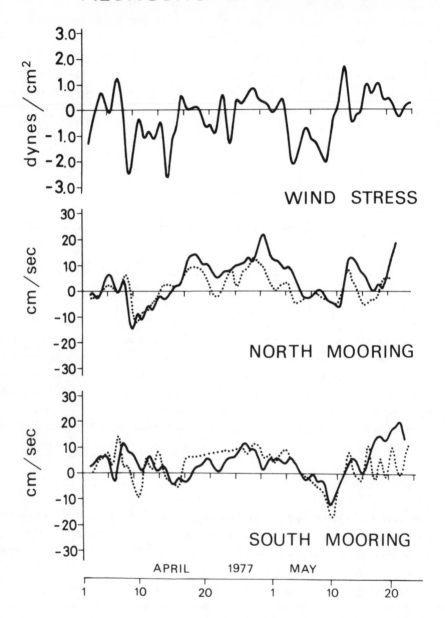

Fig. 3. Alongshore components of the low pass currents and coastal wind.
The current at 20 m is shown by a solid curve, the 80 m current by a dotted curve.

the first mode was highly correlated with the components of stress at Pisa, Civitavecchia, Olbia and Ponza but only explained 1% of the variance in the Genova record. In contrast, the second mode accounted for 95% of the Genova variance but was uncorrelated with the other locations. A similar result was obtained when the analysis was repeated for the N-S components of stress.

As a result of this partition of the wind records it was decided that the Genova wind was not representative and that the record had been unduly influenced by orographic effects; it was therefore excluded from further analysis. The remaining 4 vector series were then averaged to produce a time series of the large-scale wind. The geostrophic wind stress, calculated from atmospheric pressure, was not coherent with the observed wind and was a poor predictor of the current response.

The coastal currents

In Fig. 3 are shown the low pass alongshore components of the currents and mean wind, positive currents and wind stress being directed up the coast towards the N.W. The current records showed marked fluctuations at a time scale comparable to the record length while the wind appeared to contain more energy at the shorter time scales. The flow was northward when the wind was near zero and only reversed its direction during strong southward winds. Linear regression showed that under conditions of zero wind the flow would be towards the N.W. at both mooring locations, and that this density-driven flow which appeared to be independent of depth had a strength of about 5 cm s^{-1}.

Spectra for current, adjusted sea level and atmospheric pressure at Livorno, and the alongshore wind are shown in Fig. 4 and show that the energy was contained in the longer period motions. Sea level and wind spectra peaked at around 20 days while the current and atmospheric pressure had maxima at periods comparable to the record length. The spectra decreased steadily between 20 days and 3 days; in particular, there was no pronounced spectral peak at the 5 day time scale. However, when the coherence was computed between pairs of variables then the coherence was usually greater at 5 days than at other periods. As an example, Fig. 5 shows the coherence between the alongshore wind and each of the four current records. At both the north and south mooring the bottom currents were highly coherent with the wind at the 5 day time scale. Comparable results were found when sea level and atmospheric pressure were analysed: the coherent response appeared to be confined to a band around the 5 day period. Since the forcing variables (e.g. wind and pressure) did not contain an excess of energy at this time scale it suggests that the coastal waters were responding in an organised manner to the 5 day forcing.

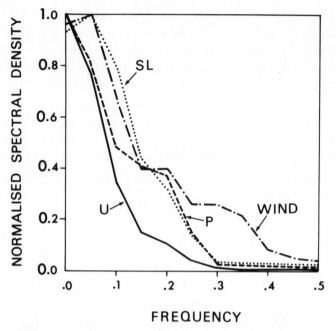

Fig. 4. Normalised spectra of alongshore current and wind, sea level and atmospheric pressure. Frequency is in cycles per day.

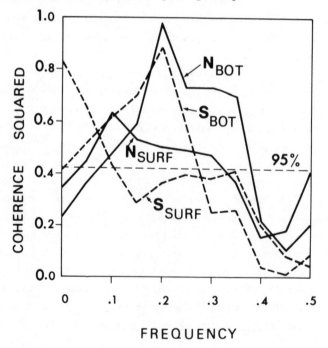

Fig. 5. Coherence squared between the alongshore currents and alongshore wind; N signifies the north mooring, S signifies the south mooring.

A MODEL OF THE LIGURIAN AND TYRRHENIAN SEAS

Since the local bottom topography, the geometry of the deep basins and rotational effects were suspected of being factors which could influence the currents, the problem is being approached numerically through the use of a hydrodynamic model. A depth-integrated model, suitable for general application to both deep and shallow water areas, was developed; the model includes density terms as well as a salinity balance and a temperature (or pollutant) balance equation. The full system of equations used in the study was

$$\frac{\partial \eta}{\partial t} = (d+\eta)\frac{\sigma}{\rho} - \frac{\partial}{\partial x_1}\left[(d+\eta)u_1\right] - \frac{\partial}{\partial x_2}\left[(d+\eta)u_2\right]$$

$$\frac{\partial}{\partial t}\left[(d+\eta)u_1\right] = \frac{\partial}{\partial x_1}\left[(d+\eta)\ N\frac{\partial u_1}{\partial x_1}\right] + \frac{\partial}{\partial x_2}\left[(d+\eta)\ N\frac{\partial u_1}{\partial x_2}\right]$$

$$+ f(d+\eta)u_2 - g(d+\eta)\frac{\partial \eta}{\partial x_1} - \frac{g}{\rho}\frac{(d+\eta)^2}{2}\frac{\partial \rho}{\partial x_1} + FS_1 - FB_1$$

$$\frac{\partial}{\partial t}\left[(d+\eta)u_2\right] = \frac{\partial}{\partial x_1}\left[(d+\eta)\ N\frac{\partial u_2}{\partial x_1}\right] + \frac{\partial}{\partial x_2}\left[(d+\eta)\ N\frac{\partial u_2}{\partial x_2}\right]$$

$$- f(d+\eta)u_1 - g(d+\eta)\frac{\partial \eta}{\partial x_2} - \frac{g}{\rho}\frac{(d+\eta)^2}{2}\frac{\partial \rho}{\partial x_2} + FS_2 - FB_2$$

$$\frac{\partial}{\partial t}(d+\eta)s = \frac{\partial}{\partial x_1}\left[(d+\eta)\ K\frac{\partial s}{\partial x_1}\right] + \frac{\partial}{\partial x_2}\left[(d+\eta)\ K\frac{\partial s}{\partial x_2}\right]$$

$$- \frac{\partial}{\partial x_1}\left[(d+\eta)u_1 s\right] - \frac{\partial}{\partial x_2}\left[(d+\eta)u_2 s\right]$$

$$\frac{\partial}{\partial t}\left[(d+\eta)c\right] = \frac{\partial}{\partial x_1}\left[(d+\eta)\ K\frac{\partial c}{\partial x_1}\right] + \frac{\partial}{\partial x_2}\left[(d+\eta)\ K\frac{\partial c}{\partial x_2}\right]$$

$$- \frac{\partial}{\partial x_1}\left[(d+\eta)u_1 c\right] - \frac{\partial}{\partial x_2}\left[(d+\eta)u_2 c\right] + (d+\eta)\sigma_c$$

$$\rho = \rho_o(1 + \alpha s + \beta c)$$

where u_1 and u_2 are the components of the horizontal velocity, η is the surface elevation above mean sea level, d is the bottom depth with respect to mean sea level, s is the salinity and c represents the concentration of a pollutant or, with some modification to the equation, the temperature distribution. N and K represent the horizontal eddy stress and diffusivity. A term representing a fresh

448

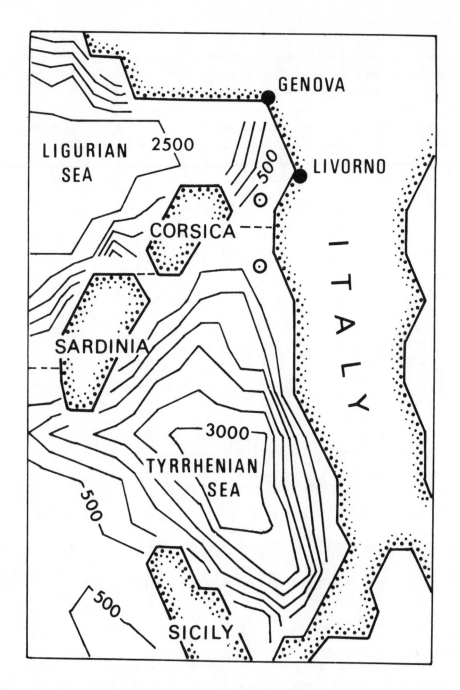

Fig. 6. Bottom topography of the Ligurian and Tyrrhenian Basins as used in the numerical study; the mooring locations are circled. The area shown is approximately 550 x 380 km^2 and the grid size is 20 km. The dashed lines represent the boundaries used to separate the two basins.

water source was included in the continuity equation to allow for river run-off.
The components of surface stress, FS_1 and FS_2, were calculated using a quadratic
drag law for the wind stress, and FB_1 and FB_2 were the components of bottom stress
— also quadratic in form. Consequently, the equations form a basically linear
system with the exception of the quadratic friction term.

The equations were solved explicitly in a standard manner using a leap-frog
scheme with centred space differences on a regular grid. The variables were
spatially staggered on the grid so that surface elevation, density and depth
were specified at the centre of each rectangular element while the two velocity
components were specified at the mid-points of adjacent sides of the grid elements
(Leendertse, 1970; Heaps, 1969; Tee, 1976). Under many circumstances (including
the present) density effects can be neglected, in which case only the first three
equations need to be solved.

The time scales of the Ligurian and Tyrrhenian Basins

As a first step towards resolving the normal modes of the entire Western Mediterranean the model was applied to the more restricted region comprising the
Ligurian and Tyrrhenian Seas as shown in Fig. 1. *In addition, attention was
directed towards the shorter time scales*, i.e. those of about one day and less.
The current observations discussed previously provided a problem of a qualitative
nature to which the model could be applied: Before the current records had been
low pass filtered, both progressive vector diagrams and spectral calculations had
revealed a marked difference in the currents at the two mooring locations, especially during the first month of measurement. Whereas the currents at the northern
mooring were generally of long period (>10 days) and flowed parallel to the coastline (having twice as much energy in the alongshore direction as in the on/offshore),
the currents at the southern mooring were mainly inertial with a period of around
17.5 hours. There was nearly an order of magnitude more inertial energy at the
southern mooring than there was in the north, the tidal energy being insignificant
at both locations.

Previous observations of inertial motions in the Mediterranean have shown that
they are rarely coherent over horizontal scales exceeding 10 km. However, in the
present case we are not investigating the coherence between inertial motions — we
are, instead, seeking an explanation for the apparent lack of such motions at the
north mooring.

To determine the spatial variability in the local response to forcing the model
was applied to the region shown in Fig. 6. For the boundary conditions there was
assumed to be no flow through the Straits of Sicily, and the surface elevations
along the left hand open boundary were specified everywhere as being the sum of
250 harmonic terms of equal amplitude but with periods ranging uniformly from 12

minutes to 50 hours. Thus the interior of the model was forced by the periodic elevations along the left hand boundary, and we shall be looking for evidence of local resonance with the applied forcing. The model was run to simulate 100 hours of currents and elevations within the system and the predicted currents at the two mooring locations and at other selected positions were stored on disc. Cross-spectral analysis was then used to isolate the coherent fluctuations. A second numerical experiment was then made by separating the two basins as shown by the dashed lines in Fig. 6 and the calculations were repeated. Similar techniques have been used to compute the periods of oscillation of the English Channel (Flather, 1976), the Gulf of Genova (Papa, 1977a) and the South Sicilian Basin (Colucci and Michelato, 1976).

Table I summarises the results for the separate and combined basins. It shows that whereas the Tyrrhenian Basin responded locally to oscillations with a near-inertial period (around 17.5 hours) these periods were locally suppressed in the Ligurian Basin. The coherent motions in the Ligurian Basin were found to be at 33 hours, 3.6 hours and between 1 to 2 hours; this agrees with previous calculations (Papa, 1977a,b) which isolated the 3.6 and 1.2 hour response but not that at 33 hours. The local suppression of near-inertial signals at the mooring position within the Ligurian Basin could explain the current characteristics discussed above. The oscillations at 3.6 and 1 to 2 hours can be identified with the seiche motions that would arise in the northern basin shown in Fig. 6. However, this mechanism cannot explain the coherence at 33 hours. For the typical shelf widths of the Ligurian Sea as resolved by the model we can compute the phase speed of the funda-mental barotropic shelf wave (Robinson, 1964) to be of the order of 400 km/day; such a wave would take about 30 hours to travel around the Ligurian coastline shown in Fig. 1. Thus if energy were being supplied over a broad spectrum by the deep water lying to the SW of the Ligurian Sea then we might expect to see a local response with a time scale of around 30 hours.

TABLE I

Periods of coherent fluctuations (for which coherence squared exceeded the 95% significance level): (a) computed for the combined system, (b) computed for the Ligurian Basin, (c) computed for the Tyrrhenian Basin, (d) from analysis of sea level at Genova and Livorno.

	Period in hours							
(a) Combined	33.1	25.0	19.9	–	–	–	3.6	1.2
(b) Ligurian	33.1	–	–	–	–	–	3.6	1.2
(c) Tyrrhenian	–	–	19.9	16.6	14.2	12.4	–	–
(d) Genova and Livorno sea level	33.4	25.4	21.4	–	–	–	3.6	–

An independent check of the calculations was made by analysing hourly unfiltered two-month long sea level records from Genova and Livorno (Fig. 1). The records were coherent at all periods longer than 4 days (due mainly to barometric effects) and also at the periods given in Table I. Good agreement was found between the periods computed by the model and those isolated from the elevation data. In particular, evidence for periodicities within the Ligurian Basin at 33 and 3.6 hours was obtained.

DISCUSSION

The observational data suggest that the coastal waters of N.W. Italy may not respond like a straight open coastline (e.g. like the west coast of the United States) but may instead have characteristics more similar to a large enclosed lake. Thus the dominant response to forcing may involve the dynamics of the entire Western Mediterranean. Consequently work is now in progress to resolve the normal modes of the western basin and to look for a possible 5 day periodicity. Alternatively, the weather systems themselves may be more spatially coherent at the 5 day time scale and may thus be more efficient at driving the coastal response: this possibility is also being investigated further.

The data analysis has made use of a 'large-scale wind' which was obtained by computing the vector mean of several coastal wind records. In view of the poor coherence found between this wind and the coastal currents it suggests that this may not be a meaningful approach. It may be necessary to specify the wind everywhere (as is done in modelling the storm surge effects in the North Sea, e.g. Heaps (1969)) in order to obtain a satisfactory prediction of the coastal response.

Returning to the problem of acoustic variability in the coastal waters, we can make some estimate of the effects of the coastal currents. From the observations the alongshore current had an amplitude of around 7.0 cm s^{-1} at the 20 day time scale: this is equivalent to a maximum displacement of the water of about 6.1 km in one day. Since the maximum observed gradient in sound speed was about 0.1 m s^{-1} per km the current could produce a change of about 0.60 m s^{-1} from one day to the next at a fixed location. In comparison, seasonal heating caused an increase in sound speed of about 0.26 m s^{-1} per day. Thus, even in this region of extremely low horizontal gradients, the currents may have contributed significantly to the day-to-day variability in sound speed. Therefore in other areas, especially near frontal locations, accurate prediction of acoustic characteristics will not be obtained until we have a better understanding of the dynamics of the coastal zone.

ACKNOWLEDGEMENTS

F. De Strobel assisted with the collection of data at sea, and P. Giannecchini, E. Nacini and G. Tognarini contributed to the analysis and presentation of the results.

REFERENCES

Cantu, V., 1977. The climate of Italy. In: C.C. Wallen (Editor), World Survey of Climatology. Elsevier, pp. 127-183.
Colucci, P. and Michelato, A., 1976. An approach to the study of the Marrubbio Phenomenon. Boll. Geof. Teor. Appl., XIX:3-10.
Flather, R.A., 1976. A tidal model of the north-west European continental shelf. Mem. Soc. Roy. Sci. Liege, 6:141-164.
Heaps, N.S., 1969. A two-dimensional numerical sea model. Phil. Trans., A265:93-137.
Leendertse, J.J., 1970. A water-quality model for well mixed estuaries and coastal seas. Rand Corporation, RM-6230-RC.
Miller, A.R., Tchernia, P. and Charnock, H., 1970. Mediterranean Sea Atlas. Woods Hole Oceanographic Institute Atlas Series, Vol. III.
Papa, L., 1977a. The free oscillations of the Ligurian Sea computed by the H-N method. Deutsch. Hydrogr. Zeitschr., 30:81-90.
Papa, L., 1977b. The free oscillations of the Ligurian Sea. A statistical investigation. Boll. Geof. Teor. Appl., XIX:269-276.
Robinson, A.R., 1964. Continental shelf waves and the response of sea level to weather systems. J. Geophys. Res., 69:367-368.
Tee, K.T., 1976. Tide-induced residual current, a 2-D nonlinear numerical tidal model. J. Mar. Res., 34:603-628.
Wallace, J.M. and Dickinson, R.E., 1972. Empirical orthogonal representation of time series in the frequency domain. J. App. Meteor., 11:887-892.

A NUMERICAL MODEL FOR SEDIMENT TRANSPORT

J.P. LEPETIT, A. HAUGUEL

E.D.F., Direction des Etudes et Recherches, Laboratoire National d'Hydraulique,
Chatou (France)

ABSTRACT

We introduce here a numerical two dimensional model for sediment transport which
permits to compute the impact of a coastal structure on the bottom evolution.
The introduction of current disturbance and some assumptions using difference of
time scale between current and bottom evolutions permits to obtain a propagation
equation driving the bottom evolution.
The model has been calibrated in the case of the local scour around a jetty. A last,
it has been applied to the bottom evolution in the vicinity of the new port of
Dunkerque.

INTRODUCTION

One of the impacts of a large coastal structure is its effect on current pattern
in the vicinity of the structure. These changes in current conditions will induce
changes in the sediment transport pattern and may disturb an existing equilibrum
thus causing large changes in bottom topography in the vicinity of the structure.
To assess the severity and extend of topographical changes induced by the structure
the interaction of the resulting fluid motion with the bottom evolution must be
properly reproduced.

The study of sediment drifting and movable bed evolution is a difficult problem
from a physical and mechanical point of view. But the sediment transport relationship
admitted, the problem is reduced to the study of a conservative phenomena.

An other problem is the difference of time scale between current and bottom
evolution. It is impossible (because of cost), to compute simultanously the bottom
evolution and the current by the classical way. Nevertheless, the interaction between
the two is fundamental for the bottom evolution.

This paper presents a two dimensional mathematical sediment transport model taking
into account the influence of the bottom evolution upon the current pattern and shows
how this particular aspect of the interaction drives the ripples propagation.

THEORETICAL ANALYSIS

Bed continuity equation and sediment transport relationship

Let $\bar{T}$ be the sediment transport vector and ξ the bottom elevation ; the bed conti
nuity equation may be expressed as

$$\frac{\partial \xi}{\partial t} + \text{div } \bar{T} = 0$$

How express $\bar{T}$ as a function of the velocity ? That is a real problem. Many relation
can be found taking into account waves or not. For ourselves we have used the Meyer-
Peter relationship for the sediment transport vector $\bar{T}$ which is supposed in the
direction of the current bottom shear stress which is evaluated using Chezy's
relationship.

So the bed continuity equation can be transformed into :

$$\frac{\partial \xi}{\partial t} + T_{Xu} \frac{\partial u}{\partial x} + T_{Xv} \frac{\partial v}{\partial x} + T_{Yu} \frac{\partial u}{\partial y} + T_{Yv} \frac{\partial v}{\partial y} = 0 \tag{1}$$

with

$$T_{Xu} = \frac{u}{W} \frac{\partial T}{\partial u} + T \frac{v^2}{W^3}, \qquad T_{Yu} = \frac{v}{W} \frac{\partial T}{\partial u} - T \frac{uv}{W^3}$$

$$T_{Xv} = \frac{u}{W} \frac{\partial T}{\partial v} - T \frac{uv}{W^3}, \qquad T_{Yv} = \frac{v}{W} \frac{\partial T}{\partial v} + T \frac{u^2}{W^3}$$

$$\left. \begin{array}{ll} T = 8 \sqrt{\dfrac{g}{\varpi} \dfrac{1}{\varpi_S - \varpi}} \ (\tau - \tau_c)^{3/2} & \text{if } \tau > \tau_c \\[2mm] T = 0 & \text{if } \tau < \tau_c \end{array} \right] \quad \text{sediment transport}$$

$\tau_c = A \ (\varpi_S - \varpi) \ D_M$ \quad (0,02 < A < 0,06 Shields). Critical bottom shear stress

$\tau = \varpi \dfrac{W^2}{C^2}$ \quad bottom shear stress

u, v are the two components of the depth averaged current
$W^2 = u^2 + v^2$
ϖ, ϖ_S specific weight of water and sediment
D_M mean diameter of sediment.

Influence of bottom evolution upon the current pattern

With the initial bottom shape ξ_0 and the new geometric conditions the depth
averaged flow pattern is (u_0, v_0). This current modifies the bottom shape which in
turn modifies the current by $(u_1 (t), v_1 (t))$.

At time t, the current pattern in given by $(u_o + u_1 (t), v_o + v_1 (t))$ and the bottom level by $\xi(t)$ $(\xi_S = \xi - \xi_o$ is the bottom evolution).

The resulting disturbance (u_1, v_1) is assumed to be without effect upon the surface elevation z_o. This assumption is equivalent to neglect the characteristic response time of the surface wave propagation compared to the characteristic response time of the bottom evolution.

The resolution of the fluid continuity equation shows that the current disturbance (u_1, v_1) can be written in two different terms :
- the first one $(\bar{u}_1, \bar{v}_1)$ comes directly from the bottom elevation ξ and expresses the flow conservation along the stream lines of the undisturbed field of currents $(u_o\ v_o)$

$$\bar{u}_1 = u_o \frac{\xi - \xi_o}{z_o - \xi} = u_o \frac{\xi_S}{h} \qquad \bar{v}_1 = v_o \frac{\xi - \xi_o}{z_o - \xi} = v_o \frac{\xi_S}{h}$$

- the second one $(\tilde{u}_1, \tilde{v}_1)$ is a deviation of the flow due to the bottom slope. It is governed by :

$$\frac{\partial}{\partial x}\left[\tilde{u}_1 (Z_o - \xi)\right] + \frac{\partial}{\partial y}\left[\tilde{v}_1 (Z_o - \xi)\right] = 0$$

Bottom equation

These two terms are introduced in the bed continuity equation (1) which can be written :

$$\frac{\partial \xi_S}{\partial t} + C \left(\frac{u}{W} \frac{\partial \xi_S}{\partial x} + \frac{v}{W} \frac{\partial \xi_S}{\partial y}\right) = - T_{Xu}\left[\frac{\partial}{\partial x} (u_o + \tilde{u}_1) + \xi_S \frac{\partial}{\partial x} (\frac{u_o}{h_o})\right]$$

$$- T_{Xv}\left[\frac{\partial}{\partial x} (v_o + \tilde{v}_1) + \xi_S \frac{\partial}{\partial x} (\frac{v_o}{h_o})\right] - T_{Yu}\left[\frac{\partial}{\partial y} (u_o + \tilde{u}_1) + \xi_S \frac{\partial}{\partial y} (\frac{u_o}{h_o})\right] \qquad (2)$$

$$- T_{Yu}\left[\frac{\partial}{\partial y} (v_o + \tilde{v}_1) + \xi_S \frac{\partial}{\partial y} (\frac{v_o}{h_o})\right]$$

with $C = \frac{1}{h} (u \frac{\partial T}{\partial u} + v \frac{\partial T}{\partial v})$

Equation (2) governs a ripples propagation in the direction of the initial current pattern with the celerity C. This phenomena comes directly from the adaptation of current disturbance $(\bar{u}_1, \bar{v}_1)$. By neglecting the disturbance it is impossible to reproduce the ripples propagation.

The second member can be divided in two differents parts :
- contribution of the initial current pattern which is conserved at time t
- contribution of the deviation of the flow $(\tilde{u}_1, \tilde{v}_1)$ which drives a ripple deformati

Fluid equation

To determine the current disturbance (u_1, v_1) an other assumption is required : a irrotational current disturbance pattern $(\bar{u}_1 + \tilde{u}_1, \bar{v}_1 + \tilde{v}_1)$ is assumed. So $\tilde{u}_1$ and $\tilde{v}_1$ are obtained from the three-dimensional stream function ψ, which yields a Poisson type equation (3).

So the actual current pattern is defined by :

$$u = u_o + u_o \frac{\xi_s}{h} + \overbrace{\frac{1}{h} \frac{\partial \psi}{\partial y}}^{\tilde{u}_1}$$

$$v = v_o + v_o \frac{\xi_s}{h} \underbrace{- \frac{1}{h} \frac{\partial \psi}{\partial x}}_{\tilde{v}_1}$$

$h = z_o - \xi$ actual depth and ψ obtained from

$$\Delta \psi = + h \frac{\partial}{\partial x} (v_o \frac{\xi_s}{h}) - \frac{\partial}{\partial y} (u_o \frac{\xi_s}{h}) + \tilde{u}_1 \frac{\partial h}{\partial y} - \tilde{v}_1 \frac{\partial h}{\partial x} \qquad (3)$$

NUMERICAL MODEL

A finite difference scheme is used to solve equations (2) and (3). The computational grids ψ and u, v, ξ are shifted. The initial conditions (u_o, v_o, z_o, ξ_o) are obtained with an other numerical model or recorded on a scale model.

Each time step involves two stages :
- computation of the bottom level ξ ; equation (2) is solved by the characteristic method. All functions are explicited but the scheme is stable.
- Computation of the new velocities ; only $\tilde{u}_1$, $\tilde{v}_1$ have to be computed. Equation (3) is solved by an iterative process.

NUMERICAL EXAMPLES

Local scour around a jetty

Several numerical examples have been computed. In figures 1 and 2, the local scou around a jetty, and the flow pattern evolution are shown. The conditions are : flat

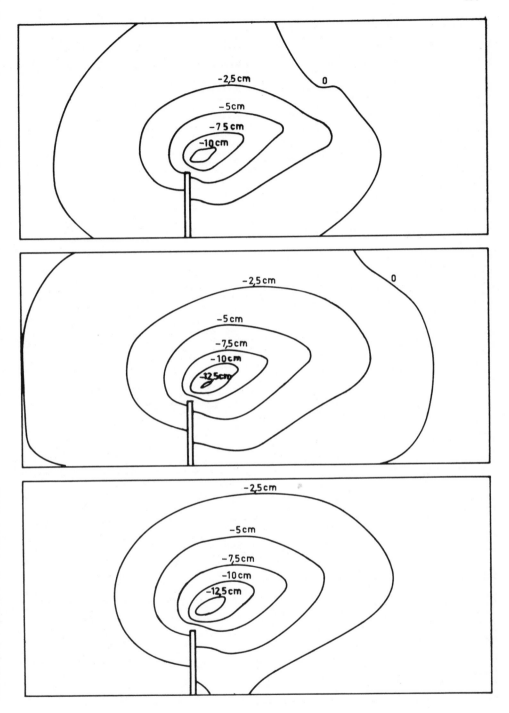

Fig.1_ EROSIONS AFTER 1, 2 AND 3 HOURS

458

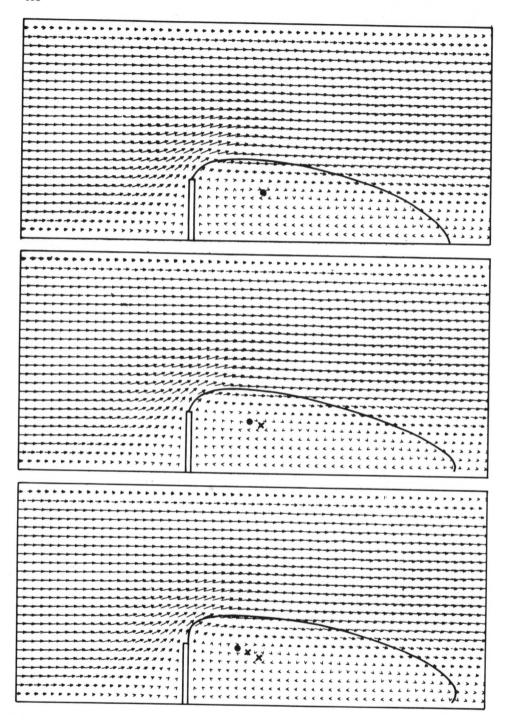

Fig.2 _ CURRENT PATTERN AFTER 1,2 AND 3 HOURS

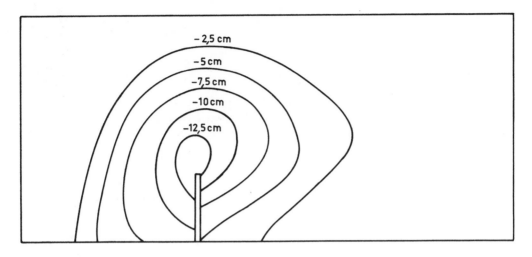

EXPERIMENT

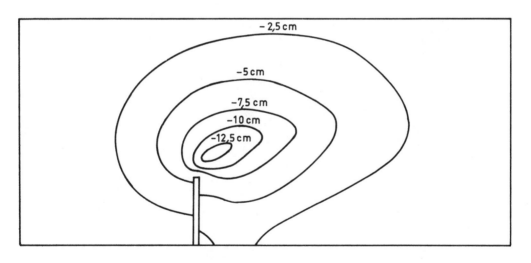

COMPUTATION

Fig.3 _COMPARISON BETWEEN MEASURED
AND COMPUTED EROSIONS

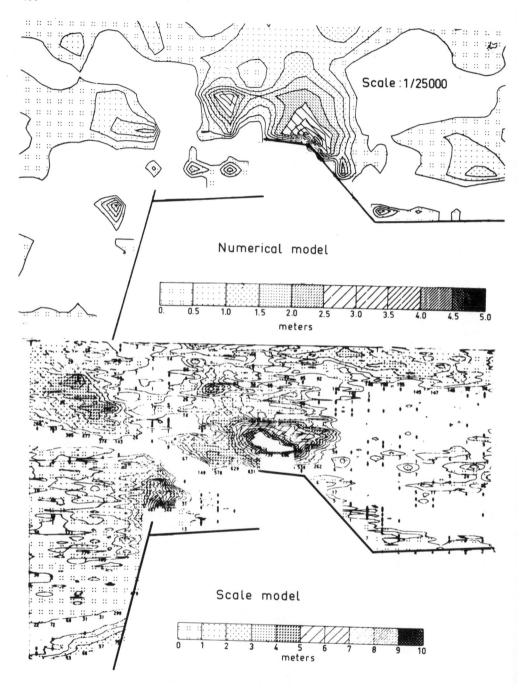

Scale : 1/25000

Numerical model

0. 0.5 1.0 1.5 2.0 2.5 3.0 3.5 4.0 4.5 5.0
meters

Scale model

0 1 2 3 4 5 6 7 8 9 10
meters

Fig.4 _ EROSIONS NEAR DUNKERQUE PORT

initial bottom, far field mean velocity = 41 cm/s, water depth = 20 cm, width = 46 cm, ratio jetty lenght over flume width = 1/3 and particle diameter 4,5 mm. The initial current pattern has been computed with an other numerical model. In figure 3, comparison between computed and measured scour is shown.

Study of new port of Dunkerque

The Port Autonome of Dunkerque has built a new port able to receive 22 meters draught ships. Many studies have been carried on during ten years. Particularly, a movable bed model have been built to study the bottom evolution due to tidal currents near the new port.

The numerical model has been used in this particular case, but to decrease the cost of computation the second kind of disturbance has been neglected. Only equation (2) was solved. The initial current pattern used for the computation was recorded on the scale model.

The comparison between the computed and mesured erosions and accretions is presented on figures 4 and 5. The main difference takes place near the jetties and it probably comes from the initial current pattern which was not conservative because of the precision of measurements on the scale model.

CONCLUSION

A simple kinematical study of the sediment transport equation has shown how can the ripples propagation be obtained. It has also allowed a numerical integration on a computer. The characteristic response time of the surface wave propagation compared to the characteristic response time of the bottom evolution put a stop to any sort of computation of the disturbed current in the classical way. The introduction of current disturbance and several assumption permits the computation of the bottom evolution during a long time.

This kinematical and mathematical aspect almost understood, studies are going on a more physical and dynamical point of view to determine the influence of the different parameters in transport relationship and to find a best dynamical approximation on the current disturbance. In the same time, a mean of averaging the tide in tidal problems is investigaded.

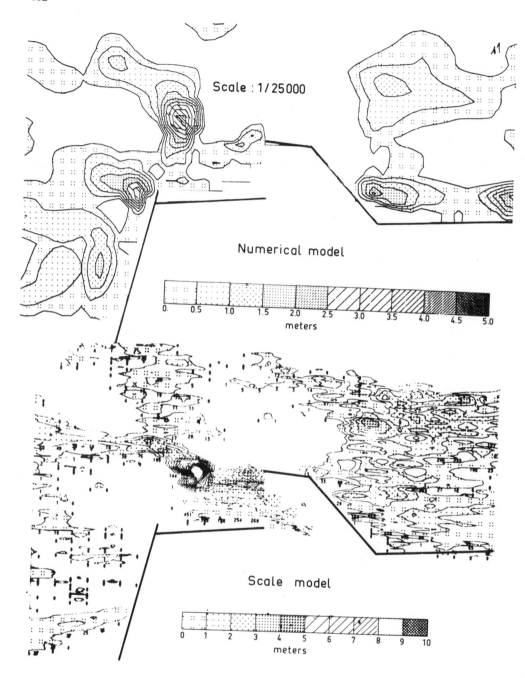

Scale : 1/25000

Numerical model

0.   0.5   1.0   1.5   2.0   2.5   3.0   3.5   4.0   4.5   5.0

meters

Scale model

0   1   2   3   4   5   6   7   8   9   10

meters

Fig. 5 _ ACCRETIONS NEAR DUNKERQUE PORT

REFERENCES

Daubert, A., Lebreton, J.C., Marvaud, P., Ramette, M., 1966. Quelques aspects du cal-
cul du transport solide par charriage dans les écoulements graduellement variés.
Bulletin du CREG n° 18.
Zaghoul, N.A., Mc Corquodale, J.A., 1975. A stable numerical model for local scour.
Journal of Hydraulic Research.
Bonnefille, R. Essai de synthèse des lois de début d'entraînement des sédiments sous
l'action d'un courant en régime continu. Bulletin du CREG n° 5.
Lepetit, J.P., 1974. Nouvel avant-port de Dunkerque, étude sur modèle réduit sédimen-
tologique d'ensemble de l'évolution des fonds au voisinage de l'avant-port.
Rapport Electricité de France, Direction des Etudes et Recherches.
Gill, M.A., 1972. Erosion of sand beds around spur-dikes. Journal of Hydraulic
Division.

SECURITY OF COASTAL NUCLEAR POWER STATIONS IN RELATION WITH THE STATE OF THE SEA

J. BERNIER , J. MIQUEL

Laboratoire National d'Hydraulique, Chatou (France)

E43/78-58

ABSTRACT

The safety of a coastal power plant is concerned with two phenomena : the wind waves, and the maximum and minimum tide levels. This paper presents methods of statistical analysis for estimating the probabilities of extreme events to be taken into account by the designer. First are recalled the definitions of these phenomena, in particular the relationships existing between the maximum of N waves and the significant wave. Then the case is approached where, because of little information available, the use of either meteorological data or uncommon events recorded in a far-off past is necessary. The paper concludes with an example of statistical study of storm durations.

INTRODUCTION

The figure below shows a vertical cross section of a power plant bordering on the sea :

The rates of flow required for the power plant cooling is pumped in the tranquillization basin, which is protected against the waves by the dike. The tranquillization basin is communicating with the sea and its level is equal to that of the tide. A maximum residual agitation of 30 cm in the basin is consistent with the operation of the pumping station.

The designer needs following complementary information :

1. Extreme wind waves probabilities, so that the stability of the dike may be ensured against centennial events at least.

2. Maximum and minimum tide level probabilities, so that protection may be ensured against the flood (maximum level) on the one hand, against failing of the pumps on the other (minimum level).

DEFINITION OF THE WIND WAVE TAKEN INTO ACCOUNT

Among the numerous statistical waves characteristics, the most frequently used for the dike design is the significant wave denoted by $H_{1/3}$ (Average upper third of the greatest waves). This is the parameter that has been selected for the estimation of the wind waves risks.

However, it should be indicated that many other parameters may be directly related to $H_{1/3}$:

1. Cartwright and Longuet-Higgins have demonstrated that in the case the wind waves follow the Gaussian model the following relation may be used : $H_{1/3} = 1,6\ \overline{H} = 0,79\ H_{1/10}$. These results have been checked on some recordings (Miquel, 1975).

2. Utilizing the same assumption, Longuet-Higgins showed that the maximum of N waves is related to $H_{1/3}$, and gave the expression of its mean value. Bernier, in an internal paper published at the "Laboratoire National d'Hydraulique", verified this expression. Besides, utilizing the results obtained by Cramer and Leadbetter (Cramer and Leadbetter, 1967) he could demonstrate that $H_{MAX}(N)$ follows a law of extreme values, the mean and the standard deviation of which are :

$$m(N) = 0,496\ H_{1/2}\ \left(\sqrt{2\ \log_e N} + \frac{0,577}{\sqrt{2\ \log_e N}} \right)$$

$$\sigma(N) = \frac{0,78\ H_{1/3}}{\sqrt{12\ \log_e N}}$$

It appears then possible to evaluate the probabilities of $H_{MAX}(N)$ from those of $H_{1/3}$, either directly by combining the probabilities of $H_{1/3}$ with those of the extreme value distribution, or through simulation by reconstituting a fictitious sample in the following way :

$$H_{MAX} (N) = m (N) - \sigma (N) . \left[0,45 + 0,78 \log_e (- \log_e p) \right]$$

where p is drawn in an uniform law on $]0, 1[$.

It is important to take into consideration not only the mean value but also the variability (figured by the standard deviation) : the neglect of this variability runs counter to safety.

An exhaustive study of waves hazards should also take into account the periods. At the present, the couple (wave-period) is being studied in a frequential way (Allen, 1977) in order to assign a "probable" period to a given wave, the waves only being probabilized.

Another important point, which is likely to be taken into account soon, is the storm duration : an incipient response is given farther.

DEFINITION OF TIDE LEVELS

Definition 1 : Observed maximum level : It is the level actually reached by the sea. It will be denoted by H_1.

Definition 2 : Predicted maximum level : It is the level that the sea would reach in the absence of atmospheric perturbation : it is determined by the position of the stars (astronomic tide). In France, this level is computed by the "Service Hydrographique et Océanographique de la Marine", by summing up the ampli tudes associated with different periods, the semi-diurnal amplitude being the principal one. This level will be denoted by H_O.

Definition 3 : Tide deviation : It is the positive or negative difference between H_1 and H_O, mainly due to meteorological conditions (pressure, wind, temperature, etc...) It will be denoted by S.

So : $H_1 = H_O + S$

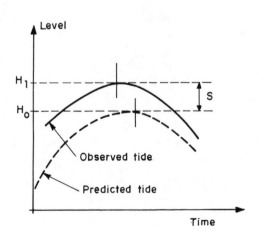

468

Generally, S is estimated by the difference between the observed H_1 and the calculated H_O. It will be shown farther that S can be sometimes estimated from meteorological conditions : $\hat{S}$ (P, V, etc.).

H_O and S being estimated, their values are somewhat uncertain. This uncertainty should be allowed for in the probabilization. It can be written : $H_1 = \hat{H}_O + \hat{S} + \mathcal{E}$, where $\mathcal{E}$ is the residue, the statistical characteristics of which must be given at the same time as the estimates $\hat{H}_O$ and $\hat{S}$.

Everything said about the maxima levels can be symmetrically extended to the minima levels.

WIND WAVES PROBABILITIES

The sample : The sample of daily waves is established, namely by choosing for each day, the surge $H_{1/3}$(i) the highest of the day. It should be made sure that all periods of the year are equally represented in the sample, otherwise a seasonal study would be necessary.

The monthly maxima method : For each month, the highest waves is selected from the sample above. The new sample $\left\{H_j\right\}$ is successively fitted to the Normal, Log. Normal, Extreme values distributions. The best of these fittings is chosen. Example :

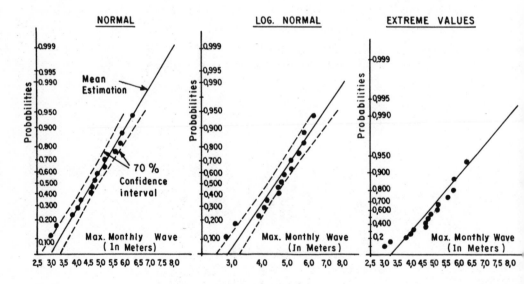

The "Renewal" method : the shortcomings of the monthly maximum method lead us
to use a method, inspired by the study of the renewal process, which is used already for about ten years to study the rates of flow of rising rivers.

Starting from the sample constituted above, the maximum wave is selected in
each storm, provided that this wave is higher than a given threshold chosen
beforehand, and that two successive waves belong undeniably to two differing
storms (independence) :

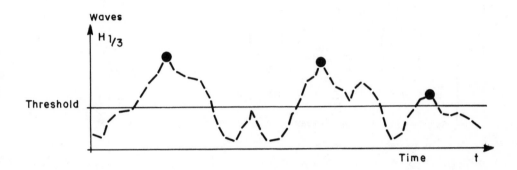

Let us take the month as a reference period. Then, two samples can be constructed :
$\{H_j\}$ is the set of the surges higher than the threshold,
$\{n_k\}$ is the catalogue of the number n_k of storms having exceeded the threshold
in the course of the kth month.

The calculation of the monthly probability of exceeding a value h, namely the
probability of the monthly maximum H^* exceeding the value h, is carried out as
follows :

Prob $\left[H^* > h\right]= 1\quad -\quad$ Prob $\left[H^* \leqslant h\right]$
Prob $\left[H^* \leqslant h\right]=$ Prob $\left[\exists\ 0\ \text{storm} \geqslant \text{threshold in the course of a month}\right]$
$\qquad\qquad + $ Prob $\left[\exists\ 1\ \text{storm} \geqslant \text{threshold and} \leqslant h\right]$
$\qquad\qquad\vdots$
$\qquad\qquad + $ Prob $\left[\exists\ r\ \text{storms} \geqslant \text{threshold and} \leqslant h\right]$
$\qquad\qquad\vdots$
$\qquad\qquad\infty$

Prob $\left[H^* \leqslant h\right] = \displaystyle\sum_{K=0}^{+\infty}$ Prob $\left[\exists\ k\ \text{storms} \geqslant \text{threshold and} \leqslant h\right]$

Prob $\left[H^* \leqslant h\right] = \displaystyle\sum_{K=0}^{+\infty}$ Prob $\left[n = k\right].\left(\text{Prob}\left[H < h\,|\,H > \text{threshold}\right]\right)$

$$\text{Prob}\left[H^* > h\right] = 1 - \sum_{K=0}^{+\infty} P(k) \cdot F^k(h)$$

where
- $P(k)$ is the probability of having k storms in the course of the month,
- $F(h)$ is the probability of a storm, higher than the threshold, being lower than or equal to h.

If h is great enough, $F(h)$ is near to 1 and this result can be simplified to :

$$\text{Prob}\left[H^* > h\right] \simeq 1 - \sum_{K=0}^{+\infty} P(k) \left\{ 1 + k(1 - F(h)) \right\}$$

$$\text{Prob}\left[H^* > h\right] \simeq \bar{n}\,(1 - F(h))$$

for
$$\sum_{K=0}^{+\infty} P(k) = 1 \text{ and } \sum_{K=0}^{+\infty} P(k) \cdot k = \bar{n}$$

$\bar{n}$ = monthly average number of storms.

From the practical point of view, the n_k catalogue enables $P(k)$ or $\bar{n}$ to be determined for the utilisation of the simplified formula ; one of the following two laws is used :

Poisson's law : $P(k) = e^{-\lambda} \dfrac{\lambda^k}{k!}$

Negative Binomial law : $P(k) = \dfrac{\Gamma(\gamma + k)}{k!\,\Gamma(\gamma)}\, p^{\gamma}\,(1 - p)^k$

Since $P(k)$ may considerably vary according to the month, it would be preferable, when sufficient information is available, to take as a reference period the year instead of the month. The probability $F(h)$ is determined by the sample of the H_j, to which are fitted the following laws :

$F(h) = 1 - e^{-\rho\,(h\, -\, \text{threshold})}$

$F(h) = 1 - e^{-\rho\,(\log_e h\, -\, \log_e \text{threshold})}$

$F(h) = 1 - e^{-\rho\,(h\, -\, \text{threshold})^p}$

$F(h) = 1 - e^{-\rho\,(h^2\, -\, \text{threshold}^2)}$

$F(h) = \theta\left[1 - e^{-\rho_1\,(h\, -\, \text{threshold})}\right] + (1 + \theta)\left[1 - e^{-\rho_2\,(h\, -\, \text{threshold})}\right]$

Example :

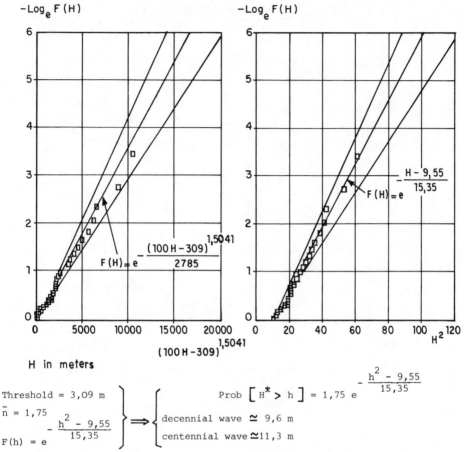

H in meters

$$\text{Threshold} = 3,09 \text{ m}$$
$$\bar{n} = 1,75$$
$$F(h) = e^{-\dfrac{h^2 - 9,55}{15,35}}$$

$$\Longrightarrow$$

$$\text{Prob}\left[H^* > h\right] = 1,75\, e^{-\dfrac{h^2 - 9,55}{15,35}}$$
$$\text{decennial wave} \simeq 9,6 \text{ m}$$
$$\text{centennial wave} \simeq 11,3 \text{ m}$$

This method has the advantage of utilizing the maximum amount of information, while warranting its homogeneity. It is possible and desirable to calculate the intervals of confidence.

TIDE LEVEL PROBABILITIES

The Observed Maximum Level : the most simple way is, like in the case of waves to constitute the sample of daily maximum levels of the high water. Then, the same methods are used as for the wave. The result is presented in the following form :

PROBABILITIES OF HIGH WATERS IN DIEPPE _

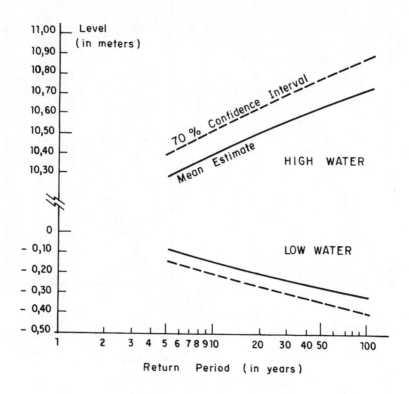

However, the question may arise of whether it will be safe to use only one statistical law for explaining the behaviour of a variable, which is made up of two phenomena entirely different : the astronomic tide and the tide deviation due to meteorological conditions. We decided therefore to study also these two phenomena.

The Predicted Maximum Level : In fact, it's a question of a random pseudovariable easy to probabilize either by constituting directly a catalogue of predicted heights, or by using the estimates based on the semi-diurnal amplitude. The two methods can be compared in the figure below.

FREQUENCIES OF PREDICTED LEVEL IN DIEPPE _

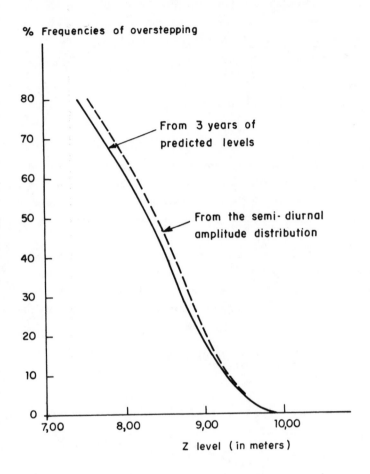

% Frequencies of overstepping

From 3 years of predicted levels

From the semi- diurnal amplitude distribution

Z level (in meters)

Tide differences : First, we constitute the catalogue of daily tide differences obtained either by means of differences $H_O - H_1$ on a series of observed tides, or by reconstitution from meteorological conditions (see farther). Then, we proceed to the same probabilistic study as for the waves.

The Sum of predicted levels and tide differences : We have $H_1 = H_O + S$. If H_O and S are independent, the probability of their sum can be easily calculated by writing :

$$\text{Prob}\left[\, H_1 > h_1 \,\right] = \int_{-\infty}^{+\infty} G\left[h_1 - x\right] \cdot f\left[x\right] \cdot dx$$

where $f\left[\,x\,\right] = \text{Prob}\left[\, x \leqslant H_O \leqslant x + dx \,\right]$

$G\left[\,y\,\right] = \text{Prob}\left[\, S > y \,\right]$

For our part, we found that if the coefficient of correlation between H_O and S could attain 0,3 during slight or medium storms, this coefficient is practically zero for heavy storms by which we are particularly concerned. This result is only indicative as it corresponds to a particular case and deserves to be tested on other sites.

If the correlation is no more zero but if there exists a relationship of the kind $S = \lambda\, H_O + S'$, where H_O and S' are independent, we can get again to the previous case by considering the independent variables : $(1 + \lambda)\, H_O$ and S'.

The figure below enables the two methods for estimating the H_1 level to be compared by studying directly H_1 or by studying the sum $H_O + S$.

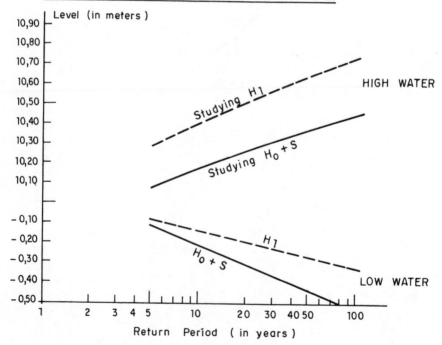

PROBABILITIES OF TIDE LEVELS IN DIEPPE _

For the design, we take the extreme limits of these estimates to which we add confidence intervals at 70%.

CASE OF POOR INFORMATION

Wave data and tide data are frequently very short, rendering the statistical estimates too uncertain : additional information should then be used.

Sometimes, it is fortunate to find another wave or tide series in the vicinity of the studied site. If the two series are closely related, the probability estimates of the long series can be easily transposed to the short one.

If this is not the case, it is necessary then to consider other possibilities.

<u>Utilization of the meteorology</u> : in the case where information, such as pressure, wind, temperature in the vicinity of the site, is available, it is possible to establish a relationship between these data and the surges or the tide fluctuations. As a test we tried multiple linear regressions of the kind :

$$H_{1/3} = f \; (P, \; V, \; V^2, \; T, \; H_O, \Delta P, \Delta V, \; ...)$$
$$S = g \; (P, \; V, \; V^2, \; T, \; H_O, \Delta P, \Delta V, \; ...)$$

where
$$\begin{cases} P & = \text{pressure} \\ \Delta P & = \text{temporary pressure variation} \\ T & = \text{temperature} \\ V & = \text{wind speed} \\ \Delta V & = \text{temporary variation of the wind} \end{cases}$$

Although the results are not yet exploitable for high events, they are incentive for low and medium events in so far as the obtained multiple correlation coefficients reached 0,8 to 0,9. Using these relationships, we reconstructed a fictitious sample of tide differences over a long period of time and we estimated then the probabilities resulting from this sample. On the figure below, the obtained results can be compared with respect to the probabilities derived from observations :

476

PROBABILITIES OF TIDE DEVIATIONS IN LE HAVRE _

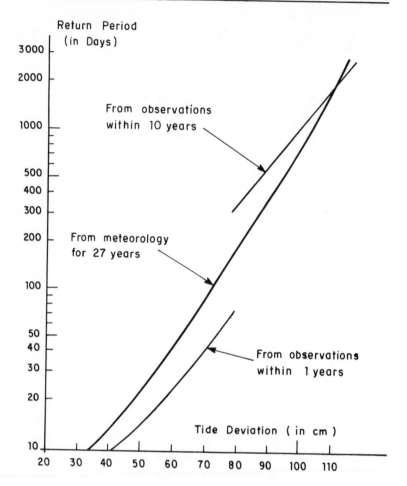

In this figure we can see that there is an acceptable compatibility between the estimates for return periods lower than 10 years. Beyond these periods, it will be necessary either to improve the statistical relationships between the meteorology and the sea states or to use mathematical prediction models.

Utilization of exceptional events : it happens that there exist recorded data on one or more exceptional events for which an estimate can be fixed, and which are known to be the highest within a long period of time (for instance, a century).

This information is precious and may be utilized, though it greatly differs from a complete catalogue of waves or tides. It allows the statistical uncertainty to be reduced and the representativity of the used sample to be proved. The detailed description of this method can be found in the references (Bernier and Miquel, 1977). It was already applied successfully to flood risk estimations :

FLOOD PROBABILITIES AT HAUCONCOURT (MOSELLE)

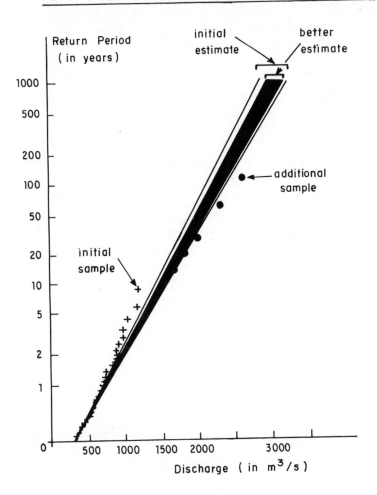

STORM DURATIONS

Recent works on random waves showed how the storm duration may affect the life-time of dikes.

Using once more the techniques applied to the study of river flow rates (Miquel and Phien Bou Pha, 1978) we can estimate, for instance, the duration probability of a storm exceeding a given surge threshold. The probabilities of the yearly sums of storm durations can be read in the figure below :

478

Durations
(in Days)

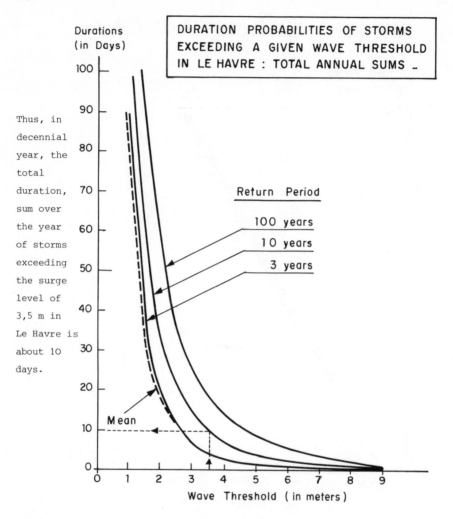

DURATION PROBABILITIES OF STORMS
EXCEEDING A GIVEN WAVE THRESHOLD
IN LE HAVRE : TOTAL ANNUAL SUMS _

Return Period

100 years

10 years

3 years

Mean

Wave Threshold (in meters)

Thus, in decennial year, the total duration, sum over the year of storms exceeding the surge level of 3,5 m in Le Havre is about 10 days.

A curve of the same kind can be obtained to describe the durations of individual storms.

Indeed, such curves will be useful to designers when they will be able to take
simultaneously into account both, the storm durations and their intensities.

REFERENCES

Allen, H., 1977. Analyse statistique des mesures de houle en différents sites du
    littoral français. Edition n° 3, rapport EDF HE 46/77.01. Chatou (France)
Bernier, J., Miquel, J., 1977. Exemple d'application de la théorie de la décision
    statistique au dimensionnement d'ouvrage hydraulique : prise en compte de l'in-
    formation hétérogène. A.I.R.H. Baden.
Cramer, Leadbetter, 1967. Stationary and related stochastic processes. Sample func-
    tion properties and their applications. John Wiley. New York.

Miquel, J., 1975. Role et importance d'un modèle statistique de la houle en vue du dépouillement et du stockage des données. A.I.R.H. Sao Paulo.

Miquel, J., Phien Bou Pha, B., 1977. Tempétiage : un modèle d'estimation des risques d'étiage. Xème Journée de l'Hydraulique. Toulouse.

SUBJECT INDEX

Tide
- High spring tide, 289, 340, 368, 400.
- Tidal high water, 286, 289.
- Tide level, 168, 301, 309, 310, 325, 328-331, 406, 412, 421, 434, 466, 467, 471.
- Tidal low water, 286, 289.
- Neap tide, 340.
- Oceanic tide, 239, 385.
- Tidal period, 251-257.
- Tidal prediction, 306, 309, 310, 315, 394, 412.
- Tidal sea, 238.
- Reversal of tide, 238, 245, 251-257, 369.
- Tidal resonance, 286.
- Tidal wave, 337, 338.

Time series analysis, 197, 200, 202, 207, 216.

Topographic effects, 91, 93.

Torrey Pines Beach, 192.

Towing resistance, 113, 114.

Tranquillization bassin, 466.

TRANSPAC, 58, 59.

Trapping scale, 67.

Turbulence
- Atmospheric turbulence, 107.
- Turbulence generation, 10, 12, 16.
- Geostrophic turbulence, 68.
- Turbulence production, 1, 2.
- Shear generated turbulence, 10, 13.
- Well-developed turbulence, 2.
- Turbulent convection, 39.
- Turbulent diffusion, 7, 20, 235, 236, 246.
- Turbulent disturbance, 7.
- Turbulent energy, 1, 6, 8, 23.
- Turbulent energy balance, 9, 24.
- Turbulent energy diffusion, 1, 9, 16, 23.
- Turbulent energy flux, 7, 9, 11, 15, 18, 21, 23, 27.
- Turbulent energy production, 5, 13, 14, 16, 19, 29, 41, 169, 177.
- Turbulent energy (rate of dissipation), 4, 41.
- Turbulent energy (time scale of dissipation), 6.
- Turbulent entrainment layer, 2-20, 23, 28, 29.
- Turbulent fluctuations, 40, 181.
- Turbulent integral length scale, 12, 162.
- Turbulent interactions, 79.
- Turbulent mixing, 9, 162.
- Turbulent operator, 245.
- Turbulent stress, 163, 169, 336.

Tyrrhenian Sea, 440, 442, 443, 448-450.

U.S. Atlantic Coast, 75, 323, 325, 328, 331.

U.S. West Coast, 451.

Veering of horizontal velocity, 255-257.

Velocity profile, 188-191, 243, 247.

Velocity shear, 1, 3, 5, 14, 16.

Velocity shear layer, 6.

Venice, 427, 428, 432, 433, 436.